Advances in Plant Ethylene Research

Advances in Plant Ethylene Research

Proceedings of the 7th International Symposium on the Plant Hormone Ethylene

Edited by

Angelo Ramina
University of Padova,
Department of Environmental Agronomy and Crop Science,
Legnaro (Padova), Italy

Caren Chang
University of Maryland,
Department of Cell Biology and Molecular Genetics,
Maryland, U.S.A.

Jim Giovannoni
Cornell University,
Boyce Thompson Institute for Plant Research,
Ithaca, U.S.A.

Harry Klee
University of Florida,
Horticultural Sciences,
Gainesville, U.S.A.

Pierdomenico Perata
Scuola Superiore Sant'Anna,
Pisa, Italy

and

Ernst Woltering
Wageningen University and Research Centre,
Agrotechnology and Food Sciences Group (AFSG),
Wageningen, The Netherlands

Springer

A C.I.P. Catalogue record for this book is available from the Library of Congress.

ISBN 978-1-4020-6013-7 (HB)
ISBN 978-1-4020-6014-4 (e-book)

Published by Springer,
P.O. Box 17, 3300 AA Dordrecht, The Netherlands.

www.springer.com

Printed on acid-free paper

PREFACE

This volume contains the main lectures and poster contributions presented at the 7th International Symposium on the Plant Hormone Ethylene held in Pisa (Italy), June 18-22, 2006. This international scientific event was organized by the University of Padova and the Scuola Superiore Sant'Anna of Pisa and took place on the premises of the Scuola Superiore Sant'Anna.

We would like to thank the Ministry of Agriculture and Forestry, the Ministry of University and Scientific Research of Italy, the University of Padova and the Scuola Superiore Sant'Anna of Pisa for partially funding this symposium. Appreciation is also extended to a number of local institutions that generously contributed to the success of this important event.

We are indebted to the members of the scientific committee. A special appreciation goes to the local organizing committee headed by Prof. Giovanni Serra.

Finally, we gratefully acknowledge Dr. Alessandro Botton for handling all the editorial aspects concerning the publication of this volume.

The Editorial Board

We dedicate the Proceedings of the 7th International Symposium on the Plant Hormone Ethylene to the memory of our dear friend and colleague, Tony Bleecker. Tony is greatly missed by the community of ethylene biologists. His unique and seminal discoveries, together with his interminable passion for science, leave a deep and lasting impact. From developing the phenotypic assay that led to the elaboration of the ethylene signal transduction pathway to identifying the ethylene receptor and providing key insights into its function, Tony continually illuminated and drove our understanding of ethylene perception. His contributions and influence were far-reaching. He brought vision and originality, and we delighted in his keen intellect. He was a champion of the Socratic method of inquiry, always challenging us to think along with him. He freely shared his time and materials with all of us and collaborated widely. He will remain an inspiration to us.

TABLE OF CONTENTS

INDEX OF AUTHORS

1. ETHYLENE BIOSYNTHESIS, PERCEPTION AND SIGNAL TRANSDUCTION

2. Materials and Methods

The *etr1-6*, *etr1-7*, *etr2-3*, *ers2-3*, and *ein4-4* mutants were obtained from Elliot Meyerowitz. The *ctr1-2* mutant was from Joseph Kieber, and the *ein2-1* mutant from Joseph Ecker. The *ers1-2* mutant was isolated and described previously (Wang *et al.* 2003; Hall and Bleecker 2003). Transgenic lines used were generated previously and are described in Wang *et al.* (2003) and Binder *et al.* (2004b).

The effects of ethylene on the growth rates and movement of hypocotyls were measured using etiolated *Arabidopsis* seedlings as described previously (Binder *et al.* 2004 a,b). For this work, seedlings were treated for 2 h with air to establish a basal growth rate and movement pattern. Ethylene was then introduced for 22 h. Images were captured every 15 min with resolutions between 33 and 90 pixels mm^{-1}. Unless otherwise specified, measurements were made in the presence of 5 μM L-α-(2-amino ethoxyvinyl)-glycine (AVG) to inhibit biosynthesis of ethylene.

To determine growth rates of hypocotyls, one of two methods were used. In one, the height in pixels of each seedling in each frame was analyzed using custom software written by Edgar Spalding in LabVIEW 5.0 (National Instruments, Austin, TX) as previously described (Parks and Spalding 1999; Folta and Spalding 2001). During nutational bending, this method was inaccurate due to the movement of the hypocotyls. In this case, the length of the hypocotyl in each frame was measured manually using the computer program Image J (http://rsb.info.nih.gov/ij/). From both methods, we calculated growth rates for each 15-min increment using Microsoft Excel. Nutation angles of hypocotyls were measured manually by drawing a line from the hypocotyl bend to the apical hook and measuring the angle with a protractor (Fig. 1A). In these plots, 90° indicates growth directly against the gravity vector while angles greater than 90° indicate the opening of the apical hook is aimed down and less than 90° that the opening is aimed up.

3. Results

3.1 Ethylene stimulates nutations in hypocotyls of dark-grown Arabidopsis *seedlings*

While examining the long-term effects of ethylene on etiolated hypocotyls, we made the unexpected observation that ethylene stimulates nutations (originally termed "circumnutations" by Darwin and Darwin in 1880) in the hypocotyls of etiolated seedlings (Fig. 1A). It is interesting to note that

Ethylene stimulates nutations of etiolated Arabidopsis hypocotyls that are dependent on the ETR1 receptor

Binder, B.M.[1,2]*, Zutz, T.C.[1], O'Malley, R.C.[1,3], Wang, W.[1,4] and Bleecker, A.B.[1,5]

[1] University of Wisconsin in Madison, Department of Botany, 430 Lincoln Drive, Madison, WI, 53706, USA; [2] Current address: U.W. Madison, Department of Horticulture, 1575 Linden Drive, Madison, WI 53706; (608) 262-1779; [3] Current address: Salk Institute for Biological Studies, 10010 North Torrey Pines Road, La Jolla, CA 92037; [4] Current address: Ceres Inc., Ceres, Inc.1535 Rancho Conejo Blvd., Thousand Oaks, CA 91320, USA; [5] Deceased. (*Corresponding author: bmbinder@wisc.edu)

1. Introduction

The phytohormone ethylene is well known to influence a number of physiological and developmental processes in plants including but not limited to seed germination, seedling growth, formation of the apical hook, senescence, fruit ripening, abscission, and gravitropism (Abeles *et al.* 1992). Among these myriad processes much attention has focused on the inhibitory effects of ethylene on the growth of etiolated seedlings. For the past 7 years this laboratory has used a high-resolution, computer-driven, time-lapse imaging system to examine the growth kinetic responses of hypocotyls of etiolated Arabidopsis seedlings to the short-term application and subsequent removal of ethylene (Binder *et al.* 2004a,b). Using this system we examined the long-term effects of ethylene and made the observation that ethylene stimulates nutations in the hypocotyls of etiolated seedlings. Nutations are oscillatory "nodding" or bending movements caused by localized differential growth. We found that the effects of ethylene on nutations could be distinguished from its inhibitory effects on hypocotyl growth. The roles of the ethylene receptors, ETR1 histidine kinase activity, and phosphorelay through the receptor receiver domain in this signaling were studied.

A. Ramina et al. (eds.), Advances in Plant Ethylene Research: Proceedings of the 7th International Symposium on the Plant Hormone Ethylene, 3–8.

nutation phenotype in *etr1-7* mutant seedlings (Fig. 2). Loss-of-function mutants in the other receptors isoforms, either singly or in combination, still nutated normally (data not shown). It should be noted that combinatorial receptor mutants that lead to constitutive growth inhibition (such as the *ers1-2 ers2-3* double mutant and the *etr2-3 ein4-4 ers2-3* triple mutant) also caused small, constitutive nutations in air (not shown). Higher-order receptor loss-of-function mutants that included either the *etr1-6* or *etr1-7* loss-of-function mutant, did not nutate; the nutation phenotype was rescued when these mutants were transformed with genomic wild-type *ETR1* supporting the importance of the ETR1 receptor (Fig. 3).

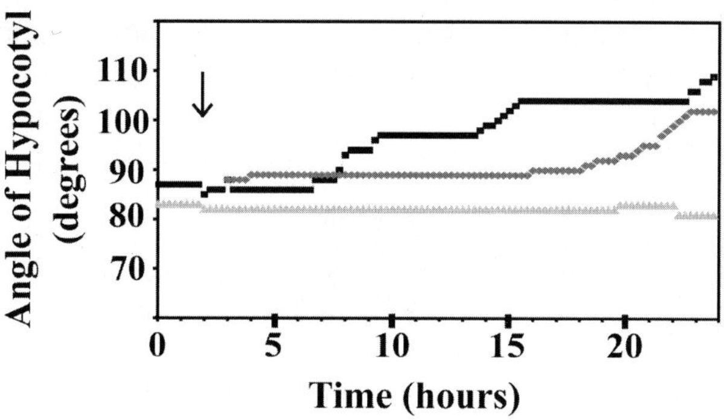

Fig. 2. Hypocotyl angle plotted as a function of time for the *etr1-7* mutant. 10 ʃL L^{-1} ethylene was added at 2 h (arrow). Three representative seedling responses are shown.

3.3 ETR1 histidine kinase activity and phosphotransfer through the receiver domain are not required for nutations

Because ETR1 contains both a functional histidine kinase and a receiver domain, we examined the role that either of these might have in the nutation phenotype. When *etr1-6 etr2-3 ein4-4* triple mutants (lacking receptors with a receiver domain) were transformed with a mutated ETR1 lacking the conserved $aspartate_{659}$ in the receiver domain required for phosphotransfer (*etr1-[D]*), four out of nine seedlings had rescue of the nutation phenotype (Fig. 3). Similarly, when *etr1-7 ers1-2* double mutants (lacking receptor histidine kinases) were transformed with a kinase-inactivated ETR1 transgene (*getr1-[HGG]*), the nutation phenotype was variably rescued with 4 out of 12 seedlings showing nutations (Fig. 3). We are currently examining the basis for the variable rescue. Nonetheless, these results suggest a noncritical role for ETR1 histidine kinase activity and this effect of ethylene was predicted over 20 years ago (Britz and Galston 1982). The growth inhibition response to ethylene was similar to that previously reported (Binder *et al.* 2004 a,b) while growth in air remained stable for about 12 h before slowly declining. When the nutation angle time-courses were plotted (examples in Fig. 1B), we found that nutations had a delay between ~3 and 12 h after the addition of ethylene with an average delay of 6.25 h. The amplitude of hypocotyl nutations at various ethylene concentrations was determined from the these time-courses by measuring the change in angle from each of the peaks to the midline of the sine wave. Nutation amplitude varied with ethylene concentration with a half maximal response at approximately 40 nL L^{-1} ethylene and saturating at approximately 100 nL L^{-1}. Nutation periodicity showed a small dependence on ethylene concentration ranging between 2.8 and 4.6 h.

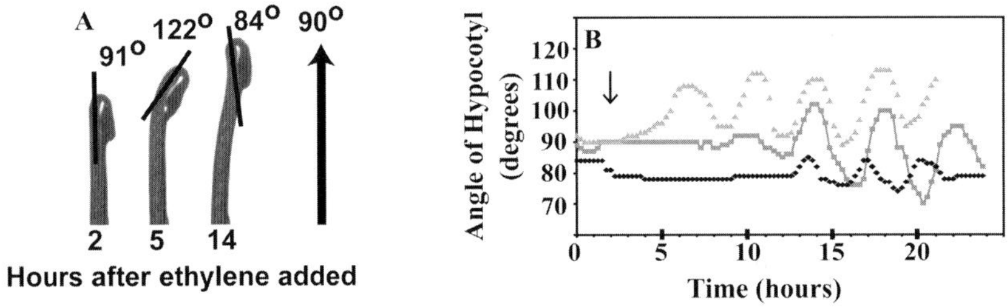

Fig. 1. Hypocotyl nutational bending in Columbia wild-type seedling hypocotyls. **(A)** Images of hypocotyls at various times after the addition of ethylene. **(B)** Hypocotyl nutation angle time-courses after the addition of ethylene in 3 representative seedlings. 10 μL L^{-1} ethylene was added at 2 h (arrow).

To confirm that these nutations were caused by ethylene, we examined nutations in air in the presence or absence of the ethylene synthesis inhibitor AVG. In the absence of AVG, small amplitude nutations were observed that were absent in the presence of AVG. Additionally, the constitutive response mutant, *ctr1-2*, showed constitutive nutations in air in the presence of AVG while the ethylene insensitive *ein2-1* mutant failed to nutate in the presence of 10 ∫L L^{-1} ethylene. Thus it appears that ethylene stimulates nutations through the well-characterized signaling pathway.

3.2 The ETR1 receptor, but not other receptor isoforms, is required for ethylene-stimulated nutations

To determine the importance of each of the ethylene receptor isoforms, we examined ethylene-stimulated nutations in receptor loss-of-function mutants. Of the five receptor isoforms, only ETR1 appears to be required for nutations stimulated by ethylene as evidenced by the lack of the phosphotransfer through the receiver domain of ETR1 in the nutation phenotype.

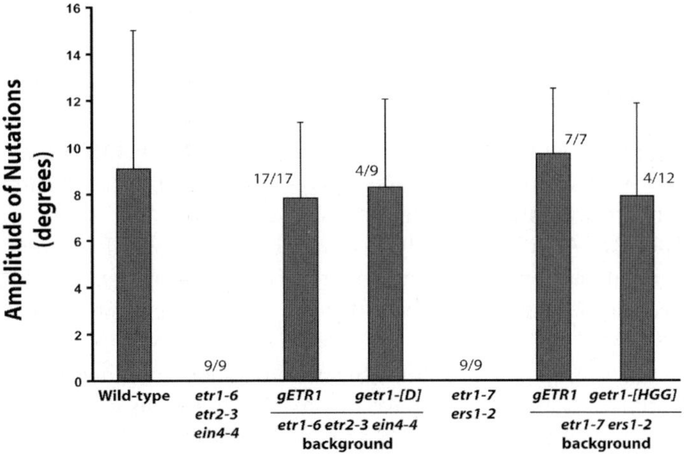

Fig. 3. Amplitude of nutations in wild-type, mutants and various transformant lines. Ratios indicate the number that show the indicated phenotype out of the total tested.

4. Summary

High-resolution, time-lapse imaging of etiolated seedlings has proven to be a useful tool for uncovering new aspects of ethylene signaling and physiology. Using this approach we found that ethylene stimulates hypocotyl nutations and involves the ethylene receptors, CTR1 and EIN2. However, of the five ethylene receptor isoforms, only ETR1 appears to be required for the nutation phenotype suggesting a nonoverlapping function for ETR1. The basis for this function does not appear to reside in either ETR1 histidine kinase activity or phosphorelay through the receiver domain. More work is required to determine the basis for this unique function of ETR1.

References

Abeles FB, Morgan PW, Saltveit ME Jr. (1992) Ethylene in Plant Biology, 2nd Edition. Academic Press, San Diego, California

Binder BM, Mortimore LA, Stepanova AN, Ecker JR, Bleecker AB (2004a) Short term growth responses to ethylene in *Arabidopsis* seedlings are EIN3/EIL1 independent. Plant Physiol. 136: 2921–2927

Binder BM, O'Malley RC, Wang W, Moore JM, Parks BM, Spalding EP, Bleecker AB (2004b) *Arabidopsis* seedling growth response and recovery to ethylene: a kinetic analysis. Plant Physiol. 136: 2913–2920

Britz SJ, Galston AW (1982) Physiology of movements in stems of seedling *Pisum sativum* L. cv. Alaska: II. The role of the apical hook and of auxin in nutation. Plant Physiol. 70: 1401–1404

Darwin C, Darwin F (1880) The Power of Movement in Plants. John Murray, London

Folta KM, Spalding EP (2001) Unexpected roles for cryptochrome 2 and phototropin revealed by high-resolution analysis of blue light-mediated hypocotyl growth inhibition. Plant J. 26: 471–478

Hall AE, Bleecker AB (2003) Analysis of combinatorial loss-of-function mutants in the arabidopsis ethylene receptors reveals that the *ers1 etr1* double mutant has severe developmental defects that are EIN2 dependent. Plant Cell. 15: 2032–2041

Parks BM, Spalding EP (1999) Sequential and coordinated action of phytochromes A and B during Arabidopsis stem growth revealed by kinetic analysis. Proc. Natl. Acad. Sci. USA 96: 14142–14146

Wang W, Hall AE, O'Malley RC, Bleecker AB (2003) Canonical histidine kinase activity of the transmitter domain of the ETR1 ethylene receptor from *Arabidopsis* is not required for signal transmission. Proc. Natl. Acad. Sci. USA 100: 352–357

A novel membrane protein conserved in plants and animals is important for ethylene receptor function in *Arabidopsis thaliana*

Resnick, J.S., Rivarola, M., Wen, C.K.[1], Shockey, J.A. and Chang, C.*

Department of Cell Biology and Molecular Genetics, University of Maryland, College Park, MD 20742, USA. Phone: (301-405-1643).
[1]National Key Laboratory of Plant Molecular Genetics, Institute of Plant Physiology and Ecology, Shanghai Institutes for Biological Sciences, Chinese Academy of Sciences, 300 Fenglin Road, Shanghai 20032, China.
(*Corresponding author: carenc@umd.edu)

1. Introduction

The known components involved in ethylene signaling generally form a linear pathway model, beginning with ethylene binding at the receptors and resulting in the activation of ethylene-response genes (Guo and Ecker, 2004). Elucidation of this pathway has been possible through the isolation of genetic mutants displaying altered responses to ethylene. *Arabidopsis* seedlings grown in the dark in the presence of ethylene exhibit a "triple response", which consists of inhibition of hypocotyl and root elongation, radial swelling of the hypocotyl, and an exaggerated curvature of the apical hook (Bleecker *et al.*, 1988; Guzman and Ecker, 1990).

Arabidopsis possesses five partially redundant ethylene receptors, which negatively regulate ethylene responses (Hua and Meyerowitz, 1998; Zhao *et al.*, 2002; Hall and Bleecker, 2003). Loss-of-function mutants of each of the receptor genes exhibit essentially a wild-type phenotype, although *etr1* loss-of-function mutants display a slightly enhanced response to ethylene (Hua and Meyerowitz, 1998; Cancel and Larsen, 2002). Downstream of the receptors, there are both positive and negative regulators in the pathway, culminating in changes in gene expression (Kieber *et al.*, 1993; Solano *et al.*, 1998; Alonso *et al.*, 1999; Guo and Ecker, 2003; Potuschak *et al.*, 2003; Gagne *et al.*, 2004). These components were primarily identified through mutants that exhibit either ethylene insensitivity or constitutive ethylene responses (Guo and Ecker, 2004). It is generally believed that such mutant screens have been saturated, but further screening may uncover additional components in the pathway.

A. Ramina et al. (eds.), Advances in Plant Ethylene Research: Proceedings of the 7th International Symposium on the Plant Hormone Ethylene, 9–14.

2. Materials and Methods

Mutagenesis, suppressor screening, map-based cloning, sequence analyses and genetic analyses are described in Resnick *et al.* (2006).

3. Results and Discussion

3.1 The RTE1 *gene is required for ethylene insensitivity in the* etr1-2 *mutant*

The *Arabidopsis etr1-2* ethylene receptor mutant shows only partial ethylene sensitivity and therefore does not display the seedling triple response phenotype when treated with ethylene. In order to identify additional components in ethylene signaling, we mutagenized *etr1-2* and screened for suppressors of the insensitivity (i.e., mutants that could now fully respond to ethylene). Three recessive extragenic mutants were isolated at the same locus, which we named *REVERSION-TO-ETHYLENE-SENSITIVITY1.* All three mutant alleles, *rte1-1*, *rte1-2*, and *rte1-3*, resulted in severe loss of *RTE1* function and caused loss of the ethylene insensitivity of *etr1-2*. In an ethylene dose response analysis of dark-grown seedlings, all three *rte1* alleles exhibited ethylene responses comparable to that of the wild type, but more similar to the *etr1* null mutant (Fig. 1A). In addition, *rte1*-suppressed *etr1-2* adult plants displayed ethylene-induced senescence, which is largely absent in the *etr1-2* single mutant (Fig. 1B). The suppression of ethylene insensitivity in both seedlings and adult plants indicated that *RTE1* is required for *etr1-2* function.

Further genetic analysis showed that rte1 does not suppress ethylene-insensitive mutations in the four other ethylene receptor genes (Resnick *et al.*, 2006), yet can suppress several other ethylene-insensitive mutations in etr1 suggesting that the suppression may be specific to mutations in ETR1 (Resnick and Chang, unpublished). Interestingly, rte1 is unable to suppress the etr1-1 mutant allele, and the basis for this is currently under investigation (Resnick *et al.*, 2006).

3.2 RTE1 *is a negative regulator of ethylene responses*

rte1 single mutants displayed an enhanced ethylene response similar to that seen in the *etr1* null mutant, *etr1-7* (Fig. 1A) (Resnick *et al.*, 2006). Further analysis indicated that null mutants of *etr1* and *rte1* both exhibit a shorter hypocotyl even in the absence of ethylene, at least partly due to an enhanced response to endogenously-produced ethylene (Cancel and Larsen, 2002; Resnick *et al.*, 2006). These similarities between *rte1* and *etr1* mutants suggested that *RTE1* might be required for the function of wild-type *ETR1*, thus acting as a negative regulator of ethylene responses (positively regulating the ETR1 receptor). A dose-response analysis of an *rte1-2 etr1-7* double mutant revealed a response virtually identical to that of either single mutant, suggesting that both genes respond to ethylene in the same pathway (Resnick *et al.*, 2006).

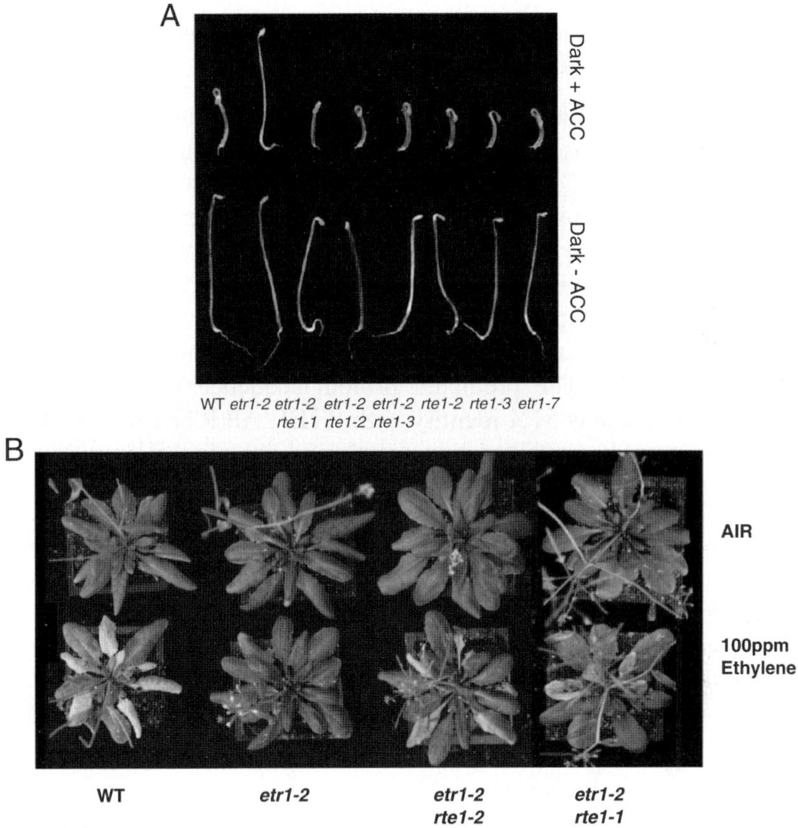

Fig. 1. (Color figure in the Annex, p.453) (**A**) Four-day old dark-grown seedlings grown in the presence or absence of 1-aminocyclopropane-1-carboxylic acid (ACC). The three *etr1-2* suppressed lines and two *rte1* single mutants all show a triple response phenotype similar to the *etr1-7* null mutant. (**B**) Ethylene-induced leaf senescence in adult plants treated with or without 100 ppm ethylene for 4 days. Ethylene-induced senescence, seen as yellowing of the leaves, occurs in wild type (WT) and the two suppressed lines (*etr1-2 rte1-2* and *etr1-2 rte1-1*), but not in *etr1-2* alone.

We isolated the *RTE1* gene by map-based cloning, and created stable transgenic plants over-expressing *RTE1* using the 35S cauliflower mosaic promoter. Over-expression of *RTE1* in wild-type seedlings resulted in partial ethylene insensitivity, consistent with *RTE1* being a negative regulator of ethylene responses (Resnick *et al.*, 2006). Interestingly, when

RTE1 was over-expressed in genotypes in which the *ETR1* gene was absent, insensitivity was much less pronounced (Resnick and Chang, unpublished). This suggested that the ETR1 receptor is required for the insensitivity phenotype, and gave further support to the hypothesis that *RTE1* is a regulator of *ETR1*. A model for *RTE1* action is shown in Fig. 2.

3.3 RTE1 *encodes a novel predicted transmembrane protein with homologs in a wide variety of organisms*

The *RTE1* gene encodes a novel protein of 250 amino acids (28 kDa), which is highly conserved in many eukaryotic organisms, including humans, *Drosophila*, *C. elegans*, and *Plasmodium*, as well as other plants. *RTE1* appears to be absent in fungi and prokaryotes. In tomato, the *RTE1* homolog is the *Green-ripe* (*Gr*) gene, which prevents tomato fruit ripening when over-expressed (Barry and Giovannoni, 2006). *RTE1* is present in a single copy in most species, except in plants, which have two to three copies. *Arabidopsis* has a second copy, which we have named *RTE1-HOMOLOG* (*RTH*). The predicted protein sequence for RTH has 231 amino acids and shares 51% identity with RTE1. All RTE1 homologs have two to four strongly predicted transmembrane domains. We are currently investigating the subcellular localization of RTE1.

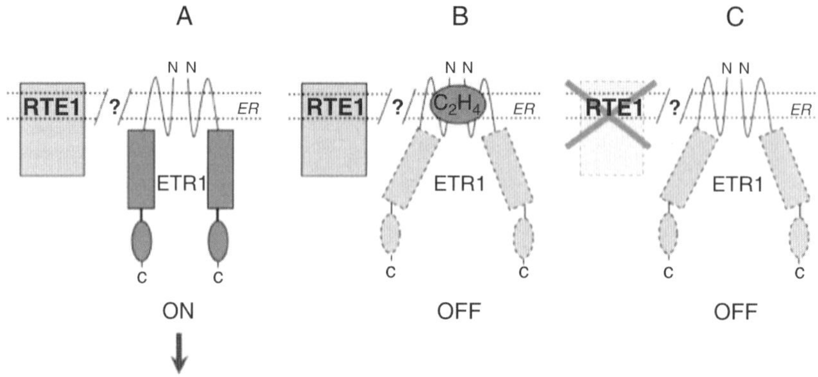

Fig. 2. Model for RTE1 action. Cartoon model showing signaling activity of the ETR1 ethylene receptor in the presence (**A**, **B**) or absence (**C**) of the *RTE1* gene. In the *rte1* loss-of-function mutant, the ETR1 receptor is nonfunctional regardless of whether ethylene is bound. RTE1 is predicted to be membrane-bound, but whether it localizes with ETR1 to the endoplasmic reticulum (ER) is unknown. Signaling "ON" means that the receptor is acting to repress ethylene responses and "OFF" indicates that ETR1 is inactive. The resulting ethylene-response phenotype (i.e., no response, wild-type response or enhanced response) is also dependent on the signaling states of the four other ethylene receptors, which are not shown.

The biochemical function of RTE1 in any organism has yet to be determined. Although the RTE1 family is defined by a conserved domain of unknown function (called DUF778), the family members contain no sequence motifs of known function. Interestingly, there are two highly conserved regions containing conserved Cys and His residues, and the *Arabidopsis rte1-1* loss-of-function allele replaces one of these residues (Cys^{161}) with a tyrosine, indicating that at least one of these conserved residues plays an important role in RTE1 function (Resnick *et al.*, 2006). Metal binding proteins often contain Cys and His residues, and since the ethylene receptor requires a copper cofactor in order to bind ethylene (Rodriguez *et al.*, 1999), one speculation has been that RTE1 is involved in delivering copper to the receptor (Barry and Giovannoni, 2006). Alternatively, RTE1 might play a direct role in the localization of ETR1, protein folding and/or ETR1 signaling. RTE1 does not appear to regulate ETR1 protein levels, since Western blot analysis shows normal levels of ETR1 in *rte1* loss-of-function mutants (Rivarola and Chang, unpublished).

Whether *RTH* plays a similar role in ethylene signaling is unclear. Analysis of an *rte1 rth* double null mutant in *Arabidopsis* may yield a better understanding of the molecular function of *RTE1* in plants and animals. Since animals do not possess ethylene receptors, *RTE1* may have a more general function, perhaps encoding a protein scaffold, cofactor, chaperone, or membrane targeting protein. The fact that *RTE1* exists in such a wide variety of organisms suggests that it performs an essential conserved function.

References

Alonso, J.M., Hirayama, T., Roman, G., Nourizadeh, S., and Ecker, J.R. (1999) *EIN2*, a bifunctional transducer of ethylene and stress responses in *Arabidopsis*. Science 284, 2148–2152

Barry, C.S., and Giovannoni, J.J. (2006) Ripening in the tomato *Green-ripe* mutant is inhibited by ectopic expression of a protein that disrupts ethylene signaling. Proceedings of the National Academy of Sciences of the United States of America 103, 7923–7928

Bleecker, A.B., Estelle, M.A., Somerville, C., and Kende, H. (1988) Insensitivity to ethylene conferred by a dominant mutation in *Arabidopsis thaliana*. Science 241, 1086–1089

Cancel, J.D., and Larsen, P.B. (2002) Loss-of-function mutations in the ethylene receptor ETR1 cause enhanced sensitivity and exaggerated response to ethylene in *Arabidopsis*. Plant Physiology 129, 1557–1567

Gagne, J.M., Smalle, J., Gingerich, D.J., Walker, J.M., Yoo, S.D., Yanagisawa, S., and Vierstra, R.D. (2004) *Arabidopsis* EIN3-binding F-box 1 and 2 form ubiquitin-protein ligases that repress ethylene action and promote growth by directing EIN3 degradation. Proceedings of the National Academy of Sciences of the United States of America 101, 6803–6808

Guo, H.W., and Ecker, J.R. (2003) Plant responses to ethylene gas are mediated by SCF (EBF1/EBF2)-dependent proteolysis of EIN3 transcription factor. Cell 115, 667–677

Guo, H.W., and Ecker, J.R. (2004) The ethylene signaling pathway: new insights. Current Opinion in Plant Biology 7, 40–49

Guzman, P., and Ecker, J.R. (1990) Exploiting the triple response of *Arabidopsis* to identify ethylene-related mutants. Plant Cell 2, 513–523

Hall, A.E., and Bleecker, A.B. (2003) Analysis of combinatorial loss-of-function mutants in the *Arabidopsis* ethylene receptors reveals that the *ers1 etr1* double mutant has severe developmental defects that are *EIN2* dependent. Plant Cell 15, 2032–2041

Hua, J., and Meyerowitz, E.M. (1998) Ethylene responses are negatively regulated by a receptor gene family in *Arabidopsis thaliana*. Cell 94, 261–271

Kieber, J.J., Rothenberg, M., Roman, G., Feldmann, K.A., and Ecker, J.R. (1993) CTR1, a negative regulator of the ethylene response pathway in *Arabidopsis*, encodes a member of the Raf family of protein kinases. Cell 72, 427–441

Potuschak, T., Lechner, E., Parmentier, Y., Yanagisawa, S., Grava, S., Koncz, C., and Genschik, P. (2003) EIN3-dependent regulation of plant ethylene hormone signaling by two *Arabidopsis* F box proteins: EBF1 and EBF2. Cell 115, 679–689

Resnick, J.S., Wen, C.-K., Shockey, J.A., and Chang, C. (2006) *REVERSION-TO-ETHYLENE SENSITIVITY1*, a conserved gene that regulates ethylene receptor function in *Arabidopsis*. Proceedings of the National Academy of Sciences of the United States of America 103, 7917–7922

Rodriguez, F.I., Esch, J.J., Hall, A.E., Binder, B.M., Schaller, G.E., and Bleecker, A.B. (1999) A copper cofactor for the ethylene receptor ETR1 from *Arabidopsis*. Science 283, 996–998

Solano, R., Stepanova, A., Chao, Q.M., and Ecker, J.R. (1998) Nuclear events in ethylene signaling: a transcriptional cascade mediated by *ETHYLENE-INSENSITIVE3* and *ETHYLENE-RESPONSE-FACTOR1*. Genes & Development 12, 3703–3714

Zhao, X.C., Qu, X., Mathews, D.E., and Schaller, G.E. (2002) Effect of ethylene pathway mutations upon expression of the ethylene receptor ETR1 from *Arabidopsis*. Plant Physiology 130, 1983–1991

New inhibitors of ethylene perception in improvement of display quality in miniature roses (*Rosa hybrida* L.)

Mibus, H.[1]*, Buanong, M.[1], Sisler, E.C.[2] and Serek, M.[1,3]

[1]Institute of Floriculture, Tree Nursery Science and Plant Breeding, Faculty of Natural Sciences, University of Hannover, Herrenhäuser Str. 2, 30419 Hannover, Germany; [2]Department of Molecular and Structural Biochemistry, North Carolina State University, Raleigh, NC 27695, USA; [3]Department of Agriculture Sciences, Laboratory for Plant and Soil Science, The Royal Veterinary and Agriculture University, Thorvaldsensvej 40, 1871 Frederiksberg C., Denmark; (*Corresponding author: mibus@zier.uni-hannover.de)

1. Introduction

The miniature potted roses are popular greenhouse crops in many parts of the world. In the 1980s the introduction of new varieties had a dramatic impact on the European and North American markets. Current world production is estimated at around 100 million pots yearly. Major centers of production include Denmark, the Netherlands, the United States, Canada and Japan with production also in France, Germany and Italy (Pemberton *et al.*, 2003). Variation in postharvest life of miniature roses is partly the result of differences in endogenous ethylene production during flower senescence, stress induced ethylene production and sensitivity to exogenous ethylene (Serek and Andensen, 1993). The sensitivity to ethylene has important implications during the transport and handling of potted roses in areas where the air is commonly contaminated with ethylene. Since ethylene inhibitors prohibit the plant tissues from responding to endogenous and exogenous ethylene, they are considered very potent for agricultural use (Sisler and Serek, 2003; Sisler and Serek, 2001; Feng *et al.*, 2004). 1-MCP and some new putative inhibitors of ethylene action, which are structural analogs of 1-MCP containing a longer side chain at the 1-position, have been found to be effective in protecting flowers against ethylene (Kenebei *et al.*, 2003). We have studied the effect of pretreatments of miniature roses with analogues of 1-MCP: 1-OCP substituted with an 8-carbon chain in the 1-position; 1-DCP with 10-carbon

A. Ramina et al. (eds.), Advances in Plant Ethylene Research: Proceedings of the 7th International Symposium on the Plant Hormone Ethylene, 15–20.

chain, compared to 1-MCP. These compounds were envisioned as a potential commercial approach for improving display quality of miniatur e potted roses. Plants of Rosa 'Lavender' of the Kordana breeding line from Rosen Kordes Sparrishoop, Germany, which is sensitive to ethylene were grown and pretreated after harvest with different levels of 1-OCP, 1-DCP, 1-MCP or air and continuously exposed to exogenous ethylene throughout the experimental period to find the best conditions for 'Lavender' as compared with 'Vanilla', which is less sensitive to ethylene.

2. Materials and Methods

2.1 Chemicals

At the Department of Molecular and Structural Biochemistry, North Carolina State University, Raleigh, USA, 1-OCP and 1-DCP were synthesized from 2-bromoalkenes and bromoform using 50% NaOH to produce a carbine and form a 1,1,2-Cyclopropane, which was then reacted with methyllithium at dry ice, to form cyclopropenes (Al Dulayymi *et al.*, 1997). The compounds were subsequently divided into smaller samples and kept at • 80°C until needed for the experiments. 1-MCP was obtained from AgroFresh Inc. (Rohm and Haas, AgroFresh Inc., Philadelphia, USA).

2.2 Plant material

Rosa hybrida L. cultivar 'Lavender', which is highly sensitive to ethylene, and cultivar 'Vanilla' with excellent postharvest performance and low sensitivity to ethylene. Both cultivars are from the Kordana breeding line of Rosen Kordes, W. Kordes' Söhne Rosenschulen GmbH & Co KG, Sparrishoop, Germany.

2.3 Optimization of concentration, treatment time and temperature of 1-OCP and 1-DCP

Plants were exposed to the appropriate concentrations of 1-OCP or 1-DCP (200, 500, 1000 and 1500 nl l^{-1}), respectively, or 200 nl l^{-1} of 1-MCP, in a 50 l glass chambers for 6 h at 20°C. Then the optimum concentrations (1000 nl l^{-1}) of 1-OCP or 1-DCP were treated at different exposure times (2, 4, 6 or 12 h) and the temperatures inside growth chambers were set 5, 10, 15 or 20°C for an exposure time of 4 h.

1-OCP and 1-DCP was dissolved in ether and then pipetted on filter paper in the glass chambers. 1-MCP was released from a powdered formulation. After the treatments, the plants were transferred and placed randomly in a new glass chamber, sealed and kept at 20°C, 60–70% RH and 12 h light (20 μmol m^{-2} s^{-1}). The ethylene treatment was provided by diffusion system (Saltveit, 1978) in a concentration of 1.25 (±0.25) μl l^{-1}. Ethylene level was controlled with a Perkin-Elmer gas chromatograph (GC Voyager, Ontario, Canada). The carrier gas was N_2 at 40 ml min^{-1} and the column temperature was 60C.

The display quality of miniature roses was evaluated every 3rd day by recording the percentage of leaf drop. 'Lavender' and 'Vanilla' plants treated with 1000 nl l^{-1} 1-OCP or 1-DCP at 20°C for 6 h or with 1-MCP (200 nl l^{-1}) for 6 h. After treatments, the plants were exposed to the same procedure as previously described.

2.3 Experimental design and statistic

Experiments were conducted in a completely randomized design using three pots (containing two cuttings) per treatment and two replications. Data obtained was subjected to analysis of variance (ANOVA) using the general linear models (Proc GLM) procedure of the Statistical Analysis System program package (SAS Institute, 2002). Multiple comparisons among means was done using the Duncan Test at $p = 0.05$.

3. Results and Discussion

3.1 Effect of concentration, treatment time and temperature of 1-OCP and 1-DCP on display quality

All tested levels of 1-OCP and 1-DCP protected the miniature potted rose cultivar 'Lavender' as compared to untreated plants. 1-OCP and 1-DCP were the most effective at concentrations 1000 and 1500 nl l^{-1} (Table 1), which was five times higher than the concentration of 1-methylcyclo-propene (1-MCP) (200 nl l^{-1}) used as a standard (Table 1). This is in contrast to the previous studies (Sisler and Serek, 2003; Kenebei *et al.*, 2003), where 1-OCP was effective at lower concentrations than 1-MCP in improving display life of kalanchoë.

The effectiveness of 1-OCP and 1-DCP was a function of time and temperature. At short (2 h) exposure times of 1-OCP and 1-DCP, the plants were highly sensitive to ethylene. Exposure time of 4 h for both 1-OCP and 1-DCP was sufficient to improve display life of miniature roses and longer exposures did not have any additional beneficial effect (Table 1). Exposure to each chemical for 12 h gave the same effect as 4 h exposure (Table 1). Similar observations were reported on other cyclo-propenes (Feng *et al.*, 2004).

Table 1. Mean of percent leaf drop of 'Lavender' pretreated with 1-OCP and 1-DCP at concentrations of 0, 200, 500, 1000, 2000 nl l^{-1}; exposure time of 1000 nl l^{-1} 1-OCP or 1-DCP 0, 2, 4, 6, 12 h or exposed to 1-OCP or 1-DCP (1000 nl l^{-1} for 4 h) at temperatures of 5, 10, 15, 20°C. 1-MCP treatment (200 nl l^{-1} for 6 h at 20°C). Leaf drop was evaluated after 3, 6, 9, 12 and 15 days). All plants were exposed to 1.25 µl l^{-1} continuous ethylene through out the experiments. Numbers marked with the same letter are not statistically different at $P < 0.05$.

Treatment		Day 3	Day 6	Day 9	Day 12	Day 15
Control		9,4^{a}	80,8^{a}	97,4^{a}	100^{a}	100^{a}
1-MCP		2,4^{b}	9,1^{f}	39,8^{cd}	55,4^{d}	60,4^{c}
1-OCP [nl^{-1}]	**200**	10^{a}	50,5^{b}	83,1^{ab}	100^{a}	100^{a}
	500	7,5^{a}	30,2^{c}	54,7^{c}	66,3^{c}	68,1^{c}
	1000	4,8^{ab}	16,3^{de}	38^{d}	55^{d}	58,1^{c}
	1500	2,9^{b}	22,4^{d}	41^{cd}	53,2^{d}	56,1^{c}
1-DCP [nl^{-1}]	**200**	4,1^{ab}	43,9^{b}	87^{a}	100^{a}	100^{a}
	500	1,9^{b}	23,7^{cd}	69,1^{b}	80,8^{b}	80,8^{b}
	1000	1,2^{bc}	18,8^{d}	68,9^{b}	80,1^{b}	81^{b}
	1500	1,5^{bc}	14,5^{e}	60,9^{bc}	74,3^{c}	77^{bc}
1-OCP [°C]	**5**	1,4^{bc}	13,2^{e}	63,4^{b}	80^{b}	83,9^{b}
	10	0,8^{bc}	19,6^{d}	67,8^{b}	81,3^{b}	84,8^{b}
	15	0,8^{bc}	46,5^{b}	54,3^{c}	74,5^{c}	80,9^{b}
	20	0,5^{c}	16^{de}	63,2^{b}	78,1^{b}	82,8^{b}
1-DCP [°C]	**5**	3,1^{b}	34,3^{bc}	65,5^{b}	75,7^{bc}	78,5^{b}
	10	1,7^{bc}	49,7^{b}	70,8^{b}	79,2^{b}	79,2^{b}
	15	3,8^{ab}	37,6^{bc}	64,3^{b}	76,3^{bc}	76,3^{bc}
	20	0,0^{c}	23,7^{cd}	57,5^{bc}	74,8^{bc}	74,8^{bc}
1-OCP [h]	**2**	1,7^{bc}	47,8^{b}	84,6^{ab}	100^{a}	100^{a}
	4	1,3^{bc}	15,3^{e}	49,7^{c}	63,7^{c}	68,2^{c}
	6	1,5^{bc}	17,9^{de}	60,2^{bc}	70,9^{c}	73,9^{bc}
	12	0,9^{c}	7^{f}	40,3^{cd}	58,3^{cd}	68,5^{c}
1-DCP [h]	**2**	3,4^{b}	58,8^{b}	85,8^{a}	100^{a}	100^{a}
	4	2,5^{b}	19,1^{d}	61,4^{b}	72,9^{b}	77,2^{bc}
	6	1,4^{bc}	22^{d}	66,3^{b}	79,6^{b}	81,1^{b}
	12	1,1^{bc}	7^{f}	42,7^{c}	66,3^{c}	79^{b}

Apparently, exposing miniature potted roses to various temperatures during treatment did not have an influence on the performance of 1-OCP and 1-DCP, while treatment at 5°C reduced the performance of both compounds; they were equally effective at temperatures of 10–20°C (Table 1). Contrary to kalanchoë (Kenebei *et al.*, 2003) and banana (Sisler *et al.*, 2001), the activity of 1-OCP was influenced by temperature, and it was more potent at 15 and 20°C than at lower (5 and 10°C) temperatures (Table 1).

3.2 Comparison of effectiveness of 1-MCP, 1-OCP and 1-DCP between 'Lavender' and 'Vanilla' cultivars

'Vanilla' control plants lost 40% of the leaves after 6 days of continuous ethylene treatment, while 'Lavender' attained 80% leaf drop during the same period (Table 2). Differences in ethylene synthesis or perception are the major factors in the process of leaf drop (Müller and Stummann, 2001; Müller and Stummann, 2003). Pretreatment of the less ethylene sensitive cultivar 'Vanilla' results after 15 days showed that 1-DCP was a more potent inhibitor of ethylene action than 1-OCP (Table 2). The percent leaf drop was similar in 'Lavender' treated with 1-OCP or 1-DCP.

Table 2. Mean of percent leaf drop cultivar 'Vanilla' and 'Lavender' pre-treated with 1-OCP, 1-DCP (1000 nl l^{-1}), respectively, or 1-MCP (200 nl l^{-1}) for 4 h at 20°C. Plants were subsequently exposed to 1.5 μl l^{-1} ethylene. Bars marked with the same letter are not statistically different ($P < 0.05$).

Cultivar	Treatment	Day 3	Day 6	Day 9	Day 12	Day 15
'Lavender'	Control	$4,8^{a}$	$76,4^{a}$	100^{a}	100^{a}	100^{a}
	1-OCP	$2,5^{a}$	$17,4^{bd}$	$45,3^{c}$	$67,3^{c}$	$75,3^{b}$
	1-DCP	4^{a}	$35,6^{c}$	$57,7^{b}$	$78,3^{b}$	$78,3^{b}$
	1-MCP	$1,1^{a}$	$3,6^{d}$	$12,8^{d}$	33,8d	$53,1^{c}$
'Vanilla'	Control	$3,4^{a}$	$38,3^{a}$	89^{a}	100^{a}	100^{a}
	1-OCP	$5,2^{a}$	$24,3^{b}$	$74,3^{b}$	$80,2^{b}$	100^{a}
	1-DCP	5^{a}	$27,9^{ab}$	$71,1^{b}$	$77,2^{b}$	$77,2^{b}$
	1-MCP	$5,1^{a}$	$15,8^{b}$	$52,6^{c}$	$76,8^{b}$	$76,8^{b}$

Pretreatment with 1-MCP was the most effective in improving the display quality of both investigated miniature roses cultivars (Table 2). It could be assumed, that ethylene exposure of 'Lavender' induces an autocatalytic rise in ethylene with constant increase in endogenous ethylene levels, similar to previously investigated ethylene sensitive cultivars (Müller and Stummann, 2001; Müller and Stummann, 2003). Therefore, for ethylene-sensitive cultivars like 'Lavender', treatment with ethylene inhibitors will be an important practical tool for increasing postharvest life. In conclusion, our results demonstrated the potency of

1-MCP and long chain CPs, 1-OCP and 1-DCP, as ethylene antagonists in potted roses. However, 1-MCP was the most potent ethylene inhibitor in improving display quality among investigated materials.

References

Al Dulayymi AR, Al Dulayymi JR, Baird MS, Koza G (1997) Simple four and five carbon cyclopropane and cyclopropenes synthetic intermediates. Russ J Org Chem 33: 798–816

Feng X, Apelbaum A, Sisler EC, Goren R (2004) Control of ethylene activity in various plant systems by structural analogues of 1-methylcyclopropene. Plant Growth Regul 42: 29–38

Kenebei Z, Sisler EC, Winkelmann T, Serek M (2003) Efficacy of new inhibitors of ethylene perception in improvement of display life of kalanchoë (*Kalanchoë blossfeldiana* Poelln.) flowers. Postharvest Biol Technol 30(2): 169–176

Müller R, Stummann BM, Andersen AS (2001) Comparison of postharvest properties of closely related miniature rose cultivars (*Rosa hybrida* L.). Sci Hortic 91: 325–338

Müller R, Stummann BM (2003) Postharvest physiology; ethylene In: Encyclopedia of Rose Science. A.V. Reberts and T.D. Serge Gudin (eds.). Elsevier Academic Press. SP. ISBN 0-12-227620-5: 557–564

Pemberton HB, Kelly JW, Ferare J (2003) Pot rose production, Introduction and world market In: Encyclopedia of Rose Science. AV Reberts TD Serge Gudin (eds.). Elsevier Academic Press. SP. ISBN 0-12-227620-5: 587–593

Saltveit ME Jr (1978) Simple apparatus for diluting and dispensing trace concentrations of ethylene in air. HortScience 13: 249–251

SAS Institute. (2002) SAS/STAT user's guide for Personal Computers, version 8. SAS Institute, Cary, NC, USA

Serek M, Andensen S (1993) AOA and BA influence on floral development and longevity of potted 'Victory Parade' miniature rose. HortScience 28(10): 1039–1040

Sisler EC, Serek M, Roh KA, Goren R (2001) The effect of chemical structure on the antagonism by cyclopropenes of ethylene responses in banana. Plant Growth Regul 33: 107–110

Sisler EC, Serek M (2001) New development in ethylene control – compounds interacting with the ethylene receptor. Acta Hortic 543: 33–40

Sisler EC, Serek M (2003) Compounds interacting with the ethylene receptor in plants. Plant biol 5: 473–480.

The melon ethylene receptor CmERS1 is localized to the endoplasmic reticulum with an N_{lum}–C_{cyt} membrane orientation

Ma, B.[1], Cui, M.L.[1], Sun, H.J.[1], Takada, K.[1], Mori, H.[2], Kamada, H.[1] and Ezura, H.[1]*

[1]Gene Research Center, University of Tsukuba, Tennodai 1-1-1, Tsukuba, Ibaraki 305-8572, Japan. Tel/Fax: 81-29-853-7263. [2]Graduate School of Bioagricultural Sciences, Nagoya University, Furo-cho Chikusa-ku, Nagoya 464-8601, Japan. Tel/Fax: 81-52-789-4167. (*Corresponding author: ezura@gene.tsukuba.ac.jp)

1. Introduction

Ethylene perception and signal transduction involves a multistep pathway, in which ethylene receptors act at the first step and play a crucial role by negatively regulating ethylene responses (Chang and Stadler, 2001). In this respect, ethylene receptors have received considerable attention, as knowledge regarding this system would provide important contributions to our understanding of the mechanisms of ethylene signaling, and also provide the means to develop effective approaches for engineering ethylene sensitivity in plants. To better understand the biochemical mechanism of ethylene receptor action, we examined the subcellular localization and membrane topology that are important determinants of the biochemical functions of multispanning membrane proteins.

Ethylene receptors have long been believed to be located at the plasma membrane (PM). However, a recent study indicated that AtETR1 is localized at the endoplasmic reticulum (ER) (Chen *et al.*, 2002). Additionally, the downstream protein kinase CTR1 was also found to be at the ER through interaction with ethylene receptors (Gao *et al.*, 2003). These findings imply a central role of the ER in ethylene perception and early signal transduction (Chen *et al.*, 2002; Gao *et al.*, 2003). However, whether the other receptor isoforms are localized at this organelle remains unknown. Using computer-based prediction, AtETR1 is thought to be located in the membrane with its N-terminus facing the extracytosolic side and its C-terminal domain facing the cytosolic side (Schaller *et al.*, 1995). Although

A. Ramina et al. (eds.), Advances in Plant Ethylene Research: Proceedings of the 7th International Symposium on the Plant Hormone Ethylene, 21–26.

the finding that the dimerization of AtETR1 is mediated by disulfide linkages at Cys4 and Cys6 residues is indicative of an extracytosolic orientation of the N-terminus (Schaller *et al.*, 1995), the overall membrane topology has yet to be clearly demonstrated.

In this study, we investigated the subcellular localization and membrane topology of the ethylene receptor CmERS1, which was previously identified from melon (Sato-Nara *et al.*, 1999; Takahashi *et al.*, 2002). Our data show that CmERS1 is an integral ER membrane protein that spans the lipid bilayer three times, with its N-terminus facing the luminal space, and the large C-terminal portion on the cytosolic side. High conservation of the topogenic sequences in all ethylene receptor homologues identified thus far suggests that these proteins may share the same membrane topology.

2. Materials and Methods

2.1 Membrane fractionation and proteolysis

Two-phase portioning was performed using a 6.4% (w/w) Dextran T500/PEG3350 mixture (Larsson *et al.*, 1987). Sucrose density gradient centrifugation was performed essentially as described by Ferrol and Bennett (1996). For proteolysis, the microsomes were incubated with 100 $\mu g\ mL^{-1}$ of proteinase K for 0–60 min at 30°C in the absence or presence of 0.1% (v/v) Triton X-100.

2.2 Microprojectile bombardment and confocal microscopy

Microprojectile bombardment was carried out using a BioRad PDS-1000/He Particle Delivery System. After 20–24 h of bombardment, GFP fluorescence was detected using a confocal microscope. All images were collected through a water-immersion objective (x63, 1.2 numerical aperture). Optical sections were 0.10 µm thick. Final image assembly was performed using Adobe Photoshop 6.0 software.

2.3 In vitro transcription/translation

The plasmids were transcribed and translated in a TNT T7 Quick Coupled Transcription/Translation System (Promega, USA) in the presence of [^{35}S] methionine according to the manufacturer's instructions. Proteinase K, Endo H, and alkali treatments were performed as described previously (Devoto *et al.*, 1999; Vilar *et al.*, 2002; McCartney *et al.*, 2004). Treated translation products were separated on SDS-PAGE gels that were subsequently fixed, dried, and analyzed using autoradiography.

3. Results and Discussion

3.1 CmERS1 is predominantly associated with the ER membrane of melon cells

Two-phase partitioning, sucrose density gradient centrifugation, and GFP imaging were used to investigate the subcellular localization of CmERS1. The results demonstrate that CmERS1 is predominantly localized to the ER (Fig. 1). Our data provide experimental evidence for the ER localization of an ERS-type receptor. These observations, together with the ER localization of AtETR1, support the deduction that the ER may play a central role in ethylene perception and early signal transduction (Chen *et al.*, 2002; Gao *et al.*, 2003).

3.2 CmERS1 spans the membrane three times with an N_{lum}–C_{cyt} orientation

A protease protection assay and *N*-glycosylation analysis were used to determine the membrane topology. The results indicate that CmERS1 integrates into the membrane with its N-terminus facing the luminal space and the large C-terminal portion lying on the cytosolic side of the ER membrane (Fig. 2; see Ma *et al.*, 2006 for detail). These observations provide the first direct experimental evidence for the membrane topology of ethylene receptor proposed by computer-based prediction (Schaller *et al.*, 1995). The N_{lum}–C_{cyt} orientation of CmERS1 provides a platform for its interaction with cytosolic signaling elements such as CTR1. Moreover, the three N-terminal transmembrane (TM) segments were found to function as topogenic sequences to determine the final topology. Our data suggest the following membrane topogenesis of CmERS1: TM1 functions as an SA-I sequence that is responsible for ER targeting and translocation of the N-terminus, TM2 integrates into the membrane as a consequence of the internal SA-I function of the following TM3, which leaves the C-terminal portion on the cytosolic side (see Ma *et al.*, 2006 for detail). High conservation of these topogenic sequences in all ethylene receptor homologues identified thus far suggests that these proteins may share the same membrane topology.

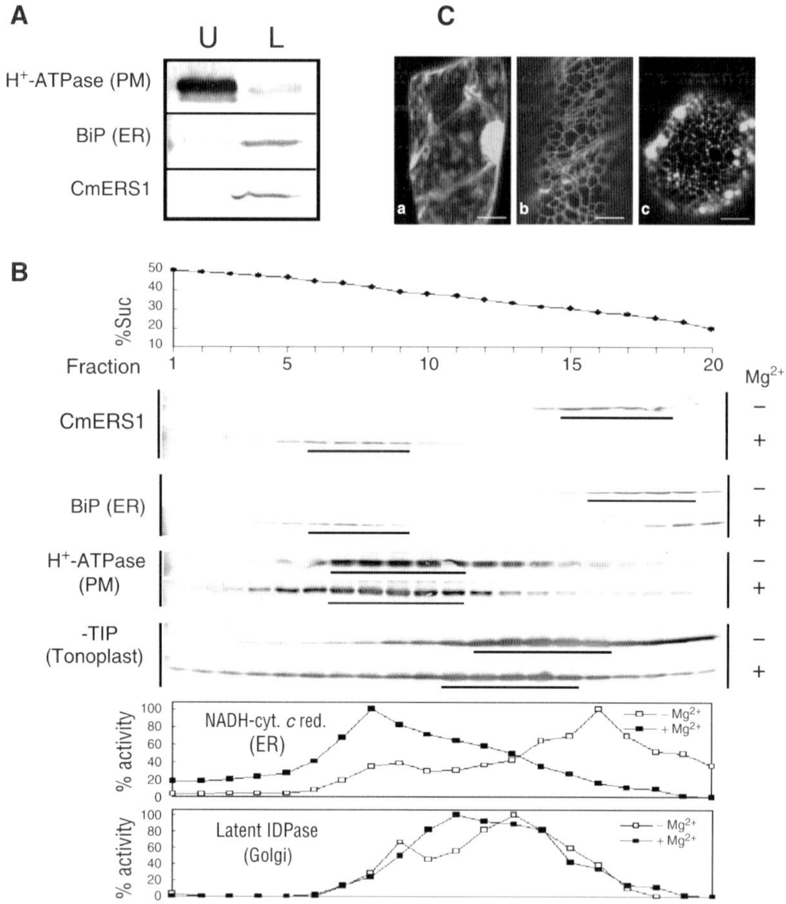

Fig. 1. (Color figure in the Annex, p.454) Subcellular localization of CmERS1. **(A)** Two-phase separation of melon leaf microsomes. Equal amounts of proteins from the upper (U) and lower (L) phases were separated using SDS-PAGE and subjected to immunoblot analysis. **(B)** Suc gradient fractionation. Microsomes from melon seedlings were separated on a linear Suc density gradient (20–50%; w/w) in the absence (–) or presence (+) of Mg^{2+}. Gradient fractions were subjected to SDS-PAGE and analyzed by immunoblotting. **(C)** GFP images. The constructs were transiently expressed in melon leaf cells by microprojectile bombardment. The images were taken from trichomes or leaf vein epidermal cells. (a) GFP, (b) GFP-CmERS1, (c) GFP-HDEL (ER marker). Bar=10 μm.

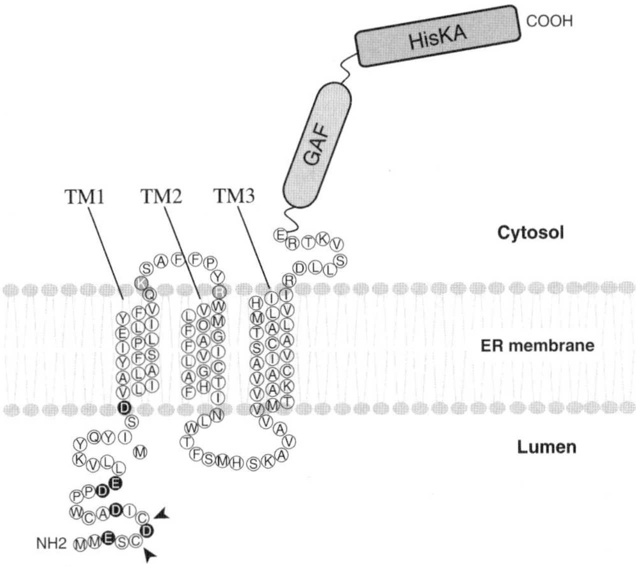

Fig. 2. A proposed model for CmERS1 topology on the ER membrane. Based on membrane fractionation, GFP imaging, protease protection assay, and glycosylation analysis, we propose that CmERS1 is predominantly localized to the ER and spans the membrane three times with its N-terminus facing the luminal space and C-terminus lying on the cytosolic side. The positively and negatively charged residues in TM1 flanking regions are shaded gray and black, respectively. Arrows indicate the residues that are possibly involved in disulfide-linked dimerization of the receptor.

References

Chang C, Stadler R (2001) Ethylene receptor action in Arabidopsis. BioEssays 23: 619–627

Chen YF, Randlett MD, Findell JL, Schaller GE (2002) Localization of the ethylene receptor ETR1 to the endoplasmic reticulum of Arabidopsis. J Biol Chem 277: 19861–19866

Devoto A, Piffanelli P, Nilsson I, Wallin E, Panstruga R, von Heijne G, Schulze-Lefert P (1999) Topology, subcellular localization, and sequence diversity of the Mlo family in plants. J Biol Chem 274: 34993–35004

Ferrol N, Bennett AB (1996) A single gene may encode differentially localized Ca^{2+}-ATPases in tomato. Plant Cell 8: 1159–1169

Gao Z, Chen YF, Randlett MD, Zhao XC, Findell JL, Kieber JJ, Schaller GE (2003) Localization of the Raf-like kinase CTR1 to the endoplasmic reticulum of Arabidopsis through participation in ethylene receptor signaling complex. J Biol Chem 278: 34725–34732

Larsson C, Widell S, Kjellbom P (1987) Preparation of high-purity plasma membranes. Methods Enzymol 148: 558–568

Ma B, Cui ML, Sun HJ, Takada K, Mori H, Kamada H, Ezura H (2006) Subcellular localization and membrane topology of the melon ethylene receptor CmERS1. Plant Physiol 141: 587–597

McCartney AW, Dyer JM, Dhanoa PK, Kim PK, Andrews DW, McNew JA, Mullen RT (2004) Membrane-bound fatty acid desaturases are inserted co-translationally into the ER and contain different ER retrieval motifs at their carboxy termini. Plant J 37: 156–173

Sato-Nara K, Yuhashi K, Higashi K, Hosoya K, Kubota M, Ezura H (1999) Stage- and tissue-specific expression of ethylene receptor homolog genes during fruit development in muskmelon. Plant Physiol 120: 321–330

Schaller GE, Ladd AN, Lanahan MB, Spanbauer JM, Bleecker AB (1995) The ethylene receptor mediator ETR1 from Arabidopsis forms a disulfide- linked dimer. J Biol Chem 270: 12526–12530

Takahashi H, Kobayashi T, Sato-Nara K, Tomita K, Ezura H (2002) Detection of ethylene receptor protein Cm-ERS1 during fruit development in melon (Cucumis melo L.) J Exp Bot 53: 415–422

Vilar M, Sauri A, Monne M, Marcos JF, von Heijne G, Perez-Paya E, Mingarro I (2002) Insertion and topology of a plant viral movement protein in the endoplasmic reticulum membrane. J Biol Chem 277: 23447–23452

Cloning and differential expression of banana genes coding for EIN3-like proteins involved in ethylene action

Mbéguié-A-Mbéguié, D.[1]*, Hubert, O.[2], Chillet, M.[2] and Fils-Lycaon, B.[3]

[1]CIRAD-FLHOR, UMR 1270 QUALITROP, Station de Neufchâteau, Sainte-Marie, 97130 Capesterre-Belle-Eau, Guadeloupe, French West Indies,[2]CIRAD-FLHOR, UR 24 TROPICAL, Station de Neufchâteau, Sainte-Marie, 97130 Capesterre-Belle-Eau, Guadeloupe, French West Indies,[3] INRA-CEPIA , UMR 1270 QUALITROP, Domaine de Duclos, Prise d'Eau, 97170 Petit-Bourg, Guadeloupe, French West Indies (*Corresponding author: mbeguie@cirad.fr)

1. Introduction

Export bananas (*Musa acuminata*, cv Cavendish) of the French West Indies are usually subjected to the "mixed-ripe" and/or "ship-ripe" processes during transit with a rapid deterioration of their quality. We hypothesized that this post harvest process is associated with the level of fruit ethylene responsiveness at harvest time. We attempted to obtain insight at the molecular level by identifying related candidate genes, which could be used as markers to improve the post harvest quality of fruit.

Here we report the isolation of four banana cDNAs (MaEIL1-4) encoding EIN3-LIKE (EIL) proteins, which are primary transacting factors that positively regulate multiple ethylene responses (Tieman *et al.*, 2001; Adams-Phillips *et al.*, 2004). Gene expression analysis of MAEIL3 and 4 shows that these genes display different expression patterns during fruit development, ripening, leaf development and after wounding.

2. Materials and Methods

Stages of harvested fruits: immature green (iMG), early mature green (eMG), late mature green (lMG), turning (Tr), Yellow (Yw) and Yellow Tiger (Yt). Tissues: flower (Fw), bract (Bc), root (Rt), intact roll leaf (iRL), wound roll leaf (WRL producing 15 μl kg^{-1} h^{-1} of C_2H_4), intact unroll leaf (iURL) and wound unroll leaf (WURL producing 3 μl $kg^{-1}h^{-1}$of C_2H_4) (Fig. 1).

A. Ramina et al. (eds.), Advances in Plant Ethylene Research: Proceedings of the 7th International Symposium on the Plant Hormone Ethylene, 27–30.

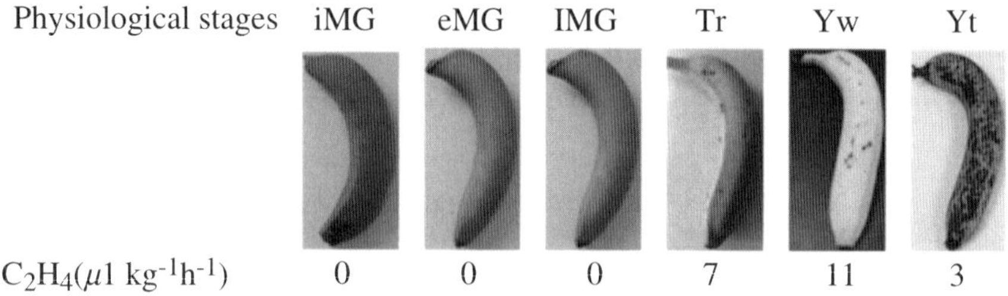

Fig. 1. (Color figure in the Annex, p.455) Fruits used in these studies.

MaEIL cDNAs were isolated from fruit peel and pulp tissues using 5'- and 3'-RACE-PCR, and degenerate primers designed from conserved regions of EIL-polypeptide sequences.

For gene-expression analysis, real-time PCR was performed using SYBR Green PCR master mix (Applied Biosystem, France), total cDNA, and specific primers designed within the 3'UTR of each MaEIL cDNA and actin (as an internal standard).

3. Results and Discussion

3.1 MaEIL cDNA sequence analysis

The partial MaEIL cDNAs, namely MaEIL1 (1817 bp), 2 (1840 bp) and 3 (1920 bp) encode proteins of 495, 517 and 517 amino acids, respectively. These clones are 51–70% and 35–70% identical at the nucleotide and polypeptide levels, respectively. The full-length cDNA (MaEIL4) is 2435 bp and encodes a protein of 635 amino acids. It is 40–70% and 32–75% identical to the three others at the nucleotide and protein levels, respectively. Phylogeny analysis shows that MaEIL1, 3 and 4 belong to the same cluster; however, they constitute a quite different subgroup. MaEIL2 appears to be the most distantly related to this cluster.

3.2 MaEIL3 and 4 expression in banana fruit and other tissues, and effect of wounding ethylene in banana leaf

MaEIL3 mRNA levels decreased in both, peel and pulp tissues from the iMG stage to late ripening stages (Figs. 2A and B). This decrease is more important in pulp than in peel tissue. MAEIL4 mRNA level increased in both, peel and pulp tissues during the early developmental stages (iMG, eMG and lMG) and decreased during the late ripening stage (Tr, Yw and Yt). As for MaEIL3, this decrease is more important in pulp that in peel tissue. In pulp tissue, both MaEIL3 and MaEIL4 mRNA levels decreased drastically at the turning stage (Tr), the beginning of ripening.

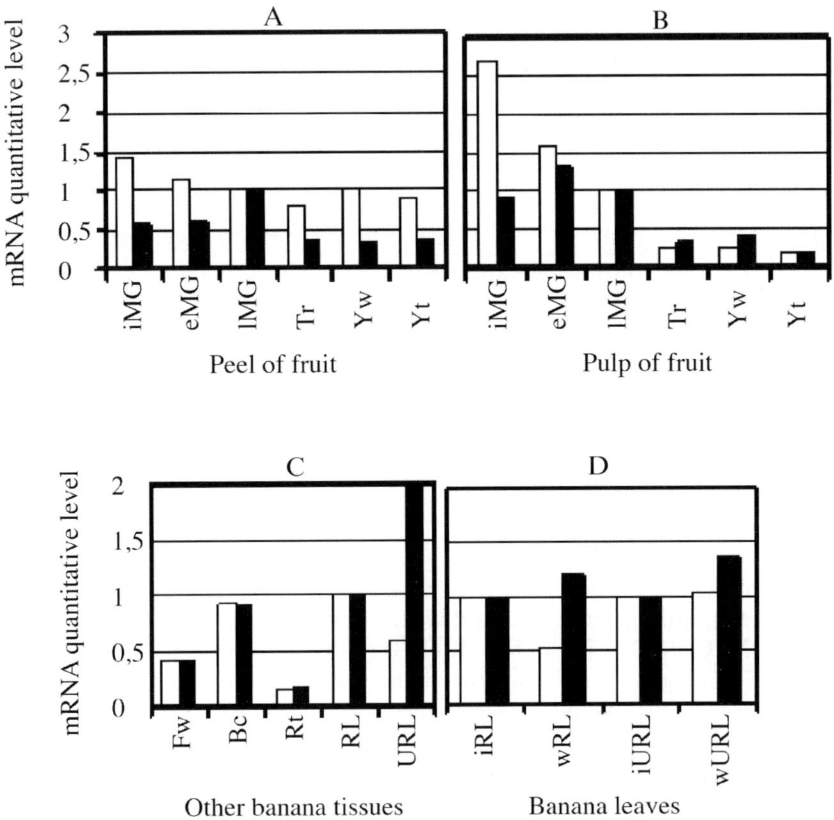

Fig. 2. Expression profiles of banana *EIL3* (□) and *EIL4* (■) genes in different banana tissues and upon ripening and wounding. *MaEIL3* and *MaEIL4* gene expression was examined during fruit ripening (iMG = immature green, eMG = early mature green, lMG = late mature green, Tr = turning, Yw = yellow, Yt = yellow tiger), in other banana tissues (Fw = flower, Bc = bract, Rt = root, RL = roll leaves, URL = unroll leaves) and in leaves after wounding (iRL = intact roll leaves, wRL = wounded roll leaves, iURL = intact unroll leaves, wURL = wounded unroll leaves).

Both MaEIL3 and MaMaEIL4 mRNA are detected at different levels in other banana tissues including flowers, bracts, roots, and young and old leaves. From young to old leaves, the MaEIL3 mRNA level was reduced by half, while that of MaEIL4 increased twofold (Fig. 2C). Wounding reduced by half the MaMaEIL3 mRNA level in roll (young) leaf but have no effect in unroll leaf. For MaEIL4, wounding induced the mRNA regardless of the developmental stage (Fig. 2D).

4. Conclusions

MaEIL3 and 4 gene expression is differentially regulated in fruit peel and pulp tissues during development and ripening, and in leaves in development and wounding.

The drastic decrease of both MaEIL3 and 4 mRNA levels at the beginning of ripening (turning stage) raises the question of their putative roles during the late ripening stage.

The gene expression patterns of both MaEIL3 and MaEIL4 are negatively correlated with ripening ethylene production. It should be interesting to assess the regulation of expression of these genes by exogenous ethylene in fruit.

References

Adams-Phillips L, Barry C, Giovannoni J (2004) Signal transduction systems regulating fruit ripening. Trends Plant Sci 9(7):331–338

Tieman DM, Ciardi JA, Taylor MG, Klee HJ (2001) Members of the tomato LeEil (EIN3-like) gene family are functionally redundant and regulate ethylene responses throughout plant development. Plant J 26(1):47–58.

Isolation and expression of ACC oxidase in *Hevea brasiliensis*

Kuswanhadi[2], Leclercq, J.[1], Sumarmadji[2], Rio, M.A.[1] and Montoro, P.[1]

[1]CIRAD, "Developmental Biology of Tree Crops" research unit, TA 80 / 03, Avenue Agropolis, 34398 Montpellier Cedex 5, France.
[2]Indonesian Rubber Research Institute, Sungei Putih & Sembawa Research Centres, Indonesia (*Corresponding author: pascal.montoro@cirad.fr)

1. Introduction

Natural rubber is mostly produced by *H. brasiliensis* and represents 33% of total consumed rubber in the world, its production reaching 7.92 million tons in 2003. Heveaculture takes advantage of the increase in demand for natural rubber and the ecological challenges (carbon sequestration) to be a major renewable source of rubber and wood.

Ethephon, an ethylene generator, is applied to the bark of rubber trees to increase rubber production by stimulating latex regeneration and flow (Jacob *et al.*, 1989). However, a good command of both the stimulation frequencies and the ethephon concentrations to be used during tree tapping is required to avoid oxidative burst that can result in tapping panel dryness, leading to a loss of production. Although little is known about the molecular response to ethylene stimulation (Pujade-Renaud *et al.*, 1994; Sumarmadji, 1999), further studies on ethylene biosynthesis and its regulation were needed to gain a better understanding of the mechanisms involved in latex production. Hence, ACO genes were isolated from the rubber tree and characterized by semiquantitative RT-PCR.

2. Results and Discussion

2.1 Isolation of three genes encoding ACC oxidase

Total RNA was extracted from various tissues of clone PB 260 in a guanidium thiocyanate solution by ultracentrifugation through a cesium chloride cushion. Using degenerate primers, partial fragments were amplified by PCR, then cloned. Full-length cDNAs were isolated either by

A. Ramina et al. (eds.), Advances in Plant Ethylene Research: Proceedings of the 7th International Symposium on the Plant Hormone Ethylene, 31–34.

cDNA library screening (clone H5) or by RACE (clones H4 and O48). *Hb*ACO-H4, *HbACO-H5*, and *HbACO-O48* encoded polypeptides of 318, 315, and 318 amino acids, respectively, having 79–92% protein identity and 75–86% nucleotide identity between them. Two genomic sequences were isolated: *HbACO-H4,* which was 1504 bp long consisting of 2 introns and 3 exons, and *Hb-ACO-H5,* which was 1456 bp long consisting of 3 introns and 4 exons. These results suggested that we had isolated three members of this multigenic family.

2.2 Expression of HbACO genes during plant development

Characterization at several stages of development revealed differential regulation of expression (Fig. 1). *HbACO-H4* was strongly expressed in all tissues, except for latex, when ethephon was applied. Conversely, *O48* expression was more restricted to juvenile material and its response to ethephon application was lower. Although *H5* was slightly expressed, it was mostly found in all tissues and did not respond to ethephon stimulation.

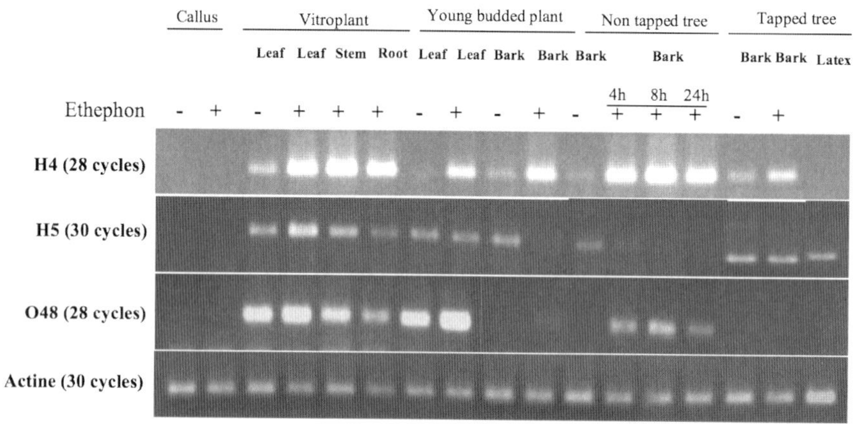

Fig. 1. Expression of *HbACO* genes during plant development. Treatments: (–) no stimulation, (+) 24 h after ethephon stimulation.

2.3 Expression of HbACO genes in response to ethephon or ethylene

Young grafted plants were treated with 2.5% ethephon or 1 ppm ethylene, and the kinetics of gene expression were analyzed from 1 to 168 h. Expression of *HbACO-H4* and *HbACO-O48* was induced by both treatments, whereas *HbACO-H5* tended to be down-regulated in bark (Fig. 2) and leaf tissues (Fig. 3). According to the number of PCR cycles, member

O48 was the most highly expressed, with expression being more transient in bark than in leaves, in which it was still expressed after 168 h.

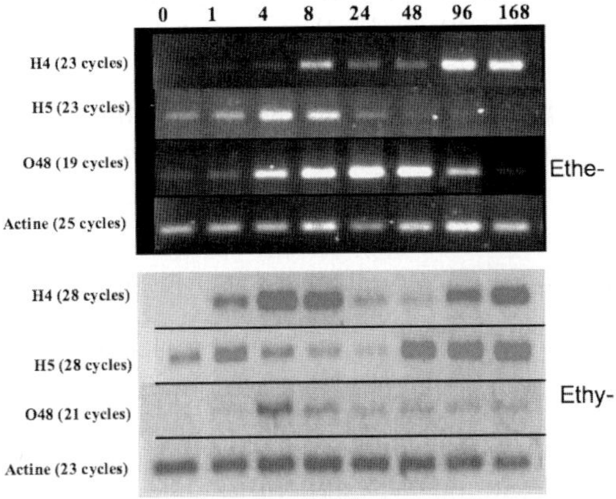

Fig. 2. Expression of *HbACO* genes in bark tissues in response to ethephon or ethylene.

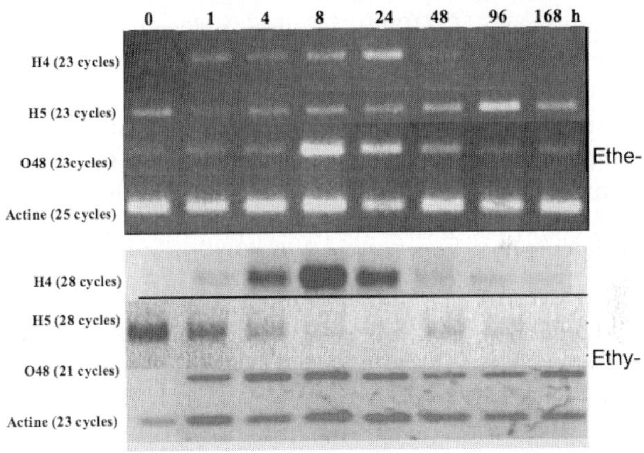

Fig. 3. Expression of *HbACO* genes leaf tissues in response to ethephon or ethylene.

3. Conclusions

The multigenic family encoding ACOs consisted of at least three members in the genome of *Hevea brasiliensis* clone PB 260. Characterization at several stages of development revealed differential regulation of expression: *HbACO-H4* was induced by ethephon in roots, leaves, and bark tissues, *HbACO-H5* displayed low but constitutive expression, and *HbACO-O48* was expressed more in juvenile tissue. None was expressed in callus.

A kinetic analysis on young grafted plants confirmed these results, but revealed very transient expression of O48 in bark, which could not be detected after 24 h. Gene expression patterns were similar in response to ethephon and ethylene, although ethylene action was faster. At that stage, *HbACO-O48* was the most strongly expressed gene in both bark and leaves. These observations correlated well with the fact that Ethrel application triggers endogenous ethylene production via autocatalytic reactions. The very weak expression found in latex tended to show that the source of ethylene biosynthesis could be in bark tissues in the vicinity of laticifer cells.

References

Jacob JL, Prévôt JC and Kekwick RG (1989) General metabolism of *Hevea brasiliensis* latex (with the exception of isoprenic anabolism). In: Physiology of rubber tree latex. Eds. d'Auzac J., Jacob J.L. and Chrestin H. CRC Press Inc. Boca Raton, Florida, pp. 101–144.

Pujade-Renaud V, Clement A, Perrot-Rechenmann C, Prevôt JC, Chrestin H, Jacob JL and Guern J (1994) Ethylene-induced increase in glutamine synthetase activity and mRNA levels in *Hevea brasiliensis* latex cells. Plant Physiology 105: 127–132.

Sumarmadji (1999) Response of Physiological Characters and Latex Production to Ethylene Stimulation on Several Rubber Tree Clones. In Advisors: Sudirman Yahya, Achmad Surkati Abidin, Alex Hartana, Bambang Sapta Purwoko, and Siswanto, Bogor.

Post-translational modification of ACC oxidase of white clover (*Trifolium repens* L.)

Du, Z., Leung, S. and McManus M.T.*

Institute of Molecular Biosciences, Massey University, Private Bag 11-222, Palmerston North, New Zealand (*Corresponding author: M.T.McManus@massey.ac.nz)

From studies on ethylene biosynthesis in the pasture legume white clover (*Trifolium repens* L.), it is clear that transcription of the two committed enzymes in the pathway, ACC synthase and ACC oxidase, is under strict developmental control, as is common in many plant species (Chen and McManus, 2006; Hunter *et al.*, 1999; Murray and McManus, 2005). In terms of post-translational control, there is now evidence from studies with the model plant *Arabidopsis thaliana* that ACC synthase activity is regulated by phosphorylation (Tatsuki and Mori, 2001). However, no similar evidence has been presented for ACC oxidase.

In white clover, four distinct ACC oxidase genes have been identified, and protein isoforms corresponding to two of these genes have been purified to homogeneity (Gong and McManus, 2000). We have shown that kinetic differences exist within the two isoforms of the gene family (TR-ACO2 and TR-ACO3), and here we speculate that the ACC oxidase activity in vitro may also be regulated by phosphorylation/dephosphorylation.

To predict putative phosphorylation sites, a bioinformatic assessment of the amino acid sequences of two translated ACC oxidase genes of white clover was undertaken [http://www.cbs.dtu.dk/services/NetPhos/]). In Table 1, residues are identified as putative kinase targets with a 'score' associated with each residue. The score denotes the probability of phosphorylation at that residue, with increasing probability as the value tends to 1.0.

In terms of higher scores (and thus increased probability), a value of .990 is recorded for Ser_{70} in TR-ACO3, and .970 for Thr_{79} in TR-ACO3. Some residues that have been identified as potential targets are unique to a particular isoform (e.g., Ser_{70} in TR-ACO3 and Ser_{98}, Thr_{264} and Tyr_{281} in TR-ACO2, while other residues with a higher score are common to both isoforms (i.e. Ser_{231} and Tyr_{229}).

A. Ramina et al. (eds.), Advances in Plant Ethylene Research: Proceedings of the 7th International Symposium on the Plant Hormone Ethylene, 35–36.

Table 1. Positions and probability scores of amino acids identified as putative kinase targets in TRACO2 and TRACO3.

Name of Amino Acid	Position		Score	
	TRACO2	TRACO3	TRACO2	TRACO3
Serine	-	70	-	0.990
	98	-	0.856	-
	231	231	0.845	0.845
Threonine	79	79	0.947	0.970
	264	-	0.955	-
Tyrosine	229	229	0.950	0.950
	281	-	0.941	-

Phosphorylation at unique residues, in particular, invites the possibility of regulation by kinases that are expressed at different leaf developmental stages, and adds to the significance of the observation that TR-ACO2 and TR-ACO3 accumulate at different developmental stages and are distinct kinetically (Gong and McManus, 2000). We are currently obtaining more direct experimental evidence, but the demonstration of the control of ACC oxidase activity by phosphorylation will add further to the evidence that this step in ethylene biosynthesis is highly controlled.

References

Chen, B.C-M. and McManus, M.T. (2006) Expression of 1-aminocyclopropane-1-carboxylate (ACC) oxidase genes during leaf development of vegetative tissues in white clover (*Trifolium repens* L.) is regulated by ontological cues. Plant Mol. Biol. 60: 451–467

Gong, D. and McManus, M.T. (2000) Purification and characterisation of two ACC oxidases expressed differentially during leaf ontogeny in white clover. Physiol. Plant 110: 13–21

Hunter, D.A., Yoo, S.D., Butcher, S.M. and McManus, M.T. (1999) Expression of 1-aminocyclopropane-1-carboxylate oxidase during leaf ontogeny in white clover. Plant Physiol. 120: 131–141

Murray, P.A. and McManus, M.T. (2005) Developmental regulation of 1-aminocyclopropane-1-carboxylate synthase gene expression during leaf ontogeny in white clover. Physiol. Plant 124: 107–120

Tatsuki, M. and Mori, H. (2001) Phosphorylation of tomato 1-aminocycloprpoane-1-carboxylic acid synthase, LE-ACS2, at the terminal region. J Biol. Chem. 276: 28051–28057

Characterisation of expression of the ACC oxidase gene family of apple (*Malus domestica*)

Binnie, J.E.,[1] Tustin, S.[2] and McManus, M.T.[1]*

[1]Institute of Molecular Biosciences, Massey University, Private Bag 11-222, Palmerston North, New Zealand; [2]Horticultural and Food Research Institute of New Zealand Ltd (HortResearch), Private Bag 1401, Havelock North, New Zealand. (*Corresponding author: M.T.McManus@massey.ac.nz)

It is now well established in plants that ACC oxidase comprises a small multigene family of typically 4–5 members. Further, in many of the species studied, this gene family displays differential expression in response to developmental, tissue-specific stimuli, and/or environmental cues (Barry *et al.*, 1996; Hunter *et al.*, 1999). In apple (*Malus domestica*), an ACC oxidase gene that is associated with fruit ripening has been identified (Ross *et al.*, 1992), as well as a fruit-associated isoform purified and characterised in terms of kinetic properties (Pirrung *et al.*, 1993). However, very little is known about the other members of the gene family in apple.

To study the ACC oxidase gene family in apple, degenerate ACC oxidase primers were used to amplify putative ACO cDNA sequences by RT-PCR from RNA isolated from leaf tissues at different developmental stages. Sequences that were shown to have identity to ACC oxidases were aligned against apple tissue EST libraries generated by HortResearch, Auckland, New Zealand (Newcomb *et al.*, 2006). Three distinct cDNAs were identified, and designated as MD-ACO1 (identical to AP4; Ross *et al.*, 1992), MD-ACO2 and MD-ACO3.

To begin to examine the differential expression of the multigene family of apple, we undertook a 'virtual' northern in which the three MD-ACO sequences were detected in EST libraries constructed from RNA isolated from different tissues of apple at different developmental stages (Table 1).

It is apparent that MD-ACO1 (clone AP4) is expressed in fruit tissue, MD-ACO2 predominantly in fruit tissue, while MD-ACO3 is expressed not only in fruit tissue but also in vegetative tissues. To dissect the pattern of expression, further Northern and RT-PCR analyses are currently underway.

A. Ramina et al. (eds.), Advances in Plant Ethylene Research: Proceedings of the 7th International Symposium on the Plant Hormone Ethylene, 37–38.

Table 1. Identification of the EST or cDNA clone indicated in EST libraries constructed to RNA isolated from the tissues at different physiological conditions. DAFB = days after full bloom.

MD-ACO1 (clone AP4)
Fruit of cv. Royal Gala fruit stored for 24 h under low oxygen/high CO_2
Cortex tissue of tree ripened fruit 150 DAFB
Peel of tree-ripened fruit 150 DAFB
Cortex of cv. Royal Gala fruit 126 DAFB
MD-ACO2 (EST 153710 – Genebank Accession: EB14037)
Fruit of cv. Royal Gala stored for 24 h under low oxygen/high CO_2
Cortex tissue of tree-ripened fruit 150 DAFB
Peel of tree-ripened fruit 150 DAFB
Cortex of cv. Royal Gala fruit 126 DAFB
Spur buds from cv. Pacific Rose trees
MD-ACO3 (EST 192136 – Genebank accession: EB151006)
Fruit of cv. Royal Gala stored @ 0.5°C for 24 h
Leaves of seedlings of cv.Royal Gala infected with *Venturia inaequalis*
Expanding leaf tissue of cv. Pinkie (X1)
Partially senescing leaf of cv. Royal Gala (X1)

Acknowledgement

We thank Dr. Robert Simpson (HortResearch, Palmerston North, New Zealand) for assistance with the EST mining and sequencing.

References

Barry, C.S., Blume, B., Bouzayen, M., Cooper, W., Hamilton, A.J. and Grierson, D. (1996) Differential expression of the 1-aminocyclopropane-1-carboxylate oxidase gene family of tomato. Plant J 9: 525–535.

Hunter, D.A., Yoo, S.D., Butcher, S.M. and McManus, M.T. (1999) Expression of 1-aminocyclopropane-1-carboxylate oxidase during leaf ontogeny in white clover. Plant Physiol 120: 131–141.

Newcomb, R.D., Crowhurst, R.N. *et al.* (2006) Analysis of expressed sequence tags from apple. Plant Physiol 141: 147–166.

Pirrung M.C., Kaiser L.M. and Chen J. (1993) Purification and properties of the apple ethylene forming enzyme. Biochemistry 32: 7445–7450.

Ross G.S., Knighton M.L. and Lay-Yee M. (1992) An ethylene-related cDNA from ripening apples. Plant Mol Biol 19: 231–238.

Characteristics of an ethylene inducible ethylene receptor *Cm-ETR2* in melon fruit

Owino, W.O., Ma, B., Sun, H.J., Shoji, T. and Ezura, H.*

Gene Research Centre, University of Tsukuba, Tennodai 1-1-1, Tsukuba, Ibaraki, 305-8572, Japan. Tel/Fax: +81-29-853-7734 (*Corresponding author: ezura@gene.tsukuba.ac.jp)

1. Introduction

Ethylene perception and signal transduction elements are conserved among various plants species where they have been examined, suggesting their importance in plant development and survival. The ethylene receptor multigene family can be divided into subfamily 1 and 2. Previous studies in our laboratory have isolated and characterized *Cm-ERS1* and *Cm-ETR1* (Sato-Nara *et al.*, 1999; Takahashi *et al.*, 2002; Ma *et al.*, 2006). These two melon ethylene receptor homologues structurally belong to the subfamily I subtype. Their mRNAs are differentially expressed during fruit development, suggesting that each of the corresponding proteins has a specific role in fruit development and ripening (Sato-Nara *et al.*, 1999; Takahashi *et al.*, 2002). In order to gain a better understanding of the role that ethylene receptor subfamilies play in melon fruit development and ripening, we isolated and characterized *Cm-ETR2,* a subfamily 2 receptor homologue from melon (*Cucumis melo* L. cv Vedrantais) fruit.

2. Materials and Methods

Melon (cv Vedrantais) fruits were grown in a greenhouse. Fruits were harvested at regular intervals as from 36 days after pollination, ethylene production was determined and the fruits were then sampled and frozen until further analyses. Total RNA was extracted from the frozen tissue and used for isolation of full length Cm-ETR2 cDNA using SMART RACE kit with primers designed from the ETR2 EST deposited in the melon EST database of Cornell University. Southern and Northern blot analyses were carried out using DIG-labeled DNA probes generated from the 3′ end of the cDNAs. A polyclonal antibody was raised against the receiver domain

A. Ramina et al. (eds.), Advances in Plant Ethylene Research: Proceedings of the 7th International Symposium on the Plant Hormone Ethylene, 39–40.

of the Cm-ETR2 protein that had been expressed in *Escherichia coli* and used for Western blot analysis of the protein. Microsomal membranes were isolated from melon seedlings, leaves and mature fruit tissues, and subjected to aqueous two-phase partitioning and sucrose gradient fractionation as described by Ma *et al.* (2006)

3. Results

Fig. 1. Schematic representation of Cm-ETR2 protein.

The putative ethylene receptor gene *Cm-ETR2* encodes a protein with a putative signal peptide, three transmembrane segments, a putative histidine kinase domain, and a putative receiver domain (Fig. 1). Northern blot analysis indicated that the levels of *Cm-ETR2* mRNA increased during melon ripening then declines with the fall in ethylene production. During ripening, transcripts of the melon subfamily II ethylene receptor gene *Cm-ETR2* exhibit earlier accumulation compared to *Cm-ETR1*, however its transcript accumulation declines when *Cm-ETR1* transcript accumulation is still high. The subcellular localization of *Cm-ETR2* seems to be tissue specific. Predominant localization to PM was observed in leaves and to both the PM and ER in mature fruits and seedlings.

References

Ma B, Cui ML, Sun HJ, Takada K, Mori H, Kamada H, Ezura H (2006) Subcellular localization and membrane topology of the melon ethylene receptor CmERS1. Plant Physiol. 141:587–597

Sato-Nara K, Yuhashi KI, Higashi K, Hosoya K, Kubota M, Ezura H (1999) Stage- and tissue-specific expression of ethylene receptor homolog genes during fruit development in muskmelon. Plant Physiol. 120:321–330

Takahashi H, Kobayashi T, Sato-Nara K, Tomita KO, Ezura H (2002) Detection of ethylene receptor protein Cm-ERS1 during fruit development in melon (*Cucumis melo* L.). J. Exp. Bot. 53:415–422

Ethylene biosynthetic and signalling genes in strawberry fruit: isolation and characterization of *ACC-synthase, -oxidase* and *CTR1*

Kiss, E.[1,3]*, Balagh, A.[1,2], Tisza, V.[1], Koncz, T.[1] and Heszky, L.[1,3]

[1]Szent István University, Department of Genetics and Plant Breeding, Gödöllő, Hungary; [2]Agricultural Biotechnological Center, Gödöllő, Hungary; [3]SIU-HAS Research Group for Molecular Plant Breeding, Gödöllő, Hungary (*Corresponding author: Kiss.Erzsebet@kk.szie.hu)

The role of ethylene in non-climacteric ripening is not clear yet. The objective of our experiments was to isolate genes involved in ethylene biosynthe sis and signal transduction in strawberry (*Fragaria* × *ananassa)*, a non-climacteric fruit. For this purpose, RT-PCR was used on RNA templates derived from different parts of strawberry plants (young and old leaves, runners and runner tips), fruits of four ripening stages and ripe fruits diseased by *Botrytis cinerea*. Using degenerate oligonucleotides described by Cazonnelli *et al.* (1998), Nakatsuka *et al.* (1998), and El-Sharkawy *et al.* (2003), the partial cDNAs for two ethylene biosynthetic genes (ACC/ 1-aminocyclopropane carboxylic acid/synthase and oxidase: *FaACS* and *FaACO*) and the partial cDNA of one component of the ethylene signalling pathway (*CTR1*: constitutive triple response: *FaCTR1*) have been obtained, cloned and sequenced. The complete cDNAs of these genes were isolated by 5' and 3' RACE. The promoters of these three genes were also isolated (by TAIL-PCR) and characteristic regulatory elements of the promoters were identified by bioinformatic analysis. The transcript accumulation of *FaACS* is the highest in green fruits and proved to be wound inducible in green leaves (Balogh *et al.* 2006). *FaACS* transcripts can be detected in old leaves, runners and runner tips. The expression of *FaACO* increased, while *FaCTR1* shows constitutive expression during ripening in fruit flesh and in vegetative tissues (Fig. 1). Several characteristics of strawberry *ACC*-synthase, -oxidase and *CTR1* genes (*FaACS, FaACO* and *FaCTR1*) are summarized in Table 1. Enhancer elements, regulatory motifs responsible for biotic–abiotic stress, auxin response and circadian control were identified in the promoter regions of *FaACS*, *FaACS* and *FaCTR* by *in silico* analysis. While expression of

A. Ramina et al. (eds.), Advances in Plant Ethylene Research: Proceedings of the 7th International Symposium on the Plant Hormone Ethylene, 41–43.

two ACO genes (*FaACO1* and *FaACO2*) in strawberry ripening has been reported previously (Trainotti *et al.* 2005), these are the first data on *FaACS* and *FaCTR1* genes to be published (Balogh *et al.* 2006).

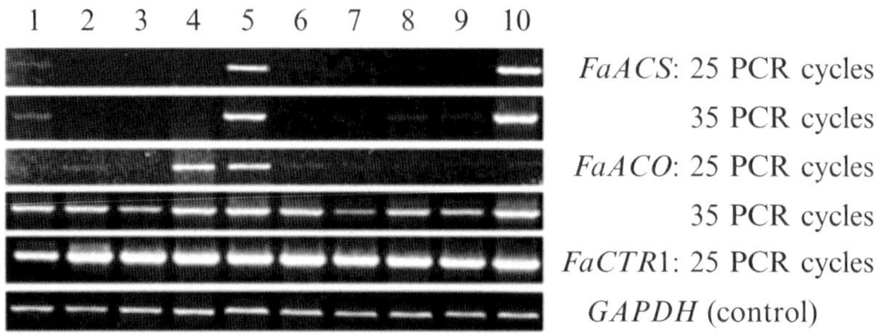

Fig. 1. Expression of *FaACS*, *FaACO* and *CTR*1 genes with semi-quantitative RT-PCR; 1, 2, 3, 4: green, white, pink and red fruit flesh; 5, 6: *Botrytis cinerea* infected ripe and overripe fruit; 7: young leaf; 8: old leaf; 9: runner; 10: runner tip.

Table 1. Characteristics of strawberry *ACS*, *ACO* and *CTR*1 genes (*FaACS*, *FaACO* and *FaCTR1*).

Gene/accession number	Highest similarity at protein level	Size of ORF (bp)	Size of CDS (bp)	Exon:intron structure
FaACS/ AY661301	Apple *MdACS-5* AB0349	2152	1473	4:3
FaACO/ AY706156	Peach *ACO* CAA5449	1163	963	3:2
FaCTR1/ AY538771	Rose *CTR1* AAK40361	2538	-	-

Acknowledgements

The research is supported by the Hungarian Scientific Research Fund (OTKA: TS 040887, T0 37861).

References

Balogh A, Kiss E, Koncz T, Dénes F, Heszky L (2006) Isolation and characterization of genes involved in the biosynthesis and signalling pathway for ethylene in strawberry. ISHS Acta Hortic 708: 541–545

Cazzonelli CI, Cavallaro AS, Botella JR (1998) Cloning and characterization of ripening-induced ethylene biosynthetic genes from non-climacteric pineapple (*Ananas comosus*) fruits. Aust J Plant Physiol 25: 513–518

El-Sharkawy I, Jones B, Li ZG, Lelievre JM, Pech JC, Latché A (2003) Isolation and characterization of four ethylene perception elements and their expression during ripening in pears (*Pyrus communis* L.) with/without cold requirement. J Exp Bot 54: 1615–162

Nakatsuka A, Murachi S, Okunishi H, Shiomi S, Nakano R, Kubo Y, Inaba A (1998) Differential expression and internal feedback regulation of 1-aminocyclopropane-1-carboxylate synthase, 1-aminocyclopropane-1-carboxylate oxidase, and ethylene receptor genes in tomato fruit during development and ripening. Plant Physiol 118: 1295–1305

Trainotti L, Pavanello A, Casadoro G (2005) Different ethylene receptors show an increased expression during the ripening of strawberries: does such an increment imply a role for ethylene in the ripening of these non-climacteric fruits? J Exp Bot 418: 2037–2046

Analysis of the protein phosphatase involved in the posttranslational regulatory mechanism of LeACS2

Kamiyoshihara, Y. and Mori, H.*

Bioagricultural Sciences, Nagoya University, Nagoya, Japan
(*Corresponding author: morihito@agr.nagoya-u.ac.jp)

1. Introduction

1-Aminocyclopropane-1-carboxylic acid synthase (ACS) is a rate-limiting enzyme in the ethylene biosynthesis pathway. Recent studies suggest that ACS is regulated posttranslationally as well as transcriptionally. We previously reported that LeACS2, a wound-inducible ACS in tomato, is phosphorylated at Ser-460 in the C-terminal region (Tatsuki and Mori, 2001). Treatments with protein kinase and phosphatase inhibitors led to the finding that phosphorylation regulates LeACS2 turnover (Mori *et al.*, 2002), which is supported by the results of experiments with *eto2* (*ethylene-overproducer2*) and *eto3* mutants (Chae *et al.*, 2003). Furthermore, ETO1 protein binds to the C-terminal region of ACS5, including the putative phosphorylation site, leading to the degradation of ACS5 via the 26S ubiquitin-proteasome pathway (Wang *et al.*, 2004). These results suggest that phosphorylation/dephosphorylation controls binding of ETO1 to ACS.

Since LeACS2 is immediately phosphorylated after translation (Mori *et al.*, 2002), the amount of LeACS2 protein in the cell would be regulated mainly at the dephosphorylation step. Little is known, however, about ACS dephosphorylation. In this study, we characterized the Ser/Thr protein phosphatase involved in LeACS2 turnover.

2. Materials and Methods

Modified phosphono peptide (biotinyl-KNNLRLXFSKRMYD-Bpa, X = phosphono-Ser, Bpa = 4-benzoyl-L-phenylalanine) was synthesized based on the LeACS2 sequence surrounding the phosphorylation site. The N-terminus was tagged with a biotin molecule and the C-terminus was tagged with L-Bpa, which acts as a benzophenone photophore. The phosphorylated

A. Ramina et al. (eds.), Advances in Plant Ethylene Research: Proceedings of the 7th International Symposium on the Plant Hormone Ethylene, 45–47.

serine residue was replaced by a phosphono serine residue to prevent hydrolysis of the phosphate group by phosphatases. The reaction mixture of the peptide (final conc. 10 μM) and tomato (*Lycopersicon esculentum* Mill.) fruit pericarp tissue extract was exposed to UV irradiation. After the reaction, proteins were applied to two-dimensional polyacrylamide gel electrophoresis, which was followed by Western blot analysis with horseradish peroxidase (HRP)-conjugated streptavidin and the antibody to the human catalytic subunit of PP2A (1D6, Upstate).

3. Results and Discussion

In a previous study, treatment with protein phosphatase inhibitors, okadaic acid and calyculin A, considerably prolonged the half-life of LeACS2 in tomato fruit pericarp tissue (Mori *et al.*, 2002). This suggests that the protein phosphatase P (PPP) family (including PP1 and PP2A), which is specifically inhibited by okadaic acid and calyculin A, has an important role in LeACS2 dephosphorylation.

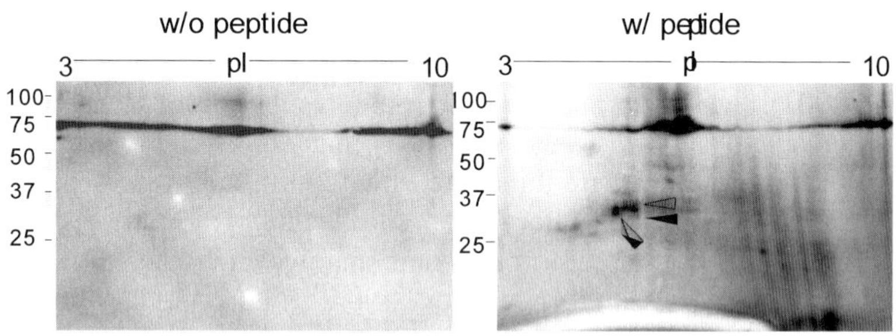

Fig. 1. Detection of biotinylated proteins after the reaction. After incubation with phosphono-peptide, proteins were applied to two dimensional-polyacrylamide gel electrophoresis, followed by detection with HRP-conjugated streptavidin (left; no peptide added, right; peptide added).

To identify the protein phosphatase involved in LeACS2 dephosphorylation, modified peptide (see "Materials and Methods") was added to the tomato fruit extract. First, reacted proteins were analyzed by HRP-conjugated streptavidin. Endogenous biotinylated proteins (e.g., 75 kDa of β-methylcrotonyl CoA carboxylase) were detected in the control experiment (no peptide added), whereas 36-kDa (pI 5.0) and 37-kDa (pI 5.2) proteins were detected when the peptide was added (Fig. 1). A competition experiment with nonlabeled peptide (without modifications of the N- and C-terminus) confirmed that these proteins were specifically detected with the modified peptide. These proteins were especially close to the molecular mass and isoelectric point of the catalytic subunit of PP2A. Western blot analysis with anti-PP2A was then performed. The 36-kDa protein detected by HRP-conjugated streptavidin corresponded to PP2A, suggesting that one of the proteins that interacted with the peptide was a catalytic subunit of PP2A.

References

Chae HS, Faure F, Kieber JJ (2003) The *eto1*, *eto2*, and *eto3* mutations and cytokinin treatment increase ethylene biosynthesis in Arabidopsis by increasing the stability of ACS protein. Plant Cell 15, 545–559

Mori H, Iwata M, Tatsuki M (2002) The regulatory mechanism of the turnover of LE-ACS2 by phosphorylation. In: Vendrell M, Klee H, Pech JC, Romojaro F (eds) Biology and Biotechnology of the Plant Hormone Ethylene β, pp. 35–40

Tatsuki M, Mori H (2001) Phosphorylation of tomato 1-aminocyclopropane-1-carboxylic acid synthase, LE-ACS2, at the C-terminal region. J Biol Chem 276, 28051–28057

Wang KLC, Yoshida H, Lurin C, Ecker JR (2004) Regulation of ethylene gas biosynthesis by the Arabidopsis ETO1 protein. Nature 428, 945–950

Hydrogen peroxide partially reverses 1-MCP-mediated inhibition of ethylene action in 'Golden Delicious' apples

Lévesque, P.G., Iyer, S., Kumar, G.N.M. and Fellman, J.K.*

Department of Horticulture and Landscape Architecture, Washington State University, Pullman, WA 99164-6414, USA (*Corresponding author: fellman@wsu.edu)

1. Introduction

In the commercial fruit industry, formulations of 1-methylcyclopropene (1-MCP) are used to inhibit the action of ethylene in apple fruits destined for long-term storage. While the use of an ethylene action inhibitor impro ves firmness retention and reduces ethylene-related disorders, an unintended consequence is the failure of the fruit to ripen normally and attain characteristic fresh-quality attributes associated with taste and aroma (Blankenship and Dole, 2003). We examined reversing 1-MCP-mediated inhibition of ethylene action by using hydrogen peroxide (H_2O_2) to initiate normal ethylene response, and studied the effects of H_2O_2 on the regulation of aminocyclopropane carboxylic acid (ACC) synthase (ACS) and ACC oxidase (ACO), the two genes of ethylene biosynthesis, using polymerase chain reaction (PCR) techniques.

2. Materials and Methods

'Golden Delicious' apples (*Malus domestica* Borkh.) harvested at 0.02 ppm internal ethylene concentration were treated with 1-MCP (0 or 3 ppm) overnight at 1°C. Subsequently, 1-MCP-treated fruits were dipped for an hour in either 0 or 30 mM H_2O_2, while the control (no 1-MCP, no H_2O_2) was dipped in water only.

A. Ramina et al. (eds.), Advances in Plant Ethylene Research: Proceedings of the 7th International Symposium on the Plant Hormone Ethylene, 49–51.

3. Results and Discussion

Ethylene production was induced in 1-MCP-treated fruit using H_2O_2 (Fig. 1); however, the small quantity of ethylene produced may not be sufficient to cause normal fruit ripening.

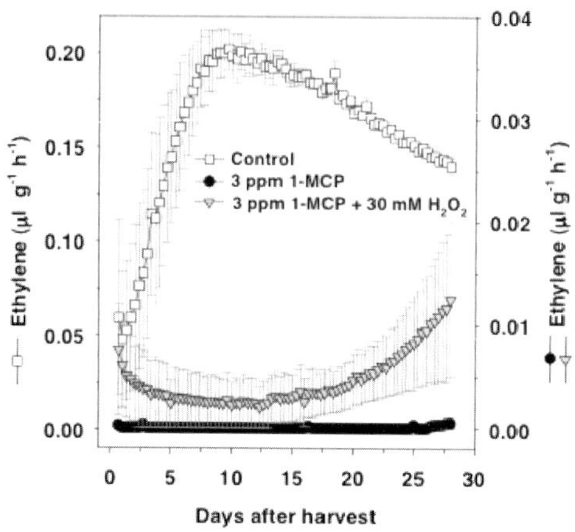

Fig. 1. Postharvest ethylene evolution from 'Golden Delicious' apples at 23°C. The experiment was a complete randomized design with three replications (five fruits per rep/chamber). In a flow-through system, gas samples from each chamber were analyzed every 8 h. Data analyzed as repeated measures. Control vs. 1-MCP treatments, P <0.0001. 1-MCP vs. 1-MCP + H_2O_2, P <0.03. Bars represent standard errors.

ACS gene expression, the rate-limiting step in ethylene biosynthesis, was detected only in fruit not treated with MCP, regardless of H_2O_2 treatment (Fig. 2), and was concurrent with the climacteric rise in ethylene production (Fig. 1). In contrast, H_2O_2 induced ACO gene expression in 1-MCP-treated fruit at postharvest days 1 and 10, while 1-MCP-treated fruit dipped in water demonstrated later ACO gene expression (postharvest days 19 and 28), but no ethylene was produced. No symptoms of superficial scald were observed in this experiment using H_2O_2 (data not shown). Further work to examine the H_2O_2 initiated ethylene production in 1-MCP-treated 'Golden Delicious' and the subsequent ability to sustain a normal ripening response is in progress.

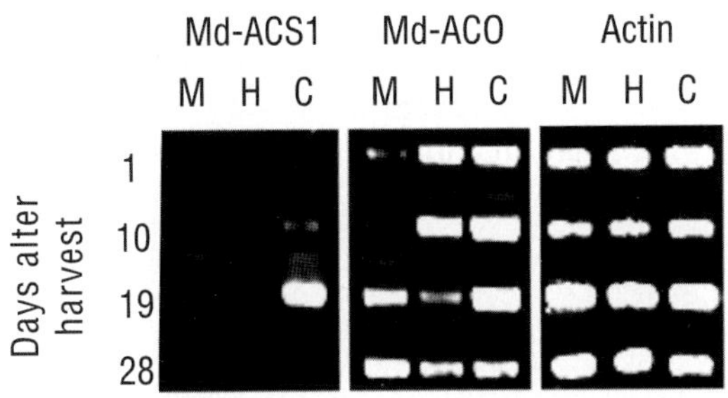

Fig. 2. RT-PCR analysis of ACS and ACO in *Malus domestica* (Md). 1-MCP-treated = M; 1-MCP + 30 mM H_2O_2 = H; and no 1-MCP + no H_2O_2 = C. 1% agarose gel stained with SYBR Gold. The PCR had 30 cycles.

References

Blankenship SM, and Dole JM. (2003) 1-Methylcyclopropene: a review. Postharvest Biol Technol 28:1–25

Characterization of three ethylene receptor genes in *Coffea canephora* Pierre

Bustamante, J.[1], Poncet, V.[2], Campa, C.[2], Noirot, M.[2], Hamon S.[2] and de Kochko A.[2]*

[1]Instituto Nacional de Investigaciones Agrícolas (INIA), Bramón, Rubio, Táchira, Venezuela. [2]Génomique et qualité du café, UMR DGPC, Centre IRD, B.P. 64501, 34394 Montpellier Cedex 5, France. (*Corresponding author: dekochko@mpl.ird.fr)

1. Introduction

The phytohormone ethylene plays a central role in physiological and developmental processes, such as germination, growth, flower initiation, senescence of leaves and flowers, organ abscission, and fruit ripening (Abeles *et al.*, 1992). It is also a major signal, mediating responses to a range of biotic and abiotic stresses. At the level of gene expression, ethylene has been shown to induce transcription of a wide range of genes involved in wound signalling and defense against pathogens. A family of five receptors mediates ethylene perception in Arabidopsis: ETR1, ERS1, ETR2, ERS2, and EIN4 (Hua and Meyerowitz, 1998; Sakai *et al.*, 1998). The ETR1 receptor is a homodimer localized in the endoplasmic reticulum membrane (Chen *et al.*, 2002).

Coffee plants are of the climacteric type, thus events during fruit maturation are tightly linked to ethylene perception, but there is little information on the response of coffee fruits to ethylene. Coffee quality depends on the stage of fruit maturation when harvested. Studies on possible relationships between ethylene receptor gene expression and fruit development and maturation should give new insights into a possible role of these receptors on coffee cup quality. Here we present results on the isolation and characterization of three genes encoding ethylene receptors in coffee (*CcETR1*, *CcETR2*, and *CcEIN4*).

A. Ramina et al. (eds.), Advances in Plant Ethylene Research: Proceedings of the 7th International Symposium on the Plant Hormone Ethylene, 53–56.

2. Results and Discussion

Table 1 gives a general description of three ethylene receptor genes from *Coffea canephora* (CAN). The *CcETR1* gene is present as a unique copy in the CAN genome. Aspects of the gene structure might indicate strong regulation at different levels of expression:

Table 1. General characterization of three ethylene receptors genes in *C. canephora.*

	CcETR1 cDNA	*CcEIN4* cDNA	*CcETR2* cDNA
Length	2,649 bp 2,683 bp 3,162 bp	2,906 bp	2,985 bp
ORF	2,223 bp	2,298 bp	2,289 bp
Putative protein	740 aa, 82.48 kDa	765 aa, 85.63 kDa	762 aa, 85.46 kDa
Identity	87.1% to *ETR1* of *Petunia × hybrida*	74.4% to *LeETR5* of *Solanum lycopersicum* 35.3% to *CcETR1*	71.5% to *LeETR4* of *Solanum lycopersicum* 37.7% to *CcETR1* 59.7% to *CcEIN4*
	GENOMIC SEQUENCE		
Intron in coding region	5	1	1
Intron size	1,240; 125; 95; 163; 1,148 bp.	2,045 bp	652 bp
Upstream Open Reading Frame (uORFs)	Yes, 35 aa	Not	Not
Intron in 5' UTR	Yes, 978 bp	Not	Not

Three independently isolated full-length cDNA clones had the same coding sequence and an identical 5′UTR, but differed in 3′UTR length, suggesting that these three clones only differed by alternative poly-adenylation sites. Indeed, some AATAAA-like motifs were found up-stream of the poly (A) tail. The three RNA forms may have a different turnover depending on the length of their 3′UTR (Meyer *et al.*, 2004).

The 5′UTR comprised a short putative open reading frame (uORF). This uORF is interrupted by one intron, which is conserved only in some *Coffea* species. In others, under the same conditions it was not possible to identify the intron by PCR amplification, indicating the absence of that intron or a strong divergence in the sequences corresponding to the primers.

RT-PCR analysis made from RNA isolated at different fruit development stages in CAN and *Coffea pseudozanguebariae* (PSE), a wild East African *Coffea* species, showed that the *ETR1* primary transcript had different alternative splice sites that take various forms (Fig. 1). They are related with retained introns at the transcript terminus or internal sections of the transcript. They include: exon skipping, unspliced introns, alternative 5′ splice donor site, and alternative 3′ splice acceptor site. In addition to the 3′ UTR, two polyadenylation sites that might be responsible for incomplete splicing at the 3′ terminus are present within the fifth intron. In some cases, these intron retentions and alternative splicing may lead to truncated proteins that interfere with the most abundant functional receptor. In other cases, the resulting protein deduced from the aberrant RNAs might be nonfunctional, but the aberrant RNAs themselves could intervene in a negative regulation of gene expression, post-transcriptional gene silencing or nonsense-mediated decay (Isshiki *et al.*, 2001).

Overexpression of *CcETR1* or *CcEIN4* in etiolated transgenic Arabidopsis plants grown on a medium without ethylene precursor (ACC) or inhibitor of ethylene synthesis (AVG) yielded a loss of gravitropic regulation of hypocotyl growth (Fig. 2), indicating a possible interference between the introduced ethylene receptors and auxin distribution (Philosoph-Hadas *et al.*, 1996).

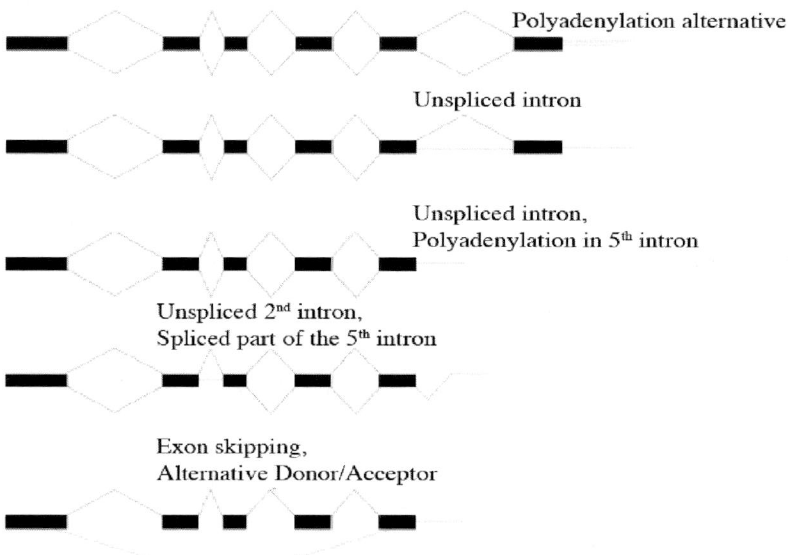

Fig. 1. Mature transcript types and alternative splicing observed in *Coffea pseudozanguebariae.*

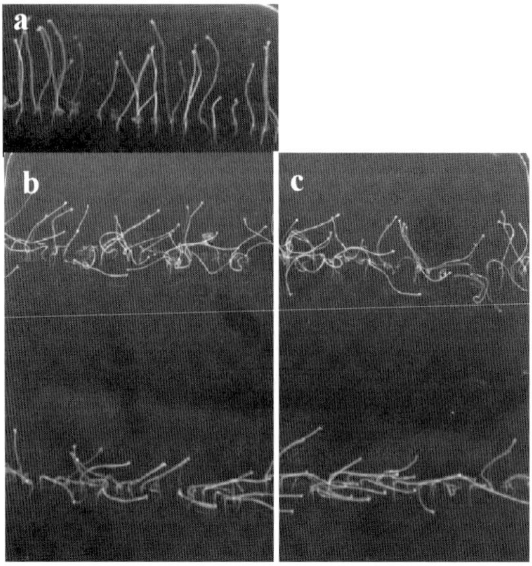

Fig. 2. (Color figure in the Annex, p.455) Effect of *CcETR1* or *CcEIN4* over-expression in *Arabidopsis* wild-type (Col-0) dark-grown seedlings. untransformed Col-0 **(a)** and two lines of transformed Col-0 with *CcETR1* **(b)** and *CcEIN4* **(c)**.

References

Abeles FB, Morgan PW, Saltveit MEJ: Ethylene in Plant Biology, 2nd ed. New York: Academic Press (1992).

Chen Y-F, Randlett MD, Findell JL, Schaller GE: Localization of the Ethylene Receptor ETR1 to the Endoplasmic Reticulum of Arabidopsis. J. Biol. Chem. 277: 19861–19866 (2002).

Hua J, Meyerowitz EM: Ethylene responses are negatively regulated by a receptor gene family in Arabidopsis thaliana. Cell 94: 261–271 (1998).

Isshiki M, Yamamoto Y, Satoh H, Shimamoto K: Nonsense-Mediated Decay of Mutant Waxy mRNA in Rice. Plant Physiol. 125: 1388–1395 (2001).

Meyer S, Temme C, Wahle E: Messenger RNA Turnover in Eukaryotes: Pathways and Enzymes. Crit Rev Biochem Mol Biol 39: 197-216 (2004).

Philosoph-Hadas S, Meir S, Rosenberger I, Halevy AH: Regulation of the Gravitorpic Response and Ethylene Biosynthesis in Gravistimulated Snapdragon Spikes by Calcium Chelators and Ethylene Inhibitors. Plant Physiol. 110: 301–310 (1996).

Sakai H, Hua J, Chen QG, Chang C, Medrano LJ, Bleecker AB, Meyerowitz EM: ETR2 is an ETR1-like gene involved in ethylene signaling in Arabidopsis. PNAS 95: 5812–5817 (1998).

Characterization of ethylene receptor subfamilies by intron position

Shoji, T., Owino, W.O., Yuhashi, K.I., Nonaka, S. and Ezura, H.*

Graduate School of Life and Environmental Science, Gene Research Center, University of Tsukuba, Tennodai 1-1-1, Tsukuba City, Ibaraki 305-8572, Japan (*Corresponding author: ezura@gene.tsukuba.ac.jp)

Ethylene receptors exist as a protein family in higher plants and ethylene receptor-like proteins have been demonstrated to exist in cyanobacteria, *Synechocystis* and *Anabaena* (Bleecker, 1999; Mount and Chang, 2002). *Synechocystis* and *Anabaena* are thought to share a common ancestor with the cyanobacterial lineage that evolved into the modern chloroplast of higher plants. There is a possibility that ethylene receptor genes may have been transferred to higher plant genomes from a plastid genome (Huang *et al.*, 2003; Stegemann *et al.*, 2003. However, it is unclear how and when the ethylene receptor genes diverged during plant evolution. The reason why higher plants have multiple ethylene receptor genes is intriguing to plant biologists. Studies on the divergence of ethylene receptor genes from evolutionary point of view might be helpful in understanding why plants have several ethylene receptors. In this study, we try to shed some light on the divergence of ethylene receptor genes in angiosperms on the basis of intron position.

We characterized the genomic DNA structure of three ethylene receptor genes in melon (*Cm-ERS1*, *Cm-ETR1* and *Cm-ETR2*), which were isolated based on the cDNA sequences reported in Sato-Nara *et al.* (1999) and Owino *et al.* (2006). The numbers of introns are four, five and one in *Cm-ERS1*, *Cm-ETR1* and *Cm-ETR2*, respectively. In the subfamily 1 genes, *Cm-ERS1* and *Cm-ETR1*, four of the five intron positions in *Cm-ETR1* are similar to those in *Cm-ERS1*. The subfamily 2 gene *Cm-ETR2* contained one intron, which differs in position from the subfamily 1 genes. The number and the position of introns between subfamilies 1 and 2 are also conserved in several plants, including *Arabidopsis*, tomato and rice. These results suggest that subfamily 1 and 2 may have derived from separate ancestors by duplication, and that the ethylene receptor subfamilies diverged before the divergence of angiosperms as shown in Fig. 1.

A. Ramina et al. (eds.), Advances in Plant Ethylene Research: Proceedings of the 7th International Symposium on the Plant Hormone Ethylene, 57–58.

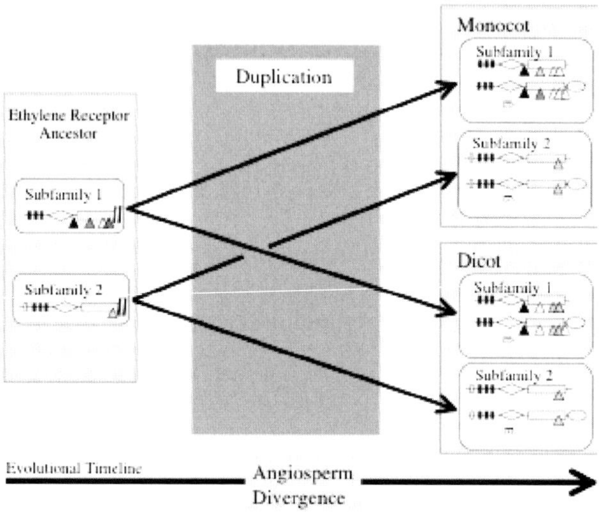

Fig. 1. Hypothetical model for divergence of ethylene receptor genes. Intron positions are indicated by arrowheads. The arrow at the bottom indicates the flow of evolution. Intron number and position between ethylene receptor subfamily 1 and 2 were conserved in several plants including monocots and dicots.

References

Bleecker AB (1999) Ethylene perception and signaling: an evolutionary perspective. Trends Plant Sci 4:269–274

Huang CY, Ayliffe MA, Timmis JN (2003) Direct measurement of the transfer rate of chloroplast DNA into the nucleus. Nature 422:72–76

Mount SM, Chang C (2002) Evidence for a plastid origin of plant ethylene receptor genes. Plant Physiol 130:10–14.

Owino W, Ma B, Sun HJ, Shoji T, Ezura H (2006) Characterization of an ethylene inducible ethylene receptor homolog, Cm-ETR2, in melon fruit. Proceedings of 7th International Symposium on the Plant Hormone Ethylene (June, 18–22, Pisa, Italy).

Sato-Nara K, Yuhashi KI, Higashi K, Hosoya K, Kubota M, Ezura H (1999) Stage- and tissue-specific expression of ethylene receptor homolog genes during fruit development in muskmelon. Plant Physiol 120:321–30

Stegemann S, Hartmann S, Ruf S, Bock R (2003) High-frequency gene transfer from the chloroplast genome to the nucleus. Proc. Natl. Acad. Sci. USA 100:8823–8833

2. INTERACTIONS BETWEEN ETHYLENE AND OTHER HORMONES

Interactions with the ethylene pathway: a puzzle yet to be completed

Van Der Straeten, D.*, Vandenbussche, F., De Grauwe, L., Dugardeyn, J. and Hagenbeek, D.

Unit Hormone Signalling and Bio-imaging, Dept. Molecular Genetics, Ghent University, K.L. Ledeganckstraat 35, B-9000 Gent; phone +3292645185/fax +32926453332) (*Corresponding author: dominique. vanderstraeten@ugent.be)

1. Introduction

Plants perceive a plethora of external signals and continuously adapt to changes in their environment. This phenotypic plasticity is obtained by integration of external signals as light and temperature, with internal, mainly hormonal signals that regulate development (Vandenbussche and Van Der Straeten, 2004). Much progress has been made in understanding how these pathways interact, and in the identification of proteins that serve as signal integrators, leading to the biologically most adequate output. Bipartite interactions are most simple to consider, but may not reflect the actual situation *in planta*, where several signals impinge on the same process. Interactions between signals can be of triple nature: signal X influencing either the metabolism (biosynthesis and degradation) or the transduction (perception/signaling or transport) of signal Y; or alternatively, the use of indiscriminate signaling intermediates, shared between the pathways of X and Y. The first two mechanisms can be regulated either at the transcriptional or at the posttranscriptional level. In recent years, it has become clear that the latter is at least equally, if not even more important than the former. The role of the 26S proteasome machinery in breakdown regulatory proteins has been extensively documented (Smalle and Vierstra, 2004).

DELLA proteins, a subfamily of the plant-specific GRAS family of putative transcription factors that regulate plant growth in response to gibberellins (GA), have been demonstrated to function as integrators of multiple plant growth regulatory signalling inputs, converging ethylene, abscissic acid, and auxin signals. It can be hypothesized that also brassinolide, jasmonate, and external signals as light ultimately interact with the GA pathway, in order to determine a developmental output. Yet,

A. Ramina et al. (eds.), Advances in Plant Ethylene Research: Proceedings of the 7th International Symposium on the Plant Hormone Ethylene, 61–68.

the GA pathway is not unique with respect to control of cell expansion (Vandenbussche *et al.*, in this book). This chapter will overview some of the interactions with the ethylene pathway, which have been studied in more detail.

2. Interactions With the Ethylene Pathway

2.1 Interactions revealed by genetic analysis

Genetic screens have enabled identification of factors involved in ethylene biosynthesis and signal transduction. The most widely used genetic screen for ethylene related mutations is the triple response assay (Roman *et al.*, 1995). Despite the fact that this screen was originally designed to identify specific components in the ethylene pathway, mutations therein often confer changes in sensitivity to other hormones as well. Alleles of mutations in ethylene signaling have also been recovered in screens using auxin transport inhibitors, in those recovering mutants resistant to cytokinin application, or in screens for suppressor and enhancer mutants of abscisic acid (ABA), or finally, in assays to uncover regulators of sugar metabolism (Zhou *et al.*, 1998; Vogel *et al.*, 1998a; Beaudoin *et al.*, 2000; Ghassemian *et al.*, 2000). These observations were indicative for a non-linear representation of hormone pathways controlling specific aspects of plant growth and development, rather supporting an interactive network wherein a given signal interferes with a variety of developmental and metabolic signals.

Models for ethylene and ABA interaction have been proposed, wherein the hormones work in the same or parallel pathways (Beaudoin *et al.*, 2000). Genetic analysis suggested that ethylene and ABA antagonize each other at the level of germination. Two independent screens designed to discover mutants involved in ABA responsiveness identified ethylene signaling mutants (Beaudoin *et al.*, 2000; Ghassemian *et al.*, 2000). *era3* mutants, which were originally identified as ABA hypersensitive, were found to be allelic to *ein2*. Furthermore, *ctr1* and *ein2* mutants were identified as enhancers and suppressors of *abi1* mutants, respectively. Hence, ethylene is a negative regulator of ABA signaling in seeds. Ethylene appears to promote seed germination by altering endogenous ABA levels and/or by decreasing the sensitivity of the seeds to ABA. Contrary to the situation in seeds, ABA and ethylene signaling act additively in roots, as supported by reduced sensitivity of *etr1* roots to exogenous ABA. However, ethylene-overproducing mutants have decreased ABA sensitivity, implying another antagonistic interaction. One suggested explanation for this apparent inconsistency is that ABA inhibits root growth by signaling through the ethylene response pathway, but is unable to use this pathway in the presence of ethylene (Ghassemian *et al.*, 2000).

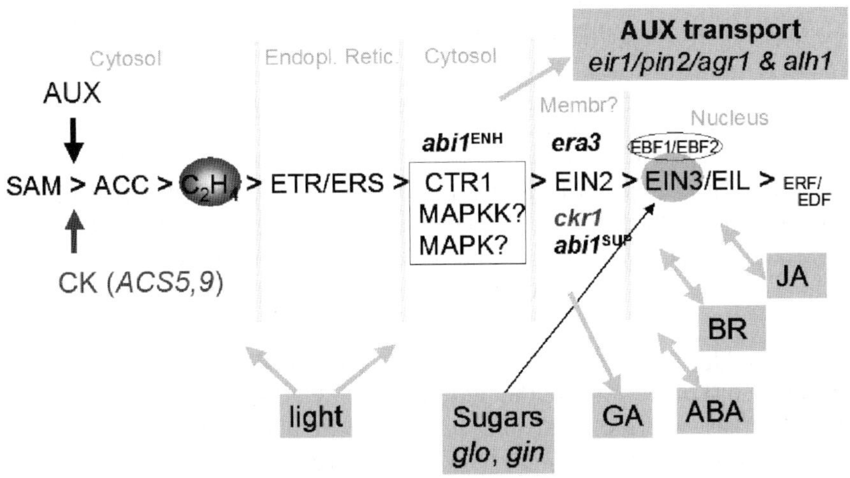

Fig. 1. Overview of interactions with the ethylene pathway. Abbreviations: SAM, S-adenosyl-L-methionine; ACC, 1-aminocyclopropane-1-carboxylic acid; ACS, ACC synthase; ETR/ERS, ethylene receptors; CTR1, constitutive triple response 1; MAPK(K), mitogen activated protein kinase (Kinase); EIN2, ethylene insensitive 2; EIN3, ethylene insensitive 3; EIL, EIN3-like proteins; EBF, EIN3 binding F-box factors; ERF, ethylene response factors; EDF, ethylene response DNA binding factors; *abi1*, ABA insensitive 1; *era3*, enhanced response to ABA 3; *ckr1*, cytokinin resistant 1; *eir1/pin2/agr1*, ethylene insensitive root 1; *alh1*, ACC-related long hypocotyl; *glo*, glucose oversensitive; *gin*, glucose insensitive.

2.2 Interactions at the level of ethylene biosynthesis: mechanisms

Several hormones can influence the biosynthesis of ethylene. Auxin has been shown to stimulate ethylene biosynthesis at the level of transcription of most Arabidopsis genes encoding ACC synthases, with *ACS7* and *ACS9* as notable exceptions (Abel *et al.*, 1995; Yamagami *et al.*, 2003). Since auxin and ethylene production follow a circadian rhythm, it is possible that rhythms in ethylene biosynthesis are caused by auxin rhythms (Thain *et al.*, 2004). The interaction auxin–ethylene also appears to work in the other direction: ethylene can regulate auxin transport and distribution (Vandenbussche *et al.*, 2003a; De Grauwe *et al.*, 2005).

Previous studies showed that cytokinin also enhances ethylene levels. Cytokinins have been proven to increase the stability of ACS5 and ACS9, genes that were respectively identified from *eto2* and *eto3* ethylene overproducer mutants (Chae *et al*., 2003; Chae and Kieber, 2005). As a consequence, many of the growth defects attributed to cytokinin are the result of ethylene overproduction. Furthermore, this explains the cytokinin insensitivity of *ein2* mutants, as mutants insensitive to ethylene obviously also are insensitive to exogenous cytokinin (Vogel *et al*., 1998b).

2.3 Interactions at the level of ethylene signaling: mechanisms

A second possibility of cross talk resides in the integration of signal transduction routes. EIN3 plays a pivotal role in ethylene signaling (Guo and Ecker, 2004). Ethylene regulates EIN3 activity by $SCF^{EBF1/EBF2}$-dependent proteolysis (Guo and Ecker, 2003; Potuschak *et al*., 2003; Yanagisawa *et al*., 2003; Gagne *et al*., 2004). Ethylene stabilizes EIN3; however, in the absence of ethylene, EIN3 is targeted to the proteasome for degradation. In contrast, glucose accelerates the degradation of EIN3 (Leon and Sheen, 2003). The antagonistic relationship between ethylene and glucose was first identified by the genetic and phenotypic analyses of *Arabidopsis* mutants with *glucose-insensitive* (*gin*) and *glucose-oversensitive* (*glo*) phenotypes (Zhou *et al*., 1998). The ethylene-insensitive *etr1* and *ein2* mutants displayed *glo* phenotypes, while *ctr1* was allelic to *gin4* (Leon and Sheen, 2003).

Ethylene signaling also interacts with the jasmonate pathway through ERF1 as a common intermediate (Lorenzo *et al*., 2003). Mobilization of defense responses to different pathogens is coordinated by ethylene, jasmonic acid (JA) and salicylic acid (SA), sometimes required individually, sometimes acting in concert (Glazebrook, 1999). ERF1 acts downstream of the intersection between ethylene and jasmonate signaling pathways and is a key element in the integration of both signals for the regulation of defense response genes. The expression of ERF1 can be activated rapidly by ethylene or jasmonate but also synergistically by both hormones. Moreover, blocking either pathway by mutations prevents *ERF1* induction by the two hormones either alone or in combination; therefore both signaling pathways are required concurrently for induction of ERF1 expression. Likewise, another member of the ERF family, *AtERF2*, was induced after ethylene and jasmonate treatment (Lorenzo *et al*., 2003).

2.4 Screening for altered ethylene responses in the light

To enable discovery of factors controlling specific developmental processes, screens on developmental stages other than etiolated seedlings are needed. Smalle *et al.*, 1997 showed that ethylene/ACC-treated light-grown *Arabidopsis* seedlings, display an elongated hypocotyl on a low nutrient medium (LNM), a response that is absent in the *etr*/*ein* mutants and constitutively present in the *ctr* mutant in the absence of ethylene. Furthermore, on LNM, leaf emergence was enhanced by the ethylene or its precursor ACC. Both the hypocotyl elongation and leaf emergence responses were exploited in search for novel mutants in ethylene response. This led to the isolation of the *alh1* (*ACC-related long hypocotyl*) (Vandenbussche *et al.*, 2003a), *slo1* (*slow*) (Dugardeyn *et al.*, in this book; Hagenbeek *et al.*, in this book) and *eer2* mutants (De Paepe *et al.*, 2005). *Alh1* proved to be involved in cross talk between auxins and ethylene. *Alh1* had an altered response to auxin, while levels of free and conjugated auxins in whole *alh1* seedlings remained unaffected. *Alh1* roots had a faster response to gravity, as demonstrated by robotized time-lapse imaging. Furthermore, the hypocotyl elongation of *alh1* and of ACC-treated wild type was reverted by auxin transport inhibitors, while auxin-upregulated genes were ectopically expressed in hypocotyls upon ACC treatment, suggesting that the ethylene response is mediated by auxins. Together, these data indicate that *alh1* is probably altered at the level of auxin transport. In addition, under particular light conditions, *alh1* displays a conspicuous leaf hyponasty and significantly longer petioles than the wild type, phenotypes characteristic of shade avoidance (Vandenbussche *et al.*, 2003b).

eer2 displays enhanced ethylene responses in the light (De Paepe *et al.*, 2005). On a low nutrient medium (LNM) light-grown *eer2* seedlings showed a significant hypocotyl elongation in response to low levels of ACC. Treatment with 1-MCP (1-methylcyclopropene), a competitive inhibitor of ethylene signalling, suppressed this hypersensitive response. The hypersensitivity of *eer2* was shown to be shoot specific. The ethylene levels in *eer2* did not differ from the wild type, indicating that ethylene overproduction is not the primary cause of the *eer2*-phenotype. In addition to its enhanced ethylene response of hypocotyls, *eer2* is also affected in the pattern of senescence and its phenotype depends on the nutritional status of the growth medium.

2.5 Influence of ethylene on the GA pathway

Cell and hence organ elongation is inhibited by the presence of a high level of RGA, a negative regulator of the GA-signalling pathway belonging to the family of DELLA proteins (Achard *et al.*, 2003; Vriezen *et al.*, 2004). Ethylene appears to modulate the GA signal by enhancing RGA stability. Root growth is repressed by DELLA proteins, which are removed from the nucleus in the presence of the growth promoting signals auxin or GA, presumably due to enhanced degradation (Fu and Harberd, 2003). This process is counteracted by ethylene, which stabilizes DELLA repressors in the nucleus, thus inhibiting root elongation. This mechanism, experimentally demonstrated for RGA, may also exist for other DELLA proteins. However, stabilization/destabilization of DELLA proteins is not the sole mechanism controlling cell elongation. Dark grown DELLA quadruple mutants, which have a constitutive gibberellin response, have a triple response, including a short hypocotyl, when treated with ACC. This indicates that another pathway apart from the GA pathway is regulating extension growth in the dark (Vandenbussche *et al.*, unpublished). In the light, ethylene controls a pathway different from the GA pathway to stimulate hypocotyl elongation (Vandenbussche *et al.*, in this book; De Grauwe *et al.*, 2005). In both light and darkness, auxins and brassinosteroids may be central players for the response (De Grauwe *et al.*, 2005).

3. Conclusions

Numerous interactions with the ethylene pathway have been described. Cross talk occurs both at the level of ethylene biosynthesis and the signal transduction route. In a number of cases, a mechanism for the interaction has been revealed. However, much remains to be done to elucidate the role of the close to 1000 ethylene response genes, some of which may play an important role in the feedback regulation of the pathway or cross-talk with other signaling routes.

References

Abel S, Nguyen MD, Chow W, Theologis A (1995) *ACS4*, a primary indoleacetic acid-responsive gene encoding 1-aminocyclopropane-1-carboxylate synthase in Arabidopsis thaliana. Structural characterization, expression in *Escherichia coli*, and expression characteristics in response to auxin. J Biol Chem 270:19093–19099

Achard P, Vriezen WH, Van Der Straeten D, Harberd, NP (2003) Ethylene regulates Arabidopsis development via modulation of DELLA protein growth repressor function. Plant Cell 15:2816–2825

Beaudoin N, Serizet C, Gosti F, Giraudat J (2000) Interactions between abscisic acid and ethylene signaling cascades. Plant Cell 12:1103–1115

Chae HS, Faure F, Kieber JJ (2003) The eto1, eto2, and eto3 mutations and cytokinin treatment increase ethylene biosynthesis in Arabidopsis by increasing the stability of ACS protein. Plant Cell 15:545–559

Chae HS, Kieber JJ (2005) Eto Brute? Role of ACS turnover in regulating ethylene biosynthesis. Trends Plant Sci 10:291–296

De Grauwe L, Vandenbussche F, Tietz O, Palme K, Van Der Straeten D (2005) Auxin, ethylene and brassinosteroids: tripartite control of growth in the Arabidopsis hypocotyl. Plant Cell Physiol 46:827–836

De Paepe A, De Grauwe L, Bertrand S, Smalle J, Van der Straeten D (2005) The Arabidopsis mutant eer2 has enhanced ethylene responses in the light. J Exp Bot 56:2409–2420

Fu X, Harberd NP (2003) Auxin promotes Arabidopsis root growth by modulating gibberellin response. Nature 421:740–743

Gagne JM, Smalle J, Gingerich DJ, Walker JM, Yoo SD, Yanagisawa S, Vierstra RD (2004) Arabidopsis EIN3-binding F-box 1 and 2 form ubiquitin-protein ligases that repress ethylene action and promote growth by directing EIN3 degradation. Proc Natl Acad Sci USA 101:6803–6808

Ghassemian M, Nambara E, Cutler S, Kawaide H, Kamiya Y, McCourt P (2000) Regulation of abscisic acid signaling by the ethylene response pathway in Arabidopsis. Plant Cell 12:1117–1126

Glazebrook J (1999) Genes controlling expression of defense responses in Arabidopsis. Curr Opin Plant Biol 2:280–286

Guo H, Ecker JR (2003) Plant responses to ethylene gas are mediated by SCF(EBF1/EBF2)-dependent proteolysis of EIN3 transcription factor.Cell 115:667–677

Guo H, Ecker JR (2004) The ethylene signaling pathway: new insights. Curr Opin Plant Biol 7:40–49

Leon P, Sheen J (2003) Sugar and hormone connections. Trends Plant Sci 8: 110–116

Lorenzo O, Piqueras R, Sanchez-Serrano JJ, Solano R (2003) ETHYLENE RESPONSE FACTOR1 integrates signals from ethylene and jasmonate pathways in plant defense. Plant Cell 15:165–178

Potuschak T, Lechner E, Parmentier Y, Yanagisawa S, Grava S, Koncz C, Genschik P (2003) EIN3-dependent regulation of plant ethylene hormone signaling by two arabidopsis F box proteins: EBF1 and EBF2. Cell 115: 679–89

Roman G, Lubarsky B, Kieber JJ, Rothenberg M, Ecker JR (1995) Genetic analysis of ethylene signal transduction in *Arabidopsis thaliana*: five novel mutant loci integrated into a stress response pathway. Genetics 139:1393–1409.

Smalle J, Haegman M, Kurepa J, Van Montagu M and Van Der Straeten, D (1997) Ethylene can stimulate Arabidopsis hypocotyl elongation in the light. Proc Natl Acad Sci USA 94:2756–2761

Smalle J, Vierstra RD (2004) The ubiquitin 26S proteasome proteolytic pathway. Annu Rev Plant Biol 55:555–590.

Thain SC, Vandenbussche F, Laarhoven LJ, Dowson-Day MJ, Wang ZY, Tobin EM, Harren FJ, Millar AJ, Van Der Straeten D (2004) Circadian rhythms of ethylene emission in Arabidopsis. Plant Physiol 136:3751–3761

Vandenbussche F, Smalle J, Le J, Saibo NJ, De Paepe A, Chaerle L, Tietz O, Smets R, Laarhoven LJ, Harren FJ, Van Onckelen H, Palme K, Verbelen JP, Van Der Straeten D (2003a) The Arabidopsis mutant *alh1* illustrates a cross talk between ethylene and auxin. Plant Physiol 131:1228–1238

Vandenbussche F, Vriezen WH, Smalle J, Laarhoven LJ, Harren FJ, Van Der Straeten D (2003b) Ethylene and auxin control the Arabidopsis response to decreased light intensity. Plant Physiol 133:517–527

Vandenbussche F, Van Der Straeten D (2004) Shaping the shoot: a circuitry that integrates multiple signals. Trends Plant Sci 9:499–506

Vogel JP, Schuerman P, Woeste K, Brandstatter I, Kieber JJ (1998a) Isolation and characterization of Arabidopsis mutants defective in the induction of ethylene biosynthesis by cytokinin. Genetics 149:417–427

Vogel JP, Woeste KE, Theologis A, Kieber JJ (1998b) Recessive and dominant mutations in the ethylene biosynthetic gene ACS5 of Arabidopsis confer cytokinin insensitivity and ethylene overproduction, respectively. Proc Natl Acad Sci USA 95:4766–4571

Vriezen WH, Achard P, Harberd NP, Van Der Straeten D (2004) Ethylene-mediated enhancement of apical hook formation in etiolated Arabidopsis thaliana seedlings is gibberellin dependent. Plant J 37:505–516

Yamagami T, Tsuchisaka A, Yamada K, Haddon WF, Harden LA, Theologis A (2003) Biochemical diversity among the 1-amino-cyclopropane-1-carboxylate synthase isozymes encoded by the Arabidopsis gene family. J Biol Chem 278:49102–49112

Yanagisawa S, Yoo SD, Sheen J (2003) Differential regulation of EIN3 stability by glucose and ethylene signalling in plants. Nature 425:521–525

Zhou L, Jang J-C, Jones TL, Sheen J (1998) Glucose and ethylene signal transduction crosstalk revealed by an *Arabidopsis* glucose-insensitive mutant. Proc Natl Acad Sci USA 95:10294–10299

Impact of jasmonate esters on ETR1 ethylene binding and the triple response

Linden, J.C.[1]* and Schaller, G.E.[2]*

[1]Department of Chemical and Biological Engineering, Colorado State University, Fort Collins, CO, USA. [2]Department of Biological Sciences, Dartmouth University, Hanover NH 03755 USA. (*Corresponding authors: jlinden@colostate.edu, george.e.schaller@Dartmouth.edu)

Abstract

It was hypothesized that methyl jasmonate (MJ) influences the binding of ethylene (C_2H_4) to membrane bound receptors. To address this question, C_2H_4 binding experiments were conducted in the presence of MJ using *Saccharomyces cerevisiae* that transgenically expresses the *Arabidopsis thaliana* ETR1 C_2H_4 receptor. Yeast were grown, washed and filtered on glass fiber filters, and then placed in airtight containers. Data obtained after injection of MJ solutions through serum caps and allowing 1 h for equilibration before $^{14}C\text{-}C_2H_4$ addition indicated that MJ did not influence C_2H_4 binding under the conditions of the experiments. However, when the yeast were grown for 18 h in the presence of MJ, more consistent improvements of between 1.5- and 2.3-fold in C_2H_4 binding, compared to controls, were observed. Similar ratios of improvement were also consistently obtained using two other jasmonate esters. Hydroxyethyl jasmonate and dihydroxypropyl jasmonate were prepared by transesterification of MJ using ethylene glycol and propylene glycol, respectively. Interactions between C_2H_4 and MJ were also examined by conducting triple response experiments using *A. thaliana* Col seeds. At all MJ concentrations, 1 ppm C_2H_4 inhibited hypocotyl extension to the same extent compared to controls (approximately 50%), but MJ suppressed the apical hook. MJ and other esters appear to improve binding of C_2H_4 to ETR1 in a nonspecific manner, as long as time and conditions allow for esters to be incorporated into the cell membranes; marginally improved C_2H_4 binding may not affect the triple effect hypocotyl extension, but appears to affect the apical hook.

A. Ramina et al. (eds.), Advances in Plant Ethylene Research: Proceedings of the 7th International Symposium on the Plant Hormone Ethylene, 69–74.

1. Background

Biophysical studies, such as those reported by Bemporad *et al.* (2005), use computer molecular dynamics simulations to determine the behavior of small solutes and large drugs in a lipid bilayer for rational drug design. Such articles are of interest because of potential impact on methyl jasmonate (MJ, Fig. 2) effect on ethylene binding in the integral membrane protein, ETR1. There is rough structural similarity between methyl jasmonate and three β-adrenoreceptor anagonists: aprenolol, atenolol, and pindolol. The aromatic portion of the molecules are named the *drug head* (commonly called drug scaffold) and the *drug tail* is commonly called drug side chain. The latter is oriented towards the membrane interior and the aromatic portion toward the membrane exterior. In addition to partitioning and diffusion coefficients, flexibility, mean orientation, reorientational correlation times, and hydrogen bonds are described.

Microscopic models of lipid–protein interactions, based on the hydrophobic matching principle, have been used to study some generic aspects of lipid membrane compartmentalization. Dispersion and activity of integral membrane proteins is found to be simulated by conformational changes governed by an external drive, such as receptor binding. Relaxation is controlled by interaction of the protein with its lipid surroundings (Seabra and Mouritsen, 1998). For instance, activation of protein kinase by the specific lipids surrounding the protein in oat root membranes was studied by Schaller *et al.* (1992). Because of previously observed ethylene (C_2H_4)/methyl jasmonate (MJ) interactions by Phisalaphong and Linden (1999) in terms of secondary metabolite formation in plant cell cultures (optimally at 7.5 μL/L C_2H_4 + 200 μM MJ), it was hypothesized that MJ might influence the binding of C_2H_4 to the membrane bound receptors, as suggested in Fig. 1.

2. Results

To address this question, ^{14}C-C_2H_4 binding experiments were conducted according to Sisler (1979) in the presence of MJ using *Saccharomyces cerevisiae* that transgenically expresses the *Arabidopsis thaliana* ETR1 C_2H_4 receptor, as described by Schaller and Bleecker (1995) in two types of experiments.

By injection of MJ solutions through serum caps and allowing 1 h for equilibration, before $^{14}C_2H_4$ was released.

Incorporation of jasmonate esters into yeast growth medium with 18 h of exposure.

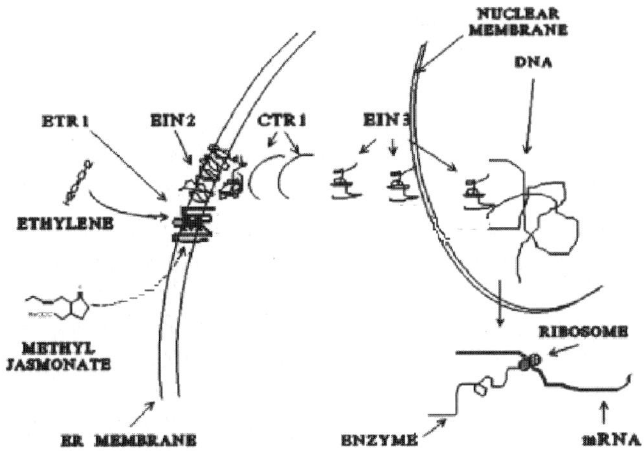

Fig. 1. Representation of hypothetical interaction between MJ and ETR1 in endoplasmic reticulum membrane. Colocalization of EIN2 is likewise hypothetical.

For the first case, the yeast were grown, washed, and filtered on glass fiber filters, and then placed in airtight containers. By injection of MJ solutions through serum caps, 1 h was allowed for equilibration. If equilibrium was established, estimated dissolved MJ concentrations (based on Henry's constant calculations) of 0.03, 0.14, 38, and 381 μM were used in two preliminary independent experiments. The MJ did not influence C_2H_4 binding under the conditions of these experiments (data not shown). It was suspected that 1 h was insufficient time for MJ equilibration in the membranes.

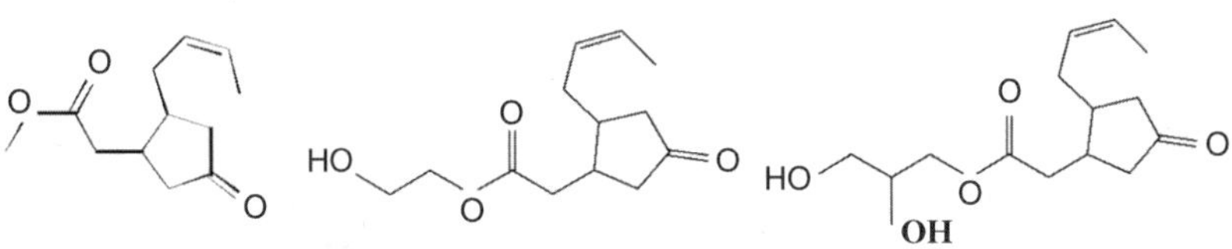

Fig. 2. Structural representations of methyl jasmonate (MJ), hydroxypropyl jasmonate (EJ), and dihydroxypropyl jasmonate, from left to right respectively.

For the second type of experiment, two other jasmonate esters were prepared by transesterification with excess alcohol and H^+ catalyst, according to the following reaction.

$$RCOOH + R'OH \leftarrow\rightarrow RCOOR' + H_2O$$

Transesterification of MJ using ethylene glycol produced hydroxyethyl jasmonate, which is abbreviated EJ. Transesterification of MJ using propylene glycol produced dihydroxypropyl jasmonate (PJ) (MJ, Fig. 2). Identity and purity were verified by GC/mass spectrometry. Quin *et al.* (2004) reported superior secondary metabolite formation using EJ.

Table 1 contains data showing that when the yeast were grown for 18 h with the three jasmonate esters dissolved at various concentrations in the defined medium, more consistent improvements of 1.2- to 2.3-fold in C_2H_4 binding, compared to controls, were observed. The effect appeared to be nonspecific.

Table 1. Ethylene binding to ETR1 in transgenic yeast grown in the presence of various concentrations of three jasmonate esters.

Conc. µM	C_2H_4 binding % of control MJ	C_2H_4 binding % of control EJ	C_2H_4 binding % of control PJ
Expt. 1			
0	100	100	100
200	160	160	190
Expt. 2			
0	100	100	100
200	190	230	120
Expt. 3			
0		100	
10		160	
50		200	
100		120	
200		150	
500		120	
1000		220	

The effect of interactions between methyl jasmonate and ethylene binding on seedling development was studied. Triple response experiments were conducted using dark-grown *A. thaliana* seedlings. The marginally improved C_2H_4 binding did not affect the triple affect hypocotyl extension, but appeared to reduce the apical hook response. Several experiments were conducted using various concentrations of ethylene and methyl jasmonate. Since MJ inhibits seed germination, it was injected in quantities to give desired dissolved equilibrium concentrations into the headspace of the vessel containing the petri dishes one day after germination (Fig. 3).

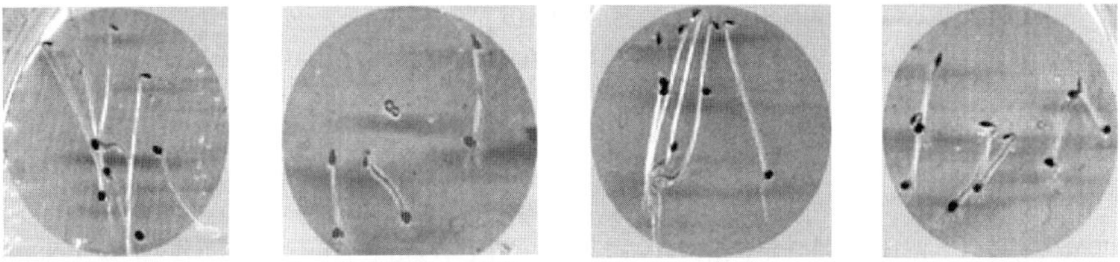

Fig. 3. Images of 5-day etiolated *A. thaliana* seedlings treated with ethylene, methyl jasmonate, and a mixture at the same concentrations are consistent with those reported by Turner *et al.* (2002).

In conclusion, MJ and other esters appear to improve binding of C_2H_4 to ETR1 in a nonspecific manner, as long as time and conditions allow for esters to be incorporated into the cell membranes. The marginally improved C_2H_4 binding does not affect the triple effect hypocotyl extension, but appears to reduce the apical hook response. Methyl jasmonate interaction results in increased binding of C_2H_4 to ETR1, which consequently increases negative regulation by CTR1. The experiments verify literature indication by Tuominin *et al.* (2004) that jasmonate interaction lies upstream of CTR1 but downstream of ETO1 in *Arabidopsis*.

References

Bemporad D, Luttmann C, Essex JW (2005) Behaviour of small solutes and large drugs in a lipid bilayer from computer simulations. Biochim Biophys Acta - Biomembr 1718: 1–21

Phisalaphong M, Linden JC (1999) Ethylene and methyl jasmonate interaction and induction models for elicited biosynthetic steps of paclitaxel in suspension cultures of *Taxus canadensis.* Biology & Biotechnology of the Plant Hormone Ethylene II. Kanellis AK *et al.* (eds) Kluwer Academic Publishers, Dordrecht pp 85–94

Quin AG, Zhao ZJ, Xu Y, Quin X, Zhong JJ (2004) Novel chemically synthesized hydroxyl-containing jasmonates as powerful inducing signals for plant secondary metabolism. Biotech Bioeng 86: 809–816

Schaller GE, Bleecker AB (1995) Ethylene-binding sites generated in yeast expressing the *Arabidopsis* ETR1 gene. Science 270: 1809–1811

Schaller GE, Harmon AC, Sussman MR (1992) Characterization of a calcium and lipid dependent protein kinase associated with the plasma membrane of oat. Biochem 31: 1721–1727

Seabra MC, Mouritsen OG (1998). Steady-state compartmentalization of lipid membranes by active proteins. Biophys J 74: 745–752
Sisler EC (1979) Measurement of ethylene binding in plant tissue. Plant Physiol 64: 538–542
Tuominin H, Overmyer K, Keinanen M, Kollist H, Kangasjarvi J (2004) Mutual antagonism of ethylene and jasmonic acid regulates ozone-induced spreading cell death in *Arabidopsis*. Plant J 39: 59–69
Turner JG, Ellis C, Devoto A (2002) The jasmonate pathway. Plant Cell Supplement on Signal Transduction: S153-S164, Figure 4.

Ethylene and abiscisic acid interaction during hibiscus (*Hibiscus rosa-sinensis* L.) flower development and senescence

Trivellini, A.[1], Ferrante, A.[2]*, Lucchesini, M.[1], Mensuali-Sodi, A.[3], Vernieri, P.[1], Tognoni, F.[1] and Serra, G.[3]

[1]Dept. Biologia delle Piante Agrarie, University of Pisa, viale delle Piagge 23, 56124, Pisa, Italy. Tel. +39 050 2216500, Fax. +39 050 2216524. [2]Dept. Produzione Vegetale, University of Milano, Via Celoria 2, 20133 Milano, Italy. Tel. +39 02 50316589, Fax. +39 02 50316575. [3]Sant'Anna School of Advanced Studies, Piazza Martiri della Libertà 33, 56100 Pisa, Italy, Tel. +39 050 883215, Fax +39 050 883325 (*Corresponding author: antonio.ferrante@unimi.it)

1. Introduction

The quality of flowering potted plants is defined by the interaction of many factors – noticeable and hides, qualitative and quantitative – that are involved in the visual appearance and the ability to preserve the ornamental characteristics. The most important parameter of quality for the flowering potted plants is obviously the flower. The natural senescence of flower is characterized by cascade events that are genetically regulated (Jones and Woodson, 1997). The hormone changes in the different flower parts may activate degeneration processes that lead the flower to wilting or death. Senescence signals seem to be mediated by flower organs. In carnation flowers, the ovary seems to be the trigger of senescence processes (Jones and Woodson, 1997). The ethylene is well known to play a primary role in the sensitive flowers, while the action of abscisic acid (ABA) is not clear. The endogenous levels of ABA and ethylene vary during flower development and senescence in the different flower organs (Jones, 2003; Ferrante *et al.*, 2006). The interaction of these two hormones may regulate the gene expression during flower senescence. In roses, the ethylene production and ABA content seem to be correlated. Exogenous treatments with ABA or ethylene increase the gene expression of ethylene receptors (Müller *et al.*, 2000).

A. Ramina et al. (eds.), Advances in Plant Ethylene Research: Proceedings of the 7th International Symposium on the Plant Hormone Ethylene, 75–79.

The aim of this work was to investigate the hormone changes, mainly ethylene and ABA, during flower development and senescence of *Hibiscus rosa-sinensis* L.

2. Materials and Methods

2.1 Plant material and treatments

Potted *Hibiscus rosa-sinensis* L. plants were grown under natural photoperiodic conditions. Plants were grown outdoor during spring–summer and indoor during autumn–winter.

Hibiscus flowers were detached at different development stages and treated with distilled water (control), 100 μM abscisic acid (ABA, Sigma Italy), 500 ppb 1-methylciclopropene (1-MCP, AgroFresh Italy). All experiments were performed in controlled room with temperature at 20°C and light intensity of 15 μmol m^{-2} s^{-1}.

2.2 ABA determination and ethylene production

Flower organ samples (80–100 mg FW) were extracted with distilled water (water:tissue ratio 10:1 v/w) for 16 h at 4°C in the dark. Quantitative analysis was performed on crude aqueous extracts using a solid-phase radioimmunoassay based on a monoclonal antibody (DBPA1) raised against free (S)-ABA, as described previously (Vernieri *et al.*, 1991).

Ethylene production was measured by enclosing flower organs in airtight containers (30–250 ml). Two ml gas samples were taken from the headspace of the containers after 1 h incubation at room temperature. The ethylene concentration in the sample was measured by a gas chromatograph (HP5890, Hewlett-Packard, Menlo Park, CA) using a flame ionization detector (FID), a stainless steel column (150 × 0,4 cm ø packed with Hysep T), column and detector temperatures of 70 and 350°C, respectively, and nitrogen carrier gas at a flow rate of 30 ml min^{-1}.

2.3 Genomic DNA isolation and ethylene receptors isolation

Genomic DNA was isolated from young Hibiscus leaves before to reach the fully expanded stage using Qiagen DNasy kit and manual instruction procedures. Ethylene receptors (ETR) isolation was performed using degenerate primers designed on highly conserved sequences that encode for ETR in other species. The semiquantitative RT-PCR was performed using specific primers. The PCR amplification cycles were 36 following standard PCR conditions.

3. Results and Discussion

3.1 Endogenous ABA content and ethylene production

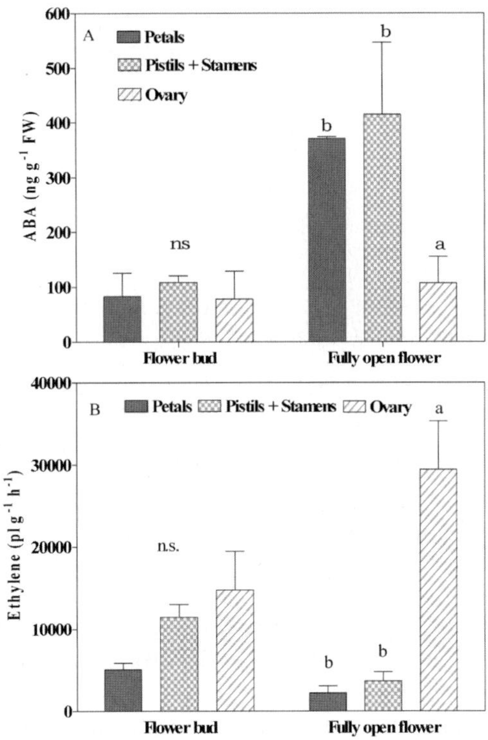

Fig. 1. Endogenous ABA content and ethylene production in petals, pistil with stamens or ovary at the bud or fully open flower stage. The values are means with standard errors (n = 5). Data were subjected to one-way ANOVA analysis. Different letters indicate statistical differences for P <0.05.

The ABA content and ethylene production were determined in petals, ovary, and pistil with stamens at bud and fully open flower stage. The endogenous ABA content was not different among flower organs at the bud stage. While in the fully open flowers, the petals and pistil with stamens showed the higher amount of ABA (Fig. 1). In petunia flowers, instead, the endogenous ABA decreases during flower development and increase again during senescence (Ferrante *et al.*, 2006). Ethylene production in the different flower organs at bud development stage was not statistically different, even if values ranged from 5 to 15 μl g^{-1} h^{-1}. On the contrary, the petals, ovary, pistil, and stamens of fully open flowers showed different ethylene production (Fig. 1). The ovary showed the highest value of ethylene production that was about tenfold higher than other organs. Analogous results were observed in carnation flowers (ten Have and Woltering, 1997).

3.2 Isolation and expression of Hibiscus ethylene receptors

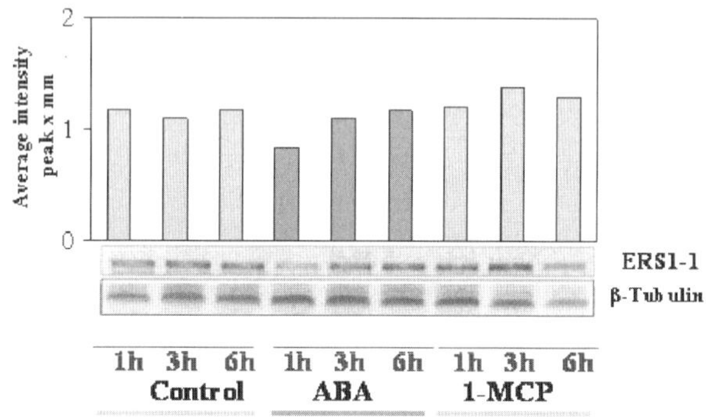

Fig. 2. Transcripts accumulation in petals of fully opened flowers treated with distilled water (control), 100 μM ABA or 500 ppb 1-MCP. The β-tubulin was used as loading internal reference. For each sampling time, 5 μg total RNA were used for RT. Each reaction was performed using 2 μl cDNA synthesized by superscript enzyme (Invitrogen).

The flower life of ethylene-sensitive flowers depends from the tissue sensitivity to ethylene. Therefore, the flower sensitivity depends from the receptors presence and their biosynthesis. Ethylene receptors were isolated using degenerate primers and genomic DNA as template. The PCR yielded DNA fragments of 532 bp that encodes for ethylene receptors. The BLAST analysis showed that the DNA amplified showed high similarity with other putative ethylene receptors (data not shown). The homologies among the DNA isolated and other receptor genes ranged from 70 to 75% at amino acids level. The isolated ethylene receptor sequence was aligned with *Arabidopsis thaliana* ethylene receptor amino acid sequences. The phylogenetic analysis showed high similarity with ethylene receptors sensor1 (ERS1), which belongs to subfamily 1. In fact, ethylene receptors isolated in *A. thaliana* are five (ETR1, ETR2, ERS1, ERS2, and EIN4) and belong to two subfamilies. Subfamily 1 includes ETR1 and ERS1, while the subfamily 2 includes ETR2, EIN4, and ERS2 (Hua *et al.*, 1998; Bleecker, 1999). Therefore, the Hrs ETR isolated is an ERS1 and belongs to subfamily 1. The semiquantitative RT-PCR approach was used for gene expression analysis of *ERS1* in petals of flowers treated with ABA or 1-MCP. Results showed that ABA down-regulated the receptor expression. A reduction of *ERS1* expression is accompanied by an increase of ethylene sensitivity as described by Bleecker (1999). On the contrary, the 1-MCP treatment, instead, increased the receptor transcript indicating a reduction of tissue sensitivity (Fig. 2).

Acknowledgment

The present work was funded by MIUR with PRIN 2006-2007 "Physiological and molecular aspects of hormonal regulation during the post-production quality of flowering potted plants."

References

Bleecker AB (1999) Ethylene perception and signalling: an evolutionary perspective. Trends in Plant Science 4: 269–274.

Ferrante A, Vernieri P, Tognoni F, Serra G (2006) Changes in abscisic acid during floral senescence. Biologia Plantarum 50(4): 581–585.

Hua J, Sakai H, Nourizadeh S, Chen QG, Bleecker AB, Ecker JR, Meyerowitz EM (1998) Ein4 and ERS2 are members of the putative ethylene receptor family in Arabidopsis. The Plant Cell 10: 1321–1332.

Jones ML, Woodson WR (1997) Pollination-induced ethylene in carnation. Role of stylar ethylene in corolla senescence. Plant Physiology 115: 205–212.

Jones ML, (2003) Ethylene biosynthetic genes are differentially regulated by ethylene and ACC in carnation styles. Plant Growth Regulation 40: 129–138.

Müller R, Stummann BM, Serek M (2000) Characterization of an ethylene receptor family with differential expression in rose (*Rosa hybrida* L.) flowers. Plant Cell Reports 19: 1232–1239.

ten Have A, Woltering EJ (1997) Ethylene biosynthetic genes are differentially expressed during carnation (*Dianthus caryophyllus* L.) flower senescence. Plant Molecular Biology 34: 89–97.

Vernieri P, Pardossi A, Tognoni F (1991) Influence of chilling and drought on water relations and abscisic-acid accumulation in bean. Australian Journal of Plant Physiology 18: 25–35.

Analysis of a novel ethylene-induced *COI1*-dependent signalling pathway in *Arabidopsis thaliana*

Adams, E.[1]*, Devoto, A.[1,2] and Turner J.G.[1]

[1]School of Biological Sciences, University of East Anglia, Norwich NR4 7TJ, UK. Phone +44-1603-592239; FAX +44-1603-592250. [2]Current Address: School of Biological Sciences, Royal Holloway, University of London, Egham, Surrey, TW20 0EX, UK. Phone +44-1784-414328; FAX +44-1784-414224. (*Corresponding author: e.adams@uea.ac.uk)

1. Introduction

Jasmonates (JAs) including jasmonic acid (JA) and methyl jasmonate (MeJA) are a family of cyclopentanone compounds synthesised from linolenic acid via the octadecanoic pathway. In *Arabidopsis*, JAs inhibit root elongation (Staswick *et al.*, 1992) and are required for pollen development, anther dehiscence (Feys *et al.*, 1994; McConn and Browse, 1996; Sanders *et al.*, 2000), and defence against insects (McConn *et al.*, 1997), necrotrophic pathogens (Thomma *et al.*, 1999) and mechanical wounding. JA biosynthesis is induced by stresses such as wounding, water deficit and pathogen attack (Creelman *et al.*, 1992; Creelman and Mullet, 1995, Penninckx *et al.*, 1998). JAs induce many defence-related genes and some of JA biosynthesis genes, not only locally but also systemically (Farmer and Ryan, 1992).

coronatine insensitive1 (*coi1*-16) is a JA-insensitive mutant, which result in unresponsiveness to growth inhibition by MeJA, male-sterility above certain temperature (Ellis and Turner, 2002), and susceptibility to insect herbivores and pathogens. (MCConn *et al.*, 1997; Thomma *et al.*, 1998). *COI1* in *Arabidopsis thaliana* was isolated by positional cloning, (Xie *et al.*, 1998), and shown to encode a 67-kD protein containing an N-terminal F-box motif and a 16 leucine rich repeats (LRRs) domain. COI1 has been shown to form a functional E3 ubiquitin ligase, SCF^{COI1} (SKP1, CDC53p/CUL1 F-box protein), in *Arabidopsis*. (Devoto *et al.*, 2002; Xu *et al.*, 2002) *COI1* is required for all JA responses studied so far.

The interaction between JA and ethylene has been studied for over a decade. Both JA and ethylene biosynthesis are induced by wounding, pests

A. Ramina et al. (eds.), Advances in Plant Ethylene Research: Proceedings of the 7th International Symposium on the Plant Hormone Ethylene, 81–87.

and pathogens (Creelman *et al.*, 1992; O'Donnell *et al.*, 1996; Kuc, 1997). The two hormones simultaneously activate defence-related genes *ETHYLENE RESPONSE FACTOR1* (*ERF1*, Lorenzo *et al.*, 2003) and synergistically induce *PATHOGENESIS-RELATED5* (*PR5*), *PLANT DEFENSIN1.2* (*PDF1.2*) and basic chitinase (*CHI-B*) (Xu *et al.*, 1994, Penninckx *et al.*, 1998; Norman-Setterblad *et al.*, 2000). Recent studies have also shown that both JA and ethylene pathways are constitutively activated in a cellulose synthase-defective mutant, *constitutive expression of VSP1* (*cev1*) (Ellis and Turner, 2001; Ellis *et al.*, 2002). On the contrary, ethylene antagonises expression of JA-responsive genes, *VEGETATIVE STORAGE PROTEINs* (*VSPs*) and a thionin (*Thi2.1*) (Rojo *et al.*, 1999, Norman-Setterblad *et al.*, 2000, Ellis and Turner, 2001).

It has been reported that *coi1*-16 shows normal sensitivity to ethylene-mediated root growth inhibition in the dark, suggesting the response is *COI1*-independent, whereas JA inhibits ethylene-induced hypocotyl hook formation in triple response in a *COI1*-dependent manner (Ellis and Turner, 2002). Also, a recent screening has suggested that light-grown *coi1*-16 appears to be more resistant to ACC-induced root growth inhibition (RGI) than wild type (A. Devoto and T. Cox, personal communication). These observations therefore suggest the interaction between ethylene response and *COI1*. We report here novel interactions between components of the ethylene and jasmonate signalling pathways.

2. Results and Discussion

Previous studies have defined plant signal pathways through mutations that suppress response to a particular hormone. Rarely have researchers tested the effect of a range of other hormones on a particular mutant. Wild type and *coi1*-16 were germinated on media containing a range of root growth inhibitors including MeJA, ethylene's precursor 1-aminocyclopropane-1-carboxylic acid (ACC), abscisic acid (ABA), epi-brassinolide (EBR), salicylic acid (SA), gibberellic acid (GA), auxin indole-3-acetic acid (IAA) and the cytokinin 6-benzylaminopurine (BAP) in continuous light (CL, 24 h light), and their root growth was compared. *coi1*-16 had significantly lower sensitivity than wild type to RGI induced by MeJA and ACC (data not shown). *coi1*-16 was also resistant to ethylene (data not shown). These results suggest that *COI1* is required for part of ethylene- and ACC-induced RGI and for most if not all of MeJA-induced RGI. We investigated the role of *COI1* in ACC-induced RGI.

Since *COI1* is required for most JA responses, it was suspected that ethylene enhanced JA biosynthesis then this would account for the reduced sensitivity of *coi1*-16 to ACC-induced RGI. To test this possibility, we examined whether ACC induced RGI in the JA biosynthesis mutants, *allene oxide synthase* (*aos*) and *12-oxophytodienoate reductase3* (*opr3*). Both mutants showed wild-type RGI by ACC (data not shown). *JASMONATE RESISTANT1* (*JAR1*) and *JASMONATE INSENSITIVE1* (*JIN1*) define genes in the JA perception-response pathway. The mutants *jar1*-1 and *jin1* showed wild-type sensitivity to ACC-induced RGI (data not shown). Together, these results indicate that ACC-induced RGI requires *COI1*, but does not require JA biosynthesis or other components of the JA perception-response pathway.

The ethylene insensitive mutant, *ethylene resistant1* (*etr1*-1), was strongly resistant to ACC-induced RGI in complete darkness (CD, 24 h dark), but it was not clearly resistant in the light (data not shown). By contrast, *coi1*-16 showed wild-type sensitivity to ACC in the dark (data not shown). To test whether *COI1* and *ETR1* contributed additively to this light-dependent, ACC-induced RGI, we constructed the *etr1*-1; *coi1*-16 double mutant. *etr1*-1; *coi1*-16 showed reduced sensitivity to MeJA in the light and in the dark, and insensitivity to ACC in the dark, as expected from the phenotypes of the single mutants. However, the double mutant showed much more resistance than *coi1*-16 or *etr1*-1 to ACC-induced RGI in the light (Table 1). Evidently, *COI1* and *ETR1* had acted additively in the light-dependent, ACC-induced RGI.

Table 1. Root growth inhibition (%) in complete darkness (CD) and complete light (CL).

	CD		CL	
	50 mM MeJA	*4 mM ACC*	*50 mM MeJA*	*4 mM ACC*
Wild type	60.0	37.8	64.5	52.5
coi1-16	23.3	47.6	8.2	38.0
etr1-1	–	–5.5	57.0	43.6
etr1-1; coi1-16	27.2	3.6	11.7	17.0

On the contrary, ACC-induced root hair formation (RHF) requires *ETR1* but not *COI1* since *coi1*-16 showed wild-type level of ACC-induced RHF but *etr1*-1 or *etr1*-1; *coi1*-16 showed much less RHF (data not shown). Additionally, we observed that MeJA inhibited germination of *etr1*-1 but not of wild type and *etr1*-1; *coi1*-16 (data not shown), indicating *ETR1* suppresses MeJA-induced, *COI1*-mediated germination inhibition.

To identify the involvement of other ethylene signalling genes in ACC-induced, *COI1*-mediated RGI, a range of mutants including *ethylene-responsive sensor1* (*ers1*-1), *ers2*-1, *etr2*-1, *ethylene insensitive2* (*ein2*-1), *ein3*-1, *ein4*, *ein5*-1, *ein6*, *constitutive triple response1* (*ctr1*-1), and a transgenic line overexpressing *ERF1*, *35S::ERF1*, was tested for light-enhanced insensitivity to ACC (Fig. 1). These results indicate that *EIN5*, *EIN6* and *ERF1* may be required for part of ACC-induced RGI in the light and in the dark, whereas the five ethylene receptors (*ETR1*, *ETR2*, *EIN4*, *ERS1* and *ERS2*), *CTR1*, *EIN2* and *EIN3* are mainly required in the dark.

Taken together, these results are summarised in Fig. 2. The *ETR1*-pathway seems to be activated in the light and in the dark since none of the ethylene perception-response mutants tested showed completely wild-type sensitivity to ACC-induced RGI (Fig. 1). In the light, on the contrary, the second pathway, the *COI1*-pathway, is activated therefore the effect of ACC seems to be divided.

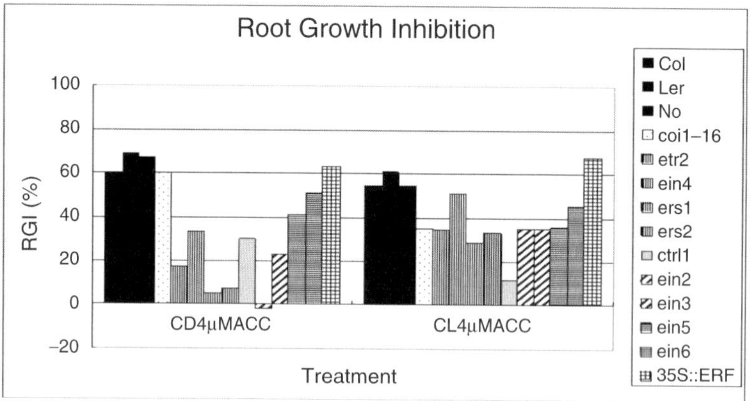

Fig. 1. ACC-induced root growth inhibition of ethylene signalling mutants.

To investigate what component of light is required to activate ACC-induced *COI1*-mediatd RGI pathway, the photoreceptor mutants including the far-red light receptor mutant, *phytochromeA* (*phyA*), the red light receptor mutant, *phyB*, and the UV-A/blue light receptor mutants, *cryptochrom1* (*cry1*), *cry2*, *phototropin1* (*phot1*), *phot2*, *cry1*; *cry2*, *phot1*; *phot2* and *cry1*; *cry2*; *phot1*; *phot2*, were tested. All of the tested mutants showed wild-type sensitivity to ACC in the light (data not shown), indicating that no particular wavelength seems to be required for the activation.

We therefore examined the effect of different photoperiods including CD, short day (SD, 8/16 h light/dark), long day (LD, 16/8 h light/dark) and

CL and different light intensity. Statistically significant differences in ACC-induced RGI between wild type and *coi1*-16 were observed in longer photoperiod and in higher light intensity (data not shown). This indicates that there may be a dose dependency of light quantity to activate the *COI1*-pathway in response to ACC.

We also examined the mechanism how the length of ACC-treated root decreased. Measurement of the length of mature root epidermal cell revealed that cell elongation inhibition accounts for 58~67% of their observed ACC-induced RGI for Col *gl* and *coi1*-16, and the rest will probably be the delayed growth, that is the lower division and elongation rate. This suggests that ACC inhibits root growth through the reduction of cell length and growth rate, and *COI1* seems to be involved in the both processes.

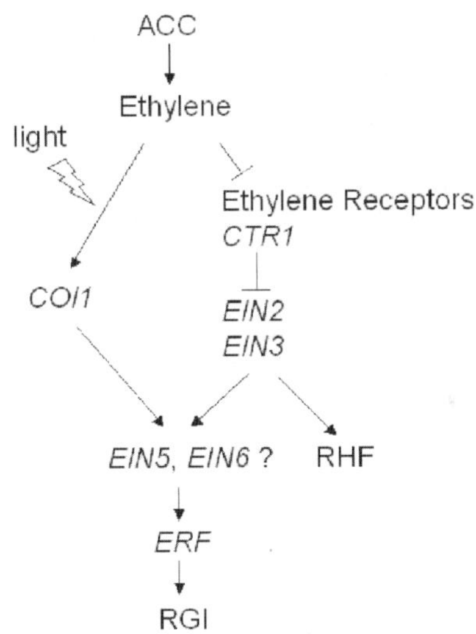

Fig. 2. Model pathway.

A mutant screen was also performed to investigate additional components of the pathway and five ACC-insensitive mutants in the light, *ethylene response insensitive1~5* (*eri1~5*), were recovered. Mapping of these genes are undergoing.

Here, we have demonstrated that a gene required for JA responses, *COI1*, is also required for ACC-induced RGI, but not for RHF, and this response is light-dependent and JA-independent. On the contrary, *ETR1* is involved in a JA response, suggesting a complicated and sophisticated regulatory mechanism of Arabidopsis in response to various environmental stimuli.

References

Creelman, R.A., and Mullet, J.E. (1995). Jasmonic acid distribution and action in plants: regulation during development and response to biotic and abiotic stress. Proc Natl Acad Sci USA 92, 4114–4119.

Creelman, R.A., Tierney, M.L., and Mullet, J.E. (1992). Jasmonic acid/methyl jasmonate accumulate in wounded soybean hypocotyls and modulate wound gene expression. Proc Natl Acad Sci USA 89, 4938–4941.

Devoto, A., Nieto-Rostro, M., Xie, D., Ellis, C., Harmston, R., Patrick, E., Davis, J., Sherratt, L., Coleman, M., and Turner, J.G. (2002). COI1 links jasmonate signalling and fertility to the SCF ubiquitin-ligase complex in Arabidopsis. Plant J 32, 457–466.

Ellis, C., and Turner, J.G. (2001). The Arabidopsis mutant cev1 has constitutively active jasmonate and ethylene signal pathways and enhanced resistance to pathogens. Plant Cell 13, 1025–1033.

Ellis, C., and Turner, J.G. (2002). A conditionally fertile coi1 allele indicates cross-talk between plant hormone signalling pathways in Arabidopsis thaliana seeds and young seedlings. Planta 215, 549–556.

Ellis, C., Karafyllidis, I., Wasternack, C., and Turner, J.G. (2002). The Arabidopsis mutant cev1 links cell wall signaling to jasmonate and ethylene responses. Plant Cell 14, 1557–1566.

Farmer, E.E., and Ryan, C.A. (1992). Octadecanoid Precursors of Jasmonic Acid Activate the Synthesis of Wound-Inducible Proteinase Inhibitors. Plant Cell 4, 129–134.

Kuc, J. (1997). Molecular aspects of plant responses to pathogens. Acta Physiol Plant 19, 551–559.

Lorenzo, O., Piqueras, R., Sanchez-Serrano, J.J., and Solano, R. (2003). ETHYLENE RESPONSE FACTOR1 integrates signals from ethylene and jasmonate pathways in plant defense. Plant Cell 15, 165–178.

McConn, M., and Browse, J. (1996). The critical requirement for linolenic acid is pollen development, not photosynthesis, in an arabidopsis mutant. Plant Cell 8, 403–416.

McConn, M., Creelman, R.A., Bell, E., Mullet, J.E., and Browse, J. (1997). Jasmonate is essential for insect defense in Arabidopsis. Proc Natl Acad Sci USA 94, 5473-5477.

Norman-Setterblad, C., Vidal, S., and Palva, E.T. (2000). Interacting signal pathways control defense gene expression in Arabidopsis in response to cell wall-degrading enzymes from Erwinia carotovora. Mol Plant Microbe Interact 13, 430–438.

O'Donnell, P.J., Calvert, C., Atzorn, R., Wasternack, C., Leyser, H.M.O., and Bowles, D.J. (1996). Ethylene as a signal mediating the wound response of tomato plants. Science 274, 1914–1917.

Penninckx, I.A., Thomma, B.P., Buchala, A., Metraux, J.P., and Broekaert, W.F. (1998). Concomitant activation of jasmonate and ethylene response pathways is required for induction of a plant defensin gene in Arabidopsis. Plant Cell 10, 2103–2113.

Rojo, E., Leon, J., and Sanchez-Serrano, J.J. (1999). Cross-talk between wound signalling pathways determines local versus systemic gene expression in Arabidopsis thaliana. Plant J 20, 135–142.

Sanders, P.M., Lee, P.Y., Biesgen, C., Boone, J.D., Beals, T.P., Weiler, E.W., and Goldberg, R.B. (2000). The arabidopsis DELAYED DEHISCENCE1 gene encodes an enzyme in the jasmonic acid synthesis pathway. Plant Cell 12, 1041–1061.

Staswick, P.E., Su, W., and Howell, S.H. (1992). Methyl jasmonate inhibition of root growth and induction of a leaf protein are decreased in an Arabidopsis thaliana mutant. Proc Natl Acad Sci USA 89, 6837–6840.

Thomma, B., Eggermont, K., Penninckx, I., Mauch-Mani, B., Vogelsang, R., Cammue, B.P.A., and Broekaert, W.F. (1998). Separate jasmonate-dependent and salicylate-dependent defense-response pathways in Arabidopsis are essential for resistance to distinct microbial pathogens. Proc Natl Acad Sci USA 95, 15107–15111.

Thomma, B.P., Eggermont, K., Tierens, K.F., and Broekaert, W.F. (1999). Requirement of functional ethylene-insensitive 2 gene for efficient resistance of Arabidopsis to infection by Botrytis cinerea. Plant Physiol 121, 1093–1102.

Xie, D.X., Feys, B.F., James, S., Nieto-Rostro, M., and Turner, J.G. (1998). COI1: an Arabidopsis gene required for jasmonate-regulated defense and fertility. Science 280, 1091–1094.

Xu, L., Liu, F., Lechner, E., Genschik, P., Crosby, W.L., Ma, H., Peng, W., Huang, D., and Xie, D. (2002). The SCF(COI1) ubiquitin-ligase complexes are required for jasmonate response in Arabidopsis. Plant Cell 14, 1919–1935.

Xu, Y., Chang, P., Liu, D., Narasimhan, M.L., Raghothama, K.G., Hasegawa, P.M., and Bressan, R.A. (1994). Plant defense genes are synergistically induced by ethylene and methyl jasmonate. Plant Cell 6, 1077–1085.

Ethylene involvement in auxin transport during apple fruitlet abscission (Malus × domestica L. Borkh)

Dal Cin, V.[1]*, Galla, G.[1], Boschetti, A.[2], Dorigoni, A.[3], Velasco, R.[3] and Ramina, A.[1]

[1]Department of Environmental Agronomy and Crop Science, University of Padova, Viale dell'Universita` 16, 35020 Legnaro (Padova), Italy. [2]CNR Institute of Photonics and Nanotechnologies, ITC, 38050 Povo Trento, Italy. [3]Experimental Institute for Agriculture, via Mach 2, San Michele all'Adige, 38010 Trento, Italy). (*Corresponding author: vdalcin@ufl.edu)

1. Introduction

Abscission is a highly coordinated process in which organ detachment occurs in specific cell layers named abscission zone (Roberts *et al.* 2002). Ethylene and auxins are the major hormones involved and their interaction is believed to be of paramount importance during abscission activation (Taylor and Whitelaw 2001).

Immature fruit abscission is the result of competition between fruitlets: the biggest and most developed ones survive whereas the smallest undergo shedding. The mechanism clearly indicates an active role of the organ subtending the abscission zone, as proposed in other systems (Alferez *et al.* 2005, Else *et al.* 2004).

In the case of apple, two weeks before shedding an increase in ethylene evolution and sensitivity occurs (Dal Cin *et al.* 2005a). This phenomenon may be a side effect or be directly connected to abscission induction. In the latter case, ethylene may trigger some specific processes at the level of seeds, where auxin is produced and subsequently translocated downstream toward the abscission zone: a decrease in auxin level or polar auxin transport across the abscission zone may desensitize it and/or modulate its activation.

The mechanisms of auxin perception and transduction have been recently described (Dharmasiri *et al.* 2005; Kepinski and Leyser, 2005) and a pivotal role is played by the AUX/IAA transcription regulators (Woodward and Bartel 2005). The AUX/IAA are proteins whose transcript amount increases following auxin application (Abel *et al.* 1995) and negatively control expression of auxin inducible genes (Worley *et al.*

A. Ramina et al. (eds.), Advances in Plant Ethylene Research: Proceedings of the 7th International Symposium on the Plant Hormone Ethylene, 89–93.

2000). These features make the amount of AUX/IAA transcripts a powerful tool to evaluate auxin quantity. In order to assess whether ethylene produced at the level of fruitlet determines a decrease in auxin level we isolated some AUX/IAA partial clones and study their expression following propylene and 1-MCP treatments. The two chemicals are known to enhance and delete abscission, respectively.

2. Materials and Methods

Intact cluster of fruitlets were collected from apple trees (cv Golden Delicious) at the beginning of abscission induction (14 days after petal fall, T0). The clusters underwent application of propylene (1000 nl h^{-1}, P) or 1-MCP (1 nl h^{-1}, M). A third bunch of clusters was left untreated and used as control (C). In order to test the responsiveness of these clones to auxin, fruitlets were immersed in buffer with a variable auxin concentration, vacuum treated for 15′ then left for 90′ in the same solution. Seed, cortex, peduncle, and abscission zone were collected and RNA extraction performed as elsewhere described (Dal Cin *et al.* 2005b). First strand RNA was performed as previously illustrated (Dal Cin *et al.* 2005a).

Table 1. Aux/IAA degenerate primers. Reverse primers are in italic.

Primer	Sequence
C1F1	5′-GGDTGGCCHCCRGTKAGATC-3′
C1R1	5′-*AACATNTCCCAWGGAACATC*-3′
C2F1	5′-GTAYKTRAAAGTBAGCATGG-3′
C2R1	5′-*TTRTAYCCHTCYSTTTCTG*-3′
C3F1	5′-AAAGTKGQCYTVAAVMTRTACA-3′
C3R1	5′-*TCRTARSTWGGHACATASTC*-3′
C4F1	5′-GTBAGYATGGATGGGGCTCC-3′
C4R1	5′-*ACATCDCCAACRAGCATCCA*-3′

DNA was extracted from 1 week old leaves with DNeasy (Qiagen) according to manufacturer instructions. Partial clones were obtained by the amplification with degenerate primers (Table 1) and the 3′ race. The amplification reaction was performed on 1 μl of cDNA in a solution containing PCR Amersham-Pharmacia Buffer (1X), dNTPs (0.25 mM), primers (2.5 μM), Amersham-Pharmacia Taq (0.5U) in a final volume of 20 μl: denaturation at 94°C, annealing for 30 s at 49°C, extension at 72°C for 45s. The total number of cycles was 40. Products were separated by electrophoresis in 2% (w/v) agarose, eluted, subcloned and sequenced.

Expression analysis was performed with specific primers (Table 2) by semiquantitative RT-PCR as previously described (Dal Cin *et al.* 2005a).

Table 2. Specific primer sequences and related annealing temperature (*T*) utilized in the semiquantitative PCR.

Primer	Sequence	*T*
IAA7	F: 5′-AATGCAAGAACAGGAGCTGA-3′ R: 5′-TCCAATCAGTCGAACAACCT-3′	64°C
IAA3	F: 5′-GGTTGAGAATCATGAAGGGT-3′ R: 5′-AAAGCCCGAGCTCTATGTCT-3′	62°C
IAA7 g	F: 5′- TGAGCTCTCTGATGCCCTA-3′ R: 5′- AGAACCGTTGAGAAACATCCA-3′	63°C

2. Results and Discussion

The partial clones identified were named accordingly to the closest similarity to Arabidopsis AUX/IAA: MdIAA3 (DQ848594), MdIAA7 (A DQ848595, B DQ848596, DQ848598), MdIAA8 (A DQ848599, B DQ848600), MdIAA 16 (d2 DQ848601, R2 DQ848602), and MdIAA27 (A PAP2 DQ848603, B PAP2 A DQ848604, B PAP2 B DQ848605, B PAP2 C DQ848606, B PAP2 D DQ848607, B PAP2 E DQ848608) and MdIAA (B1 DQ848609). Letters indicate alternative RNA processing at the 3′UTR. Moreover, by the IAA7 g primers a MdIAA7.1 genomic clone was identified (DQ848597), indicating that MdIAA7 A and B and MdIAA7 derive from the same gene through an alternative splicing. The presence of these transcripts with variable 3′UTR length was verified by comparison with other MdAUX/IAA partial clones present in database and by semiquantitative RT-PCR (results not shown). The isolation of several partial clones of MdAUX/IAA indicates that, analogously to what previously observed in several species, these proteins are encoded by a multigene family. The variability observed at the level of the 3′UTR also indicate that these transcripts undergo a complicated processing that may result in a complex post-trascriptional regulation. Several AtIAA were shown to quickly respond to auxin application (Abel *et al.* 1995). These may be the result of either a transcriptional regulation or a control at the level of RNA stability. Nevertheless, no canonical SAUR DST elements were observed (Newman *et al.* 1993) but only U-rich sequences (Mignone *et al.* 2002).

The expression analysis performed on peduncle and cortex treated with auxin indicated that transcript amount increases in a dose response manner (Fig. 1). Furthermore, propylene and 1-MCP applications indicate that the transcript accumulation is inhibited by ethylene (Fig. 2).

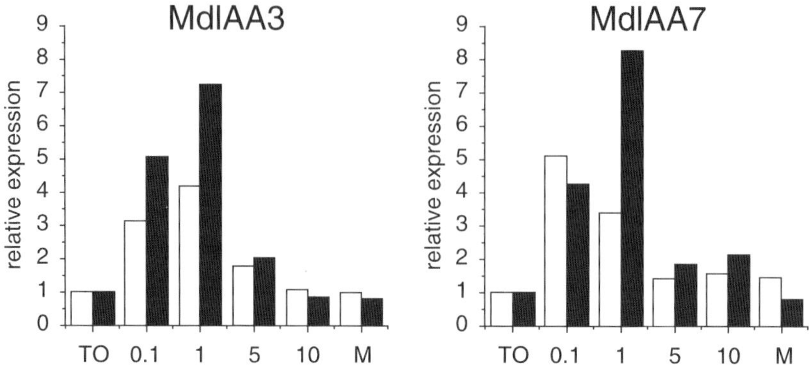

Fig. 1. Expression analysis of MdIAA3 and MdIAA7 on cortex (white bars) and peduncle (black bars) at 0.1, 1, 5, 10 mM auxin concentration. 'M' indicates the mock control.

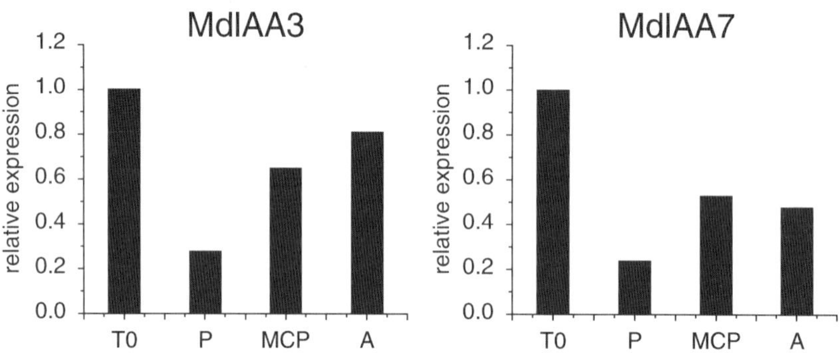

Fig. 2. Expression analysis of MdIAA3 and MdIAA7 on peduncle following propylene (P) or 1-MCP (M) applications. 'A' indicates the control in air.

These results indicate that ethylene during abscission may control directly AUX/IAA expression or determine an auxin depletion leading to a decrease in AUX/IAA transcript amount. The expression analysis of these clones as well as of auxin carriers during abscission is currently under investigation to elucidate this aspect.

References

Abel S, Nguyen MD, Theologis A (1995) The PS-IAA4/5-like family of early auxin-inducible mRNAs in *Arabidopsis thaliana*. Journal of Molecular Biology 251: 533–549.

Alferez F, Singh S, Umbach AL, Hockema B, Burns J (2005) Citrus abscission and Arabidopsis plant decline in response to 5-chloro-3-methyl-4-nitro-1H-pyrazole are mediated by lipid signaling. Plant, Cell and Environment 28: 1436–1449.

Dal Cin V, Danesin M, Boschetti A, Dorigoni A, Ramina A (2005a) Ethylene biosynthesis and perception in apple fruitlet abscission (*Malus domestica* L. Borkh). Journal of Experimental Botany 56: 2995–3005.

Dal Cin V, Danesin M, Rizzini FM, Ramina A (2005b) RNA extraction from plant tissues: the use of calcium to precipitate contaminating pectic sugars. Molecular Biotechnology 31: 113–120.

Dharmasiri N, Dharmasiri S, Estelle M. (2005) The F-box protein TIR1 is an auxin receptor. Nature 435: 441–445.

Else MA, Stankiewicz-Davies AP, Crisp CM, Atkinson C (2004) The role of polar auxin transport through pedicels of *Prunus avium* L. in relation to fruit development and retention. Journal of Experimental Botany 55: 2099–2109.

Kepinski S, Leyser O (2005) The Arabidopsis F-box protein TIR1 is an auxin receptor. Nature 435: 446–451.

Mignone F, Gissi C, Liuni S, Pesole G (2002) Untranslated regions of mRNAs. Genome Biology 3 http://genomebiology.com/2002/3/3/reviews/0004

Newman TC, Ohme-Takagi M, Taylor C, Green PJ (1993) DST Sequences, Highly Conserved among Plant SAUR Genes, Target Reporter Transcripts for Rapid Decay in Tobacco. The Plant Cell 15: 701–714.

Roberts JA, Elliott KA, Gonzales-Carranza ZH (2002) Abscission, dehiscence, and other cell separation processes. Annual Review of Plant Biology 53: 131–158.

Taylor JE, Whitelaw CA (2001) Signals in abscission. New Phytologist 151: 323–339.

Woodward AW, Bartel B (2005) A receptor for auxin. Plant Cell 17: 2425–2429.

Worley CK, Zenser N, Ramos J, Rouse D, Leyser O, Theologis A, Callis J (2000) Degradation of Aux/IAA proteins is essential for normal auxin signalling. Plant Journal 21: 553–562.

Blue light dependence of Arabidopsis seedling ethylene responses

Vandenbussche, F., Vancompernolle, B. and Van Der Straeten, D.*

Unit Hormone Signaling and Bio-Imaging, Department of Molecular Genetics, Ghent University, KL Ledeganckstraat 35, B-9000 Ghent, Belgium. Tel: +32 (0) 9 264 5186 Fax: +32 (0) 9 264 5333. (*Corresponding author: Dominique.VanDerStraeten@ugent.be)

1. Introduction

Ethylene has been well established as a hormone controlling almost all aspects of vegetative development (Smalle and Van Der Straeten, 1997). The response in dark-grown pea seedlings was the first physiological effect attributed to ethylene (Neljubow, 1901). Many years later, similar experiments with the model plant Arabidopsis yielded the triple response as a result: shorter and thicker hypocotyls, and an exaggerated apical hook with folded cotyledons and a short root, as compared with untreated plants (Bleecker *et al*., 1988). This phenotype has been successfully exploited to disect the ethylene signalling pathway and isolate various genes involved. The triple response test is the most widely used for quick analysis of ethylene phenotypes of Arabidopsis. However, the response of Arabidopsis seedlings in the light is fundamentally different of that in the dark and indicates an interaction between light and ethylene signalling pathways. Smalle *et al*. (1997) showed that in light grown seedlings, ethylene causes a short root phenotype, open but less expanded cotyledons, and most remarkably, an elongation of the hypocotyl in the presence of high exogenous ethylene or ACC levels. The opposite light versus dark phenotypes of hypocotyl growth raised the question of what mechanism may lay at the basis thereof.

The current view on hypocotyl elongation describes the process entangling external and internal factors, mainly. light and hormones. Light inhibits hypocotyl elongation as part of photomorphogenic development. White, red, far red, and blue light can inhibit elongation, with the latter two being the more effective than red light. The inhibition by white light can be counteracted by auxins, brassinosteroids, gibberellins, and ethylene. The latter hormone is generally thought of as a dwarfism inducing hormone;

A. Ramina et al. (eds.), Advances in Plant Ethylene Research: Proceedings of the 7th International Symposium on the Plant Hormone Ethylene, 95–100.

yet evidence that it is necessary for some elongation responses in non-semiaquatic plants is at hand (Smalle *et al.*, 1997; Pierik *et al.*, 2004).

Here we describe the spectral dependence of the ethylene induced hypocotyl elongation. In addition, we look in more detail at the interaction between ethylene and gibberellin, a component that negatively regulates photomorphogenesis (Alabadi *et al.*, 2004).

2. Materials and Methods

2.1 Plant material and growth conditions

Col-0, Ler-0 wild type seeds were purchased from ABRC (Ohio). The pRGA::RGA-GFP reporter line was a gift from T.P. Sun. Seeds were sown and grown on low nutrient medium as described (Smalle *et al.*, 1997), in growth chambers with a constant temperature of 22°C. Red light was provided using cool white light, filtered with appropriate hard plastic filters, combined with red filter foil. Blue light was obtained with Dragon tape LEDs (470 nm) (Osram). White light was produced by cool white lamps (Philips, the Netherlands).

2.2 Microscopy

Seedlings were grown for 3 days in blue light, harvested, and analysed within 15 min. A Zeiss axioplan (Carl Zeiss GmbH, Germany) inverted epifluorescence microscope was used.

2.3 Biometrics

Seedlings were photographed under a binocular and subsequently measured based on these pictures, using ImageJ analysis software (NIH, USA).

3. Results and Discussion

3.1 Spectral dependence of ethylene stimulated hypocotyl elongation

Contrary to the situation for dark-grown seedlings, where ethylene inhibits hypocotyl elongation, in white light, ethylene or its precursor ACC can stimulate hypocotyl elongation (Smalle *et al.*, 1997). Because not all wavelengths of the spectrum inhibit hypocotyl elongation to the same extent, the spectral requirements for counteraction of the ethylene response were investigated. In the white light experiments mentioned, usually cool white light is used, which is very poor in FR light. Therefore, the effects of ACC on hypocotyl elongation in Arabidopsis were studied in blue and red light. In blue light, the pattern found in cool white was mimicked, while in red light, there was almost no effect of the ACC (Figs. 1C and D).

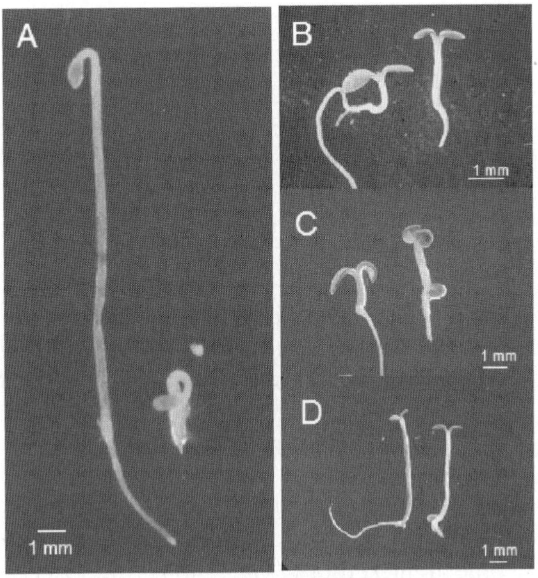

Fig. 1. (Color figure in the Annex, p.456) Response of wild-type Ler seedlings grown without (left plant) or with 50 μM ACC (right plant) in (**A**) darkness for 3 days, (**B**) white light for 7 days, (**C**) blue light for 5 days, and (**D**) red light for 5 days.

3.2 Exaggerated apical hook opening is faster in blue light than in red light

De Grauwe et al. (2005) have shown that certain aspects of the ethylene response in the light overlap with those on development, maintenance, and exaggeration of the apical hook in the dark. To study the effect of light quality and the relation with the ethylene signal in hook opening, 3 days old dark-grown seedlings were exposed to various light treatments in the presence or absence of exogenous ACC. The apical hook opened about three times faster in blue light as compared to red (Fig. 2).

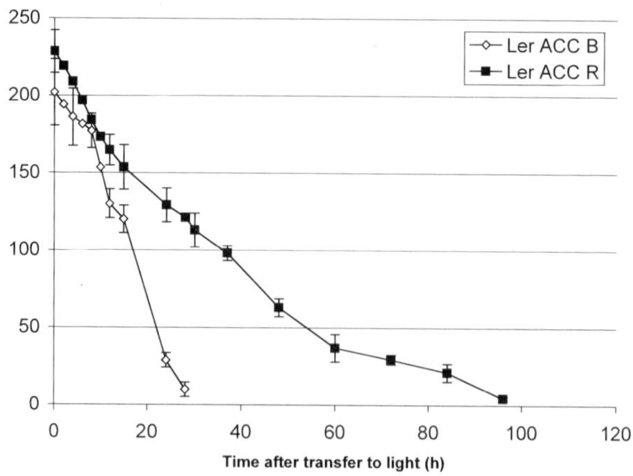

Fig. 2. Opening of the ACC induced exaggerated apical hook upon exposure to red or blue light in Ler wild-type plants. B: blue light, 30 μmol m^{-2} s^{-1}, R: red light, 30 μmol m^{-2} s^{-1}. Seedlings were grown for 3 days in darkness on a medium containing 50 μM ACC before transfer to the specific light conditions.

3.3 RGA protein accumulation in ACC treated plants

Because regulation of gibberellin signals is important for photomorphogenesis including apical hook formation, and correct inhibition of hypocotyl growth in blue light (Folta *et al.*, 2003), the effect of ethylene on gibberellin signalling compounds was investigated. The most typical and well studied gibberellin signalling gene in *Arabidopsis* is RGA. This DELLA protein negatively influences extension growth and is broken down when high levels of GA are present. It was shown earlier that in roots, ACC can stabilize RGA (Achard *et al.*, 2003). This was also the case in hypocotyls of dark-grown seedlings (Fig. 3) (Vriezen *et al.*, 2004). These results are thus consistent with ethylene mediated inhibition of elongation. To investigate whether in the light, where elongation occurs upon ethylene treatment, this response changes, pRGA::GFP–RGA plants grown in blue light with or without ACC were analysed microscopically. It was found that with ACC, more GFP–RGA protein was present in the seedlings (Fig. 3). Since RGA is a negative regulator of elongation, these results suggest that the elongation stimulated by ACC is not mediated by gibberellin signalling.

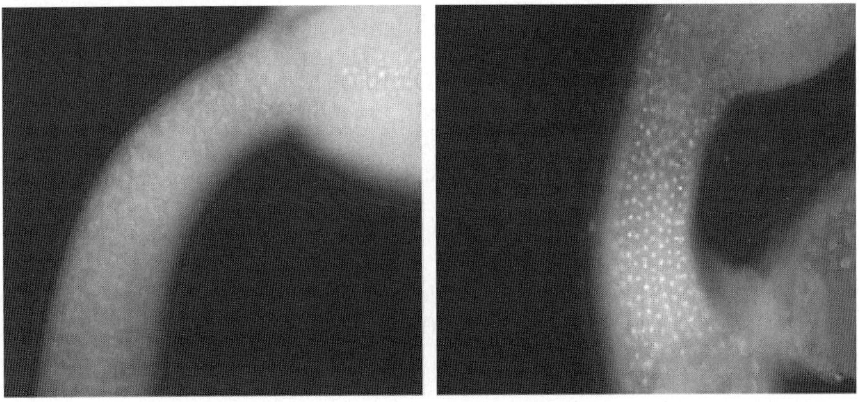

Fig. 3. Nuclear accumulation of GFP-RGA in the presence of a high ethylene signal in Arabidopsis hypocotyls. *pRGA::GFP-RGA* plants grown in 30 μmol $m^{-2}s^{-1}$ blue light for 3 days without (left panel) or with (right panel) 50 μM ACC.

4. Conclusion

Ethylene can stimulate the elongation of hypocotyls in the light. In order to shed some light on the parameters involved in this process, light quality dependence was analysed. Blue light was found to be important for the ethylene mediated hypocotyl elongation to occur. It appears that ethylene can counteract a process that is severely blue light dependent. Spectral specification is also found in the opening of the apical hook, and suggests that also in this case, blue light and ethylene are counteracting each other. The dependence of red light is far less. It is interesting to note that gibberellins, often associated with skotomorphogenesis do not play a central role in the ethylene stimulation of hypocotyl elongation. This leaves the possibility for other extension regulating signals, such as auxins or brassinosteroids (De Grauwe *et al.*, 2005), to be involved in the ethylene response. Further studies are needed to evaluate the interaction between these hormones in regulating hypocotyl growth.

References

Achard P, Vriezen WH, Van Der Straeten D, Harberd NP (2003) Ethylene regulates Arabidopsis development via modulation of DELLA protein growth repressor function. Plant Cell 15:2816–2825

Alabadi D, Gil J, Blazquez MA, Garcia-Martinez JL (2004) Gibberellins repress photomorphogenesis in darkness. Plant Physiol. 134:1050–1057

Bleecker A, Estelle M, Somerville C, Kende H (1988) Insensitivity to ethylene conferred by a dominant mutation in *Arabidopsis thaliana*. Science 241:1086–1089

De Grauwe L, Vandenbussche F, Tietz O, Palme K, Van Der Straeten D (2005) Auxin, ethylene and brassinosteroids: tripartite control of growth in the Arabidopsis hypocotyl. Plant Cell Physiol 46:827–836

Folta KM, Pontin MA, Karlin-Neumann G, Bottini R, Spalding EP (2003) Genomic and physiological studies of early cryptochrome 1 action demonstrates roles for auxin and gibberellin in the control of hypocotyl growth in blue light. Plant J 36:203–214

Neljubow DN (1901) Über die horizontale nutation der stengle von Pisum sativum und einiger anderen Pflanzen. Be ih Bot Zentralbl 10:128–139

Pierik R, Whitelam GC, Voesenek LA, de Kroon H, Visser EJ (2004) Canopy studies on ethylene-insensitive tobacco identify ethylene as a novel element in blue light and plant-plant signalling. Plant J. 38:310–319

Smalle J, Haegman M, Kurepa J, Van Montagu M, Van Der Straeten, D (1997) Ethylene can stimulate Arabidopsis hypocotyl elongation in the light. Proc Natl Acad Sci USA 94:2756–2761

Smalle J, Van Der Straeten D (1997) Ethylene and vegetative development. Physiol Plant 100:593–605

Vriezen WH, Achard P, Harberd NP, Van Der Straeten D (2004) Ethylene-mediated enhancement of apical hook formation in etiolated *Arabidopsis thaliana* seedlings is gibberellin dependent. Plant J37:505–516

Inhibition of ACC oxidase activity by melatonin and indole-3-acetic acid in etiolated lupin hypocotyls

Arnao, M.B.* and Hernández-Ruiz, J.

Department of Plant Physiology, Faculty of Biology, University of Murcia, 30100-Murcia, Spain. (*Corresponding author: marino@um.es)

1. Introduction

Melatonin (*N*-acetyl-5-methoxytryptamine)(MEL), a well-known animal hormone synthetised by the pineal gland, was originally identified in 1958. In vertebrates it plays a key role in various physiological processes, including circadian rhythmicity, sleep and seasonal photoperiod regulation. Since 1995, MEL has been detected in the roots, leaves, fruits and seeds of a considerable variety of plant species. Studies in plants have mainly focused on the quantification of MEL levels in different plant organs and species. However, very little is known about the physiological role played by MEL in plants. Studies have mainly been directed at establishing the role of melatonin as a circadian and photoperiodic regulator, a cellular protector and a plant growth regulator. An extensive review of the possible functions of MEL in plants has recently been published (Arnao and Hernández-Ruiz 2006). In the present work, we describe a novel aspect, the effect of MEL on the ethylene production in etiolated lupin.

2. Materials and Methods

Sections (3 mm) of hypocotyl or the primary root of 3-day-old etiolated lupin (*Lupinus albus* L.) were incubated in buffered media (pH 6.2) containing 0.5 mM ACC and different concentrations of MEL or IAA (0.01, 0.1 and 1 mM) for 2 h at 25°C in darkness. The treated sections were transferred to sealed vials with continuous shaking for 1 h, after which ethylene production was determined in 1 mL of the vial-atmosphere by GC-FID. ACC oxidase activity (ACCO) was expressed as nmoles ethylene·g^{-1} FW h^{-1}.

A. Ramina et al. (eds.), Advances in Plant Ethylene Research: Proceedings of the 7th International Symposium on the Plant Hormone Ethylene, 101–103.

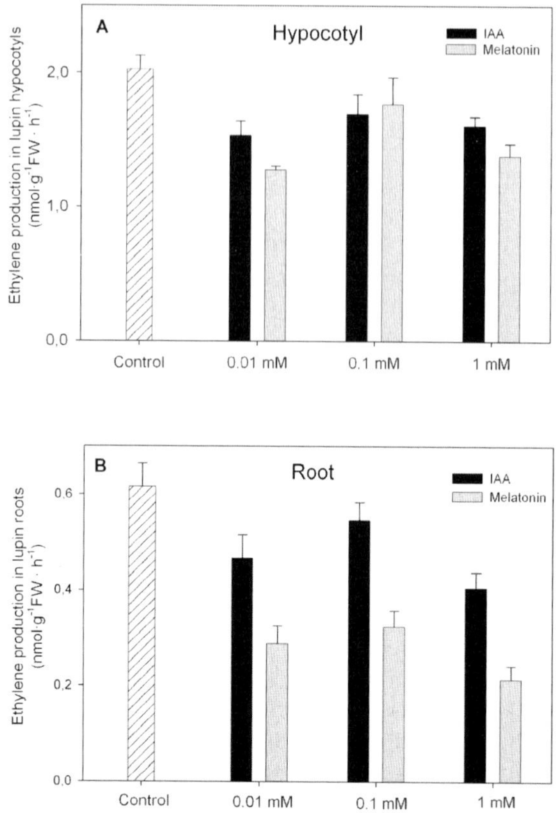

Fig. 1. ACCO activity of lupin hypocotyls and roots in presence of melatonin or IAA.

3. Results and Discussion

The ACCO activity in lupin hypocotyls and roots was inhibited by the presence of IAA or MEL (Fig. 1). In roots, inhibition was greater, especially with MEL, which produced a 65% inhibition at 1 mM. The inhibitory effect of IAA has been described by Kim *et al.* (2001) who postulated a model. However, this is the first time that a similar effect has been described for MEL, although other effects of MEL on growth have been described in several species (Hernández-Ruiz *et al.* 2004; Hernández-Ruiz *et al.* 2005).

Acknowledgements

This work was supported by FEDER-MEC-AGL2003-00638 and by Fundación Séneca (Región de Murcia) 502/PI/04. JHR has a contract with the University of Murcia.

References

Arnao MB, Hernández-Ruiz J (2006) The physiological function of melatonin in plants. Plant Signal Behav 1:89–95

Hernández-Ruiz J, Cano A, Arnao MB (2004) Melatonin: a growth-stimulating compound present in lupin tissues. Planta 220:140–144

Hernández-Ruiz J, Cano A, Arnao MB (2005) Melatonin acts as a growth-stimulating compound in some monocot species. J Pineal Res 39:137–142

Kim JE, Kim WT, Kang BG (2001) IAA and BA inhibit ethylene-regulated expression of ACCO and ACCS genes in mungbean hypocotyls. Plant Cell Physiol 42:1056–1061

Auxin and ethylene interaction during fruit growth and ripening of *Actinidia deliciosa*

Bregoli, A.M.*, Fabbroni, C., Costa, F., Raimondi, V. and Costa, G.

Dipartimento di Colture Arboree, University of Bologna, Viale Fanin 46, 40127 Bologna, Italy. (*Corresponding author: bregolia@agrsci.unibo.it)

1. Introduction

Relative to the current knowledge about ethylene, much less is known on the interaction between this hormone and auxins during fruit growth and ripening in some species as for example kiwifruit (*Actinidia deliciosa*). The 3,5,6-trichloro-2-pyridyloxyacetic acid (3,5,6-TPA), a synthetic auxin, is able to increase fruit size in peach affecting some fruit quality traits (Augustí *et al.*, 1999). During fruit growth and ripening of peaches, changes in auxin levels were related to changes in ethylene (Masia *et al.*, 1992), while ethylene emission and ACO expression showed an opposite trend (Tonutti *et al.*, 1997). It is well known that auxin influenced ethylene emission by inducing ACS and, recently, an interaction between ethylene and IAA on the expression of ACO genes was reported (Rasori *et al.*, 2003). Moreover, recent data suggest that expression of endo-1,4-β-D-glucanase (EGase), expansin and polygalacturonase (PGase) isoforms are regulated by auxins during the cell expansion phase of growth, and by ethylene during ripening (Alexander and Grierson, 2002). In the present work, we investigated the effect of the 3,5,6-TPA on fruit growth and quality in *Actinidia deliciosa* (cv Hayward) and the interaction between the chemical and ethylene biosynthesis. In addition, expression studies of genes encoding protein involved in cell wall disassembly (PGase and expansin) were performed.

2. Materials and Methods

Experiments were performed on Hayward (*Actinidia deliciosa*) orchard located at the experimental farm of the Bologna University. Different concentrations (from 2 to 15 ppm) and application times (from 20 to 100 days after full bloom) of the 3,5,6-TPA were tested. Fruit dimensions

A. Ramina et al. (eds.), Advances in Plant Ethylene Research: Proceedings of the 7th International Symposium on the Plant Hormone Ethylene, 105–107.

(diameter and length) were measured throughout the growing season. At harvest (middle of October), productive parameters (yield, fruit number and fresh weight) were also determined. Fruit quality (flesh firmness, soluble solid and dry matter content, titratable acidity), biochemical (ethylene emission, ACO activity) and molecular parameters were collected at harvest and during the 3-month cold storage period in air. For RT-PCR analyses, primers were designed on a cDNA sequence of Hayward (*Actinidia deliciosa*) homologous to ACO (GeneBank Acc. N: M97961), on a polygalacturonase cDNA sequence from *Actinidia chinensis* (Wang *et al.*, 2000), and a sequence of an expansin of *Malus domestica* (cv Mondial Gala; Stella *et al.*, 2005).

3. Results and Conclusion

Table 1. Expansin and PGC expression and ACO expression and activity of Hayward fruit treated with a synthetic auxin.

	ACO activity (nl/h x mg protein)	ACO expression (ACO/18S band intensity)	Exp expression (Exp/18S band intensity)	PGC expression (PGC/18S band intensity)
Control at harvest	0.01	1.30	0.42	0.33
Control after 1 month of storage	2.35	0.98	0.43	0.17
Control after 3 months of storage	0.3	0.05	0.31	0.03
Treated at harvest	0.01	0.58	0.17	0.19
Treated after 1 month of storage	0.16	0.59	0.32	0.34
Treated after 3 months of storage	42.5	0.88	0.73	0.55

The 3,5,6-TPA application caused a significant increase in Hayward fruit size and slowed down flesh softening and the natural increase in soluble solids content (data not shown). Taken together, these results could indicate a ripening delay probably related to the anti-senescence effect induced by auxin, as previously reported in other species. This delay on ripening might be mediated by ethylene. In fact, in control fruits ACO expression, high at harvest and after 1 month of storage, decreased to nearly undetectable levels after 3 months, while the enzyme activity was detectable at significant level only after 1 month of storage (Table 1). In treated fruits, ACO transcripts accumulation and activity were both low at harvest and after 1 month of storage, and dramatically increased at the end of the experimental period. Accumulation of expansins and PGC transcripts showed the same pattern of those of ACO in both control and treated fruits, reconfirming that the expression of these genes is ethylene controlled (Table 1). These results confirmed the data reported in literature on tomato and *Actinidia chinensis*: in the former species, an expansin gene (LeExp1) specifically expressed at the ripening stage is up-regulated by ethylene and, in the latter, a CkPGC gene expression increases concurrently with fruit softening (Rose *et al.*, 1997; Wang *et al.*, 2000).

References

Alexander, L., Grierson, D. 2002. Ethylene biosynthesis and action in tomato: a model for climacteric fruit ripening. J. Exp. Bot. 53: 2039–2055

Augustí, M., Almela, V., Andreu, I., Juan M., Zacarias, L. 1999. Synthetic auxin 3,5,6-TPA promotes fruit development and climacteric in *Prunus persica* L. Batsch. J. Hort. Sci. Biotechnol. 74: 556–560

Masia, A., Zanchin, A., Rascio, N., Ramina, A. 1992. Some biochemical and ultrastructural aspects of peach fruit development. J. Am. Soc. Hort. Sci. 117: 808–815

Rasori, A., Bertolasi, B., Furini, A., Bonghi, C., Tonutti, P., Ramina, A. 2003. Functional analysis of peach ACC oxidase promoters in transgenic tomato and in ripening peach fruit. Plant Sci. 165: 523–530

Rose, J.K.C., Lee, H.H., Bennet, A.B. 1997. Expression of a divergent expansin gene is fruit specific and ripening-regulated. PNAS 94: 5955–5960

Stella, S., Costa, F., Bregoli, A.M., Ziosi, V., Sansavini, S. 2005. Study on expression of genes involved in ethylene biosynthesis and fruit softening in apple and nectarine. 5th International Postharvest Symposium, Verona, 6–11 Giugno 2004. Acta Hort. 682: 141–148

Tonutti, P., Bonghi, C., Ruperti, B., Tornelli, G.B., Ramina, A. 1997. Ethylene evolution and 1-aminocyclopropane-1-carboxylate oxidase gene expression during early development and ripening of peach fruit. J. Am. Soc. Hort. Sci. 122: 642–647

Wang, Z., MacRae, E.A., Wright, M.A., Bolitho, M., Ross, G.S., Atkinson, R.G. 2000. Polygalacturonase gene expression in kiwifruit: relationship to fruit softening and ethylene production. Plant Mol. Biol. 42: 317–328

Jasmonates delay ripening by interfering with ethylene biosynthesis and perception and with polyamine accumulation in peach fruit

Ziosi, V.[1], Bregoli, A.M.[1], Bonghi, C.[2], Rasori, A.[2], Biondi, S.[3], Costa, G.[1] and Torrigiani, P.[3]*

[1]University of Bologna, Dip. Colture Arboree, Via Fanin 46, Bologna, Italy; [2]University of Padova, Dip. Agronomia Ambientale e Produzioni Vegetali, Agripolis, Padova, Italy; [3]University of Bologna, Dip. Biologia e.s., Via Irnerio 42, Bologna, Italy. (*Corresponding author: patrizia.torrigiani@unibo.it)

Jasmonates (JA) are cyclopentanonic compounds synthesized from linolenic acid. Allene oxide synthase (AOS) is the first specific enzyme and the major control point of their biosynthetic pathway. JA are involved in plant responses to stress and in several developmental processes, including fruit development and ripening (Franceschetti et al., 2004, Redman et al., 2001 and Westernack and House, 2002). Exogenously applied at the end of S3 stage on 'Stark Red Gold' nectarines (*Prunus persica* L. Batsch), 100 ppm of methyl jasmonate (MJ) and 50 ppm of the synthetic analogous *n*-propyl dihydrojasmonate (PDJ) dramatically inhibited ethylene production at harvest and slowed down fruit softening (Torrigiani et al., 2004) (Table 1). Both chemicals also reduced *PpACO1* transcript accumulation at harvest (day 21); on day 27, the message was recovered to almost control levels, paralleling the rise of ethylene production in treated fruit. At the same day, the ethylene receptor *PpERS1* was up-regulated by MJ and PDJ, probably as a consequence of the recovery of ethylene production in treated fruit. MJ and PDJ stimulated *PpAOS1* gene expression, confirming a positive feedback regulation of AOS transcription by JA and suggesting that exogenous JA may increase their endogenous levels. JA also induced a slight accumulation of polyamines (PA), which could contribute to delay ripening: in fact, exogenous PA cause a slowing down of peach fruit ripening. A comparative transcriptome profiling of MJ-treated versus control fruit, performed by using the peach microarray μPEACH1.0 (Trainotti et al., 2006), showed that MJ down-regulated several ripening-related genes, while stimulated mainly genes involved in stress responses. Moreover, the

A. Ramina et al. (eds.), Advances in Plant Ethylene Research: Proceedings of the 7th International Symposium on the Plant Hormone Ethylene, 109–110.

expression pattern of several cell wall–related genes differentially affected by MJ well correlated with the delay in flesh softening. In conclusion, field JA application in S3 stage inhibited climacteric ethylene production and delayed flesh softening, as confirmed by the expression pattern of ethylene biosynthetic and cell wall–related genes. JA stimulation of defence-related genes probably caused a temporary 'suspension' or slowing down of the ripening process. In fact, it has been reported that defence responses inhibited plant growth or reproduction due to different resource allocation.

Table 1. Ethylene emission (EE) and flesh firmness (FF) of untreated (C) and MJ- or PDJ-treated nectarines.

Days after treatment		17	21	27
	C	1.5 ± 0.3	11.0 ± 2.1	2.3 ± 0.3
EE	PDJ	1.1 ± 0.2	0.7 ± 0.06	2.7 ± 0.4
	MJ	0.7 ± 0.09	1.1 ± 0.2	2.5 ± 0.2
	C	50.8 ± 2.0	5.0 ± 1.0	2.5 ± 1.0
FF	PDJ	45.0 ± 8.0	25.0 ± 4.0	13.9 ± 1.6
	MJ	47.0 ± 8.0	28.0 ± 5.0	15.6 ± 2.6

References

Franceschetti M, Perry B, Thompson B, Hanfrey C, Michael AJ (2004) Expression proteomics identifies biochemical adaptations and defence responses in transgenic plants with perturbed polyamine metabolism. FEBS Lett., 576: 477–480

Redman AM, Cipollini DF, Schultz JC (2001) Fitness costs of jasmonic acid-induced defense in tomato, *Lycopersicon esculentum*. Oecologia, 126: 380–385

Torrigiani P, Bregoli AM, Ziosi V, Scaramagli S, Ciriaci T, Rasori A, Biondi S, Costa G (2004). Preharvest polyamines and aminoethoxyviniylglycine (AVG) applications control fruit ripening in Stark Red Gold nectarines (*Prunus persica* L. Batsch). Postharv. Biol. Technol., 33: 293–308

Trainotti L, Bonghi C, Ziliotto F, Zanin D, Rasori A, Casadoro G, Ramina A, Tonutti P (2006) The use of microarray μPEACH1.0 to investigate transcriptome changes during transition from pre-climacteric to climacteric phase in peach fruit. Plant Sci., 170: 606–61

Wasternack C, House B (2002) Jasmonates and octadecanoids: signals in plant stress responses and development. Prog. Nucl. Acid Res. Mol. Biol., 72: 165–221

A novel growth modulator interconnects ethylene, ABA, and sugar signaling

Dugardeyn, J. [1], Hagenbeek, D. [1], Zhang, C. [1,2], and Van Der Straeten, D. [1]*

[1]Unit Plant Hormone Signaling and Bio-imaging, Department of Molecular Genetics, Ghent University, K.L. Ledeganckstraat 35, B-9000 Gent, Belgium, Jasper.Dugardeyn@UGent.be [2]Biotechnology Research Institute, Chinese Academy of Agricultural Sciences, Beijing 100081, China. (*Corresponding author: dominique.vanderstraeten@ugent.be)

1. Introduction

Leaf emergence is ethylene dependent (Smalle and Van Der Straeten, 1997). The *slo1-1* mutant was isolated based on a delay of leaf emergence (Van Der Straeten *et al.* 1999). *SLO1* was shown to encode a pentatricopeptide repeat (PPR) protein.

PPR proteins are characterized by the presence of a 35 amino acid repeat. In Arabidopsis, the PPR protein family contains 441 members (Lurin *et al.* 2004). Based on their structure, these proteins are thought to bind RNA. Most of them are predicted to be present in chloroplasts or mitochondria (Lurin *et al.* 2004).

Ethylene and ABA often play antagonistic roles. *Ctr1-10* for instance is less sensitive to ABA than the wild type, while *ein2-45* is more sensitive (Beaudoin *et al.* 2000). EIN3 is stabilized by ethylene, whereas glucose destabilizes this protein (Yanagisawa *et al.* 2003). Furthermore, it is known that sugar and ABA act synergistically. Several sugar-insensitive mutants are also known as ABA-biosynthesis or ABA-insensitive mutants (Léon and Sheen, 2003).

2. Results

slo1-1 was proven to have an altered sensitivity to ethylene. To test the interaction with ABA, a dose-response test was performed on germination in the presence/absence of ABA. A delay in germination of *slo1-1* versus the wild type could be observed on control medium. This delay was enhanced by adding ABA. We further investigated whether the ABA

A. Ramina et al. (eds.), Advances in Plant Ethylene Research: Proceedings of the 7th International Symposium on the Plant Hormone Ethylene, 111–112.

hypersensitivity was reflected in sugar sensitivity. *Slo1-1* showed a stronger developmental delay on high sucrose concentrations as compared to the wild type.

Using a promoter GUS fusion, the expression of *SLO1-1* was analyzed. A strong induction of expression by sugar in the root tips of dark- and light-grown plants was observed. This up-regulation was counteracted by ACC, confirming the antagonistic role of SLO1-1 in the sugar and ethylene signaling pathway.

3. Conclusion

Our data strongly support an important role for SLO1-1 in the cross talk between ethylene and sugar/ABA.

References

Beaudoin N, Serizet C, Gosti F, Giraudat J (2000) Interactions between abscissic acid and ethylene signaling cascades. Plant Cell 12(7): 1103–1115

Léon P, Sheen J (2003) Sugar and hormone connections. TIPS 8(3): 110–116

Lurin C, Andres C, Aubourg S, Bellaoui M, Bitton F, Bruyere C, Caboche M, Debast C, Gualberto J, Hoffmann B, Lecharny A, Le Ret M, Martin-Magniette ML, Mireau H, Peeters N, Renou JP, Szurek B, Taconnat L, Small ID (2004) Genome-wide analysis of Arabidopsis pentatricopeptide repeat proteins reveals their essential role in organelle biogenesis. Plant Cell 16(8): 2089–2103

Smalle J, Van Der Straeten D (1997) Ethylene and vegetative development. Physiol Plant 100: 593–605

Van Der Straeten D, Smalle J, Bertrand S, De Paepe A, De Pauw I, Vandenbussche F, Haegman M, Van Caeneghem W, Van Montagu M (1999) Ethylene signaling: more players in the game. In: A.K. Kanellis, C. Chang, H. Klee, A.B. Bleecker, J.C. Pech, and D. Grierson (eds) *Biology and Biotechnology of the Plant Hormone Ethylene II.* Kluwer Academic Publishers, Dordrecht, pp.71–75

Yanagisawa S, Yoo SD, Sheen J (2003) Differential regulation of EIN3 stability by glucose and ethylene signaling in plants. Nature 425(6957): 521–525

Ethylene and ABA cross-communication and plant growth response to salt stress in tomato (*Solanum lycopersicum* L.)

Picarella, M.E.[1], Antonelli, M.G.[1], Astolfi, S.[1], Zuchi, S.[1], Vernieri, P.[2] and Soressi, G.P.[1]

[1]Dipartimento di Agrobiologia e Agrochimica, Sezione Genetica, Università della Tuscia, Via C. De Lellis, I-01100 Viterbo, Italy.
[2]Dipartimento di Biologia delle Piante agrarie, Università di Pisa, V.le delle Piagge, 23–56124 Pisa. (*Corresponding author: soressi@unitus.it)

Ethylene (ET) is a signal molecule involved in the regulation of gene expression during adaptation to abiotic stresses. The relationship between ET and ABA production in vegetative tissues under abiotic stresses appears mostly controversial. In order to ascertain the interplay between ET and ABA in young plants (5th leaf stage), grown hydroponically in control (3 mS cm^{-1}) and salt stress (10 mS cm^{-1}) conditions, ET emission, ABA content and dry weight (DW) of roots, basal and apical leaves were measured in five different tomato genotypes: cv Edkawi (EDK), salt tolerant; cv Gimar (GIM), relatively salt sensitive and its near isogenic line for the nor gene (NOR) defective in ET synthesis; the ABA mutants sitiens (SIT) and notabilis (NOT), both with different ABA biosynthetic capacity (8% vs 47%).

The recorded data (Fig. 1) showed that, under salt stress, ET production was either depressed or significantly unaffected in all the genotypes with the exception of SIT where meanly a fivefold increase of this hormone was evidenced in basal and apical leaves. ABA concentration in roots was significantly higher in EDK, GIM, SIT and in apical leaves of NOR and NOT. Interestingly the ABA mutants displayed ABA values comparable to those of the other genotypes. The salt treatment affected root and leaf DW (data not shown) of GIM, NOT, SIT and NOR (excluding apical leaf), but not of EDK.

In leaves of EDK and GIM, the salt induced an ET decrease not correlated with the ABA concentrations which remain in a steady state, while a negative correlation is evident in their root system. In the two ABA-deficient mutants, a clear general view of an interaction between

A. Ramina et al. (eds.), Advances in Plant Ethylene Research: Proceedings of the 7th International Symposium on the Plant Hormone Ethylene, 113–114.

ABA and salt stress cannot be addressed except for root of SIT similarly to that of EDK and GIM. Despite the extremely reduced ET emission in the *nor* mutant, a negative interaction with ABA is given for apical leaf. According to our experimental data a negative counteract between the two hormones cannot be ruled out. Under salt stress, the physiological pathways of the two hormones and consequently their interaction are likely influenced by the plant genetic background and/or the growth conditions.

The unexpected ABA accumulation in NOT and SIT could be explained by alternative pathways acting behind the mutation steps. This fact together with our data on ABA concentrations unaffected by salt stress, suggest the existance of a complex process regulating ABA homeostasis.

According to the DW data, the salt treatment induced growth inhibition of GIM, NOR (except for apical leaf), NOT and SIT; only EDK, costitutively salt tolerant, was unaffected. The hypothesis that the ET high level of SIT in both control and stress conditions be responsible of its reduced growth does not fit the other genotypes.

Possibly other mechanisms independent of ET and ABA in the salt/osmotic stress adaptation process are responsible for the growth reduction of plants.

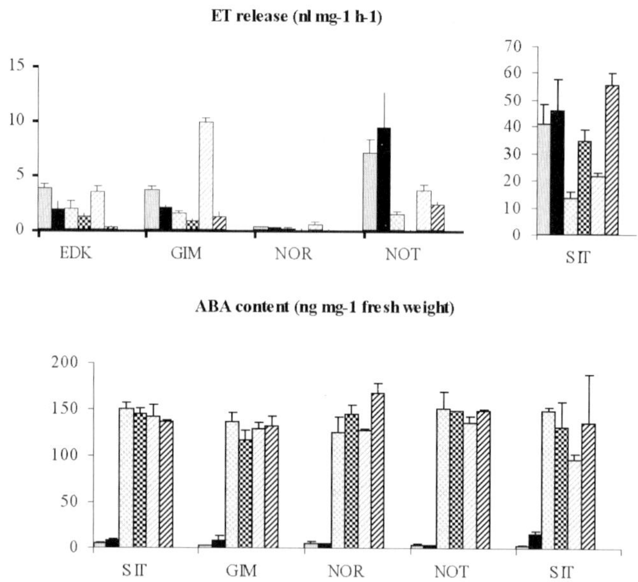

Fig. 1. Ethylene release and ABA content in five tomato genotypes. (Grey colour 3 mS cm^{-1}; black colour 10 mS cm^{-1}; solid case: root; squared case: basal leaf; barred case: apical leaf).

Auxin, ethylene and brassinosteroids: cross talk in the *Arabidopsis thaliana* hypocotyl

De Grauwe, L.[1,3], Vandenbussche, F.[1,3], Tietz, O.[2], Palme, K.[2] and Van Der Straeten, D. [1]*

[1]Unit Plant Hormone Signaling and Bio-Imaging, Ghent University, K.L.Ledeganckstraat 35, B-9000 Gent, Belgium; [2]Institut für Biologie II-Botanik, Universität Freiburg, Germany; [3]These authors contributed equally to this work. (*Corresponding author: Dominique.vanderstraeten@ugent.be)

Apical hook establishment and maintenance is an important developmental process driven by multiple hormonal cross talk. The importance of ethylene and asymmetric auxin-distribution, the requirement of gibberellins and the role of brassinosteroids have been demonstrated previously (Bleecker *et al.* 1988; Lehman *et al.* 1996; Chory *et al.* 1991; Achard *et al.* 2003; Vriezen *et al.* 2004). In the light, a similar complexity of hormonal regulation has been revealed wherein all aforementioned hormones can stimulate hypocotyl elongation.

A model of interaction between ethylene, auxins and BRs in hook maintenance and exaggeration has been proposed (Fig. 1). It is clear that the three hormones influence each other and are all necessary for the formation of the apical hook. Upon ethylene accumulation, the auxin distribution is altered, either directly or through a change in BR biosynthesis. The differential auxin distribution might be responsible for an altered ethylene production or sensitivity at the inner side of the apical hook (Raz and Ecker 1999; Vriezen *et al.* 2004), thereby causing an inhibition of cell elongation, resulting in an exaggerated apical hook.

Moreover, there are some interesting parallels between hook maintenance/ exaggeration in the dark and hypocotyl elongation in the light: they both depend upon ethylene and auxin; they both have a time window of sensitivity and a number of proteins are common to both processes (De Grauwe *et al.* 2005). The compensation of BRs for the insensitivity to ethylene in *hls* mutants, suggests a downstream role for BRs.

Although these responses appear to be largely controlled by the same mechanisms, some differences were noticed in light and darkness: IAA could only induce a BR-reporter gene in darkness; auxin transport appears

A. Ramina et al. (eds.), Advances in Plant Ethylene Research: Proceedings of the 7th International Symposium on the Plant Hormone Ethylene, 115–117.

to be involved only in hypocotyl elongation in the light, and light may also have a direct negative effect on BR biosynthesis (De Grauwe *et al.* 2005).

In conclusion, it is clear that seedling growth is dependent on a highly interacting network of hormone signals, all contributing to the most appropriate phenotype in an ever changing environment, especially upon alterations in light conditions.

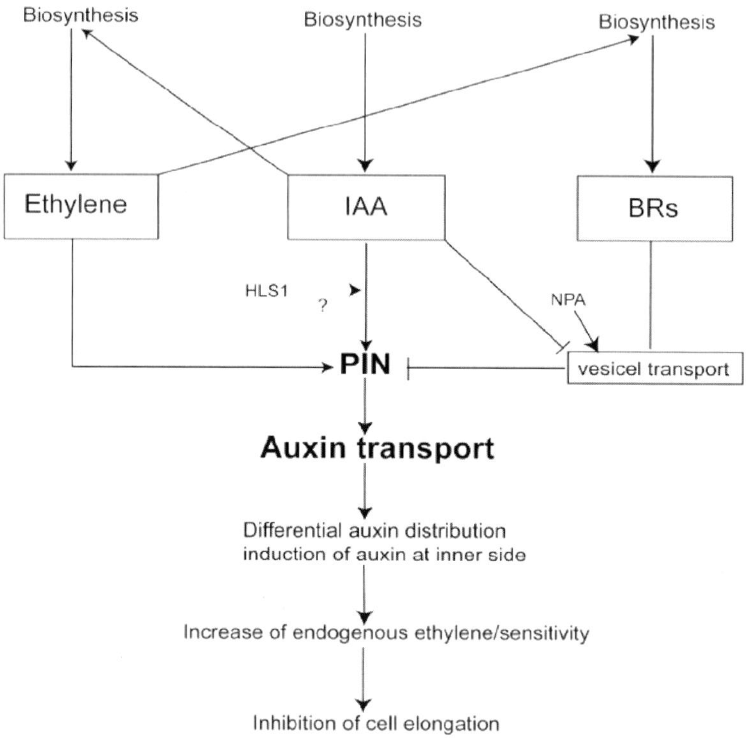

Fig. 1. Model of the interaction between ethylene, auxin and brassinosteroids in the apical hook of etiolated seedlings (De Grauwe *et al.* 2005).

References

Achard P, Vriezen WH, Van Der Straeten D, Harberd NP (2003) Ethylene regulates Arabidopsis development via the modulation of DELLA protein growth repressor function. Plant Cell 15:2816–2825

Bleecker A, Estelle M, Somerville C, Kende H (1988) Insensitivity to ethylene conferred by a dominant mutation in *Arabidopsis thaliana*. Science 241:1086–1089

Chory J, Nagpal P, Peto CA (1991) Phenotypic and genetic analysis of det2, a new mutant that affects light-regulated seedling development in *Arabidopsis*. Plant Cell 3:445–459

De Grauwe L, Vandenbussche F, Tietz O, Palme K, Van Der Straeten D (2005) Auxin, Ethylene and Brassinosteroids: Tripartite control of growth in the *Arabidopsis* hypocotyl. Plant Cell Physiol 46(6):827–836

Lehman A, Black R, Ecker JR (1996) HOOKLESS1, an ethylene response gene, is required for differential cell elongation in the *Arabidopsis* hypocotyl. Cell 85:183–194

Raz V, Ecker JR (1999) Regulation of differential growth in the apical hook of *Arabidopsis*. Development 126:3661–3668

Vriezen WH, Achard P, Harberd NP, Van Der Straeten D (2004) Ethylene-mediated enhancement of the apical hook formation in etiolated *Arabidopsis thaliana* seedlings is gibberellin dependent. Plant J 37:505–516

A PPR protein, required for normal plant development, may be involved in control of the ethylene pathway at the posttranscriptional level

Hagenbeek, D.[1], Dugardeyn, J.[1], Zhang, C.[1,2] and Van Der Straeten, D.[1]*

[1]Unit Plant Hormone Signalling and Bio-imaging, Department of Molecular Genetics, Ghent University, K.L. Ledeganckstraat 35, B-9000 Gent, Belgium, Dik.Hagenbeek@UGent.be. [2]Biotechnology Research Institute, Chinese Academy of Agricultural Sciences, Beijing 100081, China. (*Corresponding author: Dominique.VanDerStraeten@UGent.be)

1. Introduction

To enable discovery of factors controlling specific developmental processes, screens on developmental stages other than etiolated seedlings are needed. Smalle *et al.* (1997) showed that leaf emergence was enhanced by the ethylene precursor ACC. This was exploited as an assay for mutants in ethylene response.

2. Results

In order to confirm that leaf emergence is an ethylene-regulated process, we compared leaf emergence of the first two true leaves of several ethylene signalling mutants. *ctr1-1* (*constitutive triple response 1-1*) mutant seedlings had an enhanced response, close to the maximal response of wild type Col0 on ACC, while the ethylene insensitive mutants *etr1-3* (*ethylene resistant 1-3*) and *ein2-1* (*ethylene insensitive 2-1*) showed delayed leaf emergence as compared to Col0 in the absence of ACC and did not respond to ACC. This firmly proved that leaf emergence is an ethylene-regulated process.

A mutant with delayed first leaf pair emergence was identified. Map-based cloning revealed a short deletion in a gene encoding a family member of the pentatricopeptide repeat proteins (PPR; Small and Peeters, 2000; Lurin *et al.*, 2004). The mutant trait is recessive, indicating a loss of function, and not allelic to known ethylene mutants. Besides the delay in first leaf pair emergence, a general delay at different developmental stages,

A. Ramina et al. (eds.), Advances in Plant Ethylene Research: Proceedings of the 7th International Symposium on the Plant Hormone Ethylene, 119–120.

including germination, leaf emergence, and flowering, was observed. Therefore the mutant was named *slow1* (*slo1-1*). Root growth, hypocotyl growth, and rosette expansion were reduced in the mutant, rendering it semi dwarf.

Quantitative RT-PCR indicated that *SLO1* expression is ethylene-regulated.

3. Conclusion

Our data suggest a role for *SLO1* throughout the plant's life cycle. PPR proteins possibly function by binding RNA species, thus recruiting them into complexes for processing by other proteins associated to that complex. Recently, in vivo binding of an mRNA target has been proved by RIP-CHIP for CRP1, a PPR family member (Schmitz-Linneweber *et al.*, 2005). SLO1 may define a new mechanism of control of ethylene signal transduction at the posttranscriptional or translational level.

References

Lurin C, Andres C, Aubourg S, Bellaoui M, Bitton F, Bruyere C, Caboche M, Debast C, Gualberto J, Hoffmann B, Lecharny A, Le Ret M, Martin-Magniette ML, Mireau H, Peeters N, Renou JP, Szurek B, Taconnat L, Small I (2004) Genome-wide analysis of Arabidopsis pentatricopeptide repeat proteins reveals their essential role in organelle biogenesis. Plant Cell 16:2089–2103

Schmitz-Linneweber C, Williams-Carrier R, Barkan A (2005) RNA immunoprecipitation and microarray analysis show a chloroplast pentatricopeptide repeat protein to be associated with the 5' region of mRNAs whose translation it activates. Plant Cell 17:2791–2804

Small ID, and Peeters, N (2000). The PPR motif—a TPR-related motif prevalent in plant organellar proteins. Trends Biochem Sci 25:46–47

Smalle J, Haegman M, Kurepa J, Van Montagu M, Van Der Straeten D (1997). Ethylene can stimulate Arabidopsis hypocotyl elongation in the light. Proc Natl Acad Sci USA 94:2756–2761

3. ROLE OF ETHYLENE IN PLANT GROWTH AND DIFFERENTIATION

Changes in ethylene sensitivity by regulated expression of the tomato ethylene receptor family

Kevany, B.M. and Klee, H.J.*

University of Florida, Horticultural Sciences, Gainesville, FL 32611-0690, USA. (*Corresponding author: hjklee@ifas.ufl.edu)

1. Introduction

It has been well established that plants are able to fine tune phytohormone responses by control of synthesis and/or perception. Ethylene biosynthesis has been widely studied in a number of species but none more than Arabidopsis and tomato. Tomato provides a unique system to study ethylene biosynthesis because of the large, climacteric-associated increase in production at the onset of ripening. The ethylene biosynthetic pathway has two committed steps with the conversion of S-adenosylmethionine to 1-aminocyclopropane-1-carboxylic acid (ACC) by ACC synthase (ACS) being the rate-limiting step. The enzymes for ethylene biosynthesis are encoded by small gene families in tomato with each gene showing distinct expression patterns (Barry *et al.* 1996, Nakatsuka *et al.* 1998). Ethylene biosynthesis proceeds through two distinct stages during development with System I being associated with vegetative and immature fruit tissues and System II being linked with the climacteric burst at the onset of ripening. While the biosynthetic pathway has been well characterized in tomato the mechanisms by which plants perceive ethylene and transduce its signal are less understood. It is clear that immature and mature fruit respond differently to ethylene. Treatment of immature fruits with ethylene does not initiate ripening, as in mature fruit, but does hasten the onset of ripening (Yang, 1987). Fruits in which System I ethylene production is suppressed ripen later than fruits that produce these low levels of ethylene. Immature fruits appear to have a mechanism that measures cumulative ethylene exposure and uses this information to determine the timing of ripening. The manner in which fruits measure this cumulative exposure is completely unknown but research done on the ethylene receptor gene family in tomato may provide clues for how fruits achieve this end. Work is being done to gain an understanding of how ethylene signaling proceeds

A. Ramina et al. (eds.), Advances in Plant Ethylene Research: Proceedings of the 7th International Symposium on the Plant Hormone Ethylene, 123–128.

through the receptors and how the system in tomato may be fundamentally different to that of Arabidopsis.

2. Results and Discussion

2.1 The tomato ethylene receptor gene family

The ability of a plant to measure the amount of a particular hormone obligately requires the presence of hormone-specific receptors. The tomato ethylene receptor gene family is comprised of at least six members (*LeETR1-6*), all showing significant homology to ethylene receptor family members in Arabidopsis (Zhou *et al.* 1996a, Zhou *et al.* 1996b, Lashbrook *et al.* 1998,Tieman and Klee 1999). Within the tomato gene family there is significant divergence between family members with some pairings having only 50% identity to each other. All of the receptors additionally show considerable homology to two-component systems found in prokaryotic organisms (Chang *et al.* 1993). Ethylene receptors were the first genes identified in plants that showed homology to these prokaryotic systems but it is unclear whether they function in a similar manner.

The Arabidopsis receptor family consists of five members with ETR1 and ERS1 belonging to Subfamily I and ETR2, ERS2 and EIN4 belonging to Subfamily II. All of the initial receptor mutants identified where gain-of-function mutants exhibiting dominant ethylene insensitivity. Single loss-of-function (LOF) receptor mutants in Arabidopsis show no visible phenotypes likely due to functional redundancy. The presence of triple, quadruple or *etr1/ers1* double mutants results in a constitutive ethylene response in the absence of increased ethylene production (Wang *et al.* 2003). Thus, these multiple mutation lines can be said to exhibit ethylene hypersensitivity. This evidence is consistent with a model where the receptors are negative regulators of ethylene response and that in the absence of ethylene wild-type receptors actively suppress the response (Hua and Meyerowitz, 1998).

The first tomato ethylene receptor cloned was *Never-ripe* (*NR*) (Wilkinson *et al.* 1995, Yen *et al.* 1995). *Nr* was first observed as a single spontaneous non-ripening fruit in a commercial field in the late 1940's. The mutant exhibits semidominant ethylene insensitivity with defects in flower wilting, leaf and petal senescence and fruit ripening. Using *NR* as a probe, additional genes were subsequently identified. The receptor family in tomato is broken down into two similar subfamilies with Subfamily I being comprised of LeETR1, 2 and NR and Subfamily II containing *LeETR4–6*. Subfamily I members contain three transmembrane domains and all of the conserved residues of known histidine kinases. Subfamily II members contain four transmembrane domains and degenerate histidine kinase domains with LeETR5 containing none of the conserved residues. NR is the only member of the family that does not contain the carboxy-terminal receiver domain whose function in ethylene signaling is unknown. While the presence of a histidine kinase domain would suggest that it is directly involved in signaling no such relationship has ever been demonstrated. Antisense lines of *LeETR1*, *2*, *5* and *NR* are indistinguishable from wild-type plants but transgenic lines with reduced *LeETR4* or *LeETR6* expression have phenotypes consistent with a constitutive ethylene response (Lashbrook *et al.* 1998, Tieman *et al.* 2000). Antisense *ETR4* and *ETR6* lines display epinastic growth, premature flower senescence and a reduction in the time from anthesis to the first signs of ripening of up to ten days. The ripening phenotype of these transgenic lines again represents the power of using the tomato system to uncover novel functions of receptor proteins that would be otherwise impossible while using Arabidopsis. Transgenic plants with reduced expression of *NR* do not display any visible phenotypes but it is not the consequence of direct functional redundancy but in fact the increase in expression of *ETR4* that rescues the phenotype. This type of rescue has been termed functional compensation (Tieman *et al.* 2000). The overall model in tomato appears to be quite different to that of Arabidopsis with results suggesting that individual family members may be more important than others. Furthermore, loss-of-function of either of two Subfamily II members leads to altered ethylene responses. In Arabidopsis, the data support a system in which the Subfamily I members (ETR1 and ERS1) are far more important than the Subfamily II members; over-expression of a Subfamily II member cannot rescue a Subfamily I double mutant (Wang *et al.* 2003). In contrast, the ethylene receptor family in tomato does not seem to be functionally segmented. Over-expression of *NR* fully suppresses the constitutive ethylene response in *LeETR4* antisense lines.

2.2 Does receptor expression fit the model?

The ability of a plant to have strict control over a phytohormone response is key to the overall health of the plant. Quite often, hormone responses are meant to be transient reactions to internal cues or external stimuli resulting in altered growth or even death of target tissues. The plant must have mechanisms for shutting down the response as rapidly as is appropriate. In the case of ethylene, its involvement in both developmental processes and responses to biotic and abiotic stresses has clearly been demonstrated. Ethylene often causes a growth arrest in response to an external stress.

Once the stress is removed, normal growth must resume. Expression of key enzymes in the ethylene biosynthetic and signal transduction pathway is under strict control so that the plant can quickly alter its growth.

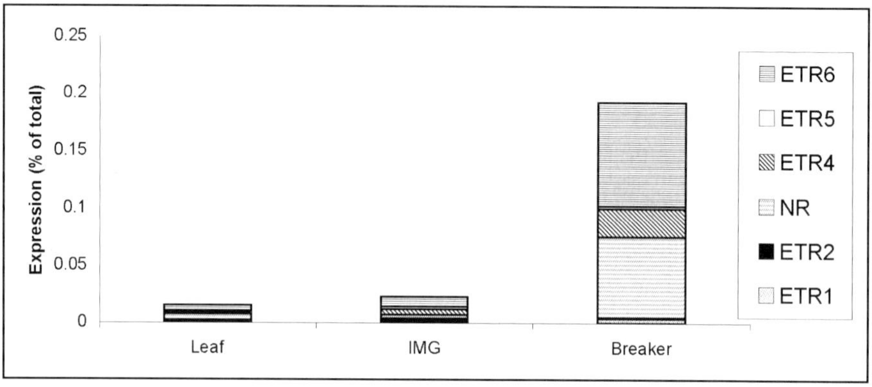

Fig. 1. Expression of each receptor family member in leaf, immature green and breaker fruit assayed by quantitative RT-PCR.

Each member of the tomato ethylene receptor gene family has a unique expression pattern. Are these expression patterns consistent with the model of how receptors mediate the plant's response to ethylene? The current model states that receptors are negative regulators of ethylene response. Increased expression would reduce sensitivity to the hormone and conversely, decreased expression would increase sensitivity. Transgenic plants with increased expression of *NR* are less sensitive to ethylene. Plants with reduced expression of either *LeETR4* or *ETR6* are more sensitive to ethylene, displaying constitutive ethylene phenotypes in the absence of an increase in ethylene production. But does this tell the whole story about what is happening in a normal plant? The expression pattern of all six receptors has been assayed throughout fruit development and in vegetative tissues and the results are somewhat paradoxical. *LeETR1*, *2* and *5* show relatively low, constitutive expression in all vegetative tissues and throughout fruit development. In contrast *LeETR4*, *6* and *NR* show low expression in vegetative and immature fruit tissue but upon the initiation of ripening there is a drastic increase in expression of these three receptors (Fig. 1). The ethylene inducibility of each receptor in immature fruit has also been assessed and only *LeETR4*, *6* and *NR* are ethylene inducible, suggesting that the increase of each of these receptors at the onset of ripening is the result of increased ethylene production. The ripening-associated increase represents a tenfold increase in total receptor content in the ripening fruit and begs the question: why would there be increased expression of a negative regulator in a tissue that requires ethylene for an essential developmental process? This paradox suggests that there is more to ethylene regulation than we currently understand.

With the notable exception of fruit ripening, receptor mRNA expression fits the proposed model of how receptors initiate ethylene signal transduction. With that said, a wealth of current data indicate that there is extensive post-transcriptional and post-translational control of hormone responses. Ethylene receptor proteins have been shown to bind ethylene in yeast and that this binding has a half-life of approximately 12 h (Schaller and Bleecker, 1995). This estimate is likely an underestimate because it does not factor in turnover of the protein. It is very possible that ethylene binding brings about irreversible changes in receptor protein activity. Protein turnover is likely a key regulatory point in the signal transduction pathway and an understanding of how ethylene binding affects turnover needs to be addressed.

3. Conclusion

Ethylene is a key regulator of multiple developmental processes and stress responses. The ethylene signal transduction pathway was first described in Arabidopsis and is now being addressed in agronomically important crops like tomato and rice. Tomato is an interesting model to study the involvement of ethylene in a myriad of responses but also allows the unique opportunity to study its role in fleshy fruit development. Work already done has shown that fruit of different developmental stages respond to ethylene in different ways. Immature fruit possess the capacity to measure how much ethylene they have encountered and use this information to determine when to begin ripening. The way this tissue accomplishes this is unknown but likely lies somewhere in the signal transduction system and the receptors appear to be good candidates to fit this role. Further work needs to be done to address how the receptors fit into the regulatory network but current data suggests they are a key point of control.

Acknowledgments

We wish to acknowledge the support of the U.S.D.A. National Research Initiative and the Dickman family endowment to the Klee lab for support of our tomato ethylene research.

References

Barry, C.S., Blume, B., Bouzayen, M., Cooper, W., Hamilton, A.J. and Grierson, D. (1996). Differential expression of the 1-aminocyclopropane-1-carboxylate oxidase gene family of tomato. Plant J 9: 525–535.

Chang, C., Kwok, S.F., Bleecker, A.B. and Meyerowitz, E.M. (1993). Arabidopsis ethylene-responsive gene *ETR1*: similarity of products to two-component regulators. Science 262:539–544

Hua, J. and Meyerowitz, E.M. (1998). Ethylene responses are negatively regulated by a receptor gene family in *Arabidopsis thaliana*. Cell 94:261–271

Lashbrook, C.C., Tieman, D.M. and Klee, H.J. (1998). Differential regulation of the tomato ETR gene family throughout plant development. Plant J 15(2): 243–252

Nakatsuka, A., Murachi, S., Okunishi, H., Shiomi, S., Nakano, R., Kubo, Y. and Inaba, Y. (1998). Differential expression and internal feedback regulation of 1-aminocyclopropane-1-carboxylate synthase, 1-aminocyclopropane-1-carboxylate oxidase, and ethylene receptor genes in tomato fruit during development and ripening. Plant Physiol 118:1295–1305

Schaller, G.E. and Bleecker, A.B. (1995). Ethylene-binding sites generated in yeast expressing the *Arabidopsis ETR1* gene. Science 270:1809–1811

Tieman, D.M. and Klee, H.J. (1999). Differential expression of two novel members of the tomato ethylene-receptor family. Plant Physiol 120:165–172.

Tieman, D.M., Taylor, M.G., Ciardi, J.A. and Klee, H.J. (2000). The tomato ethylene receptors NR and LeETR4 are negative regulators of ethylene response and exhibit functional compensation within a multigene family. PNAS 97:5663–5668

Wang, W., Hall, A.E., O'Malley, R. and Bleecker, A.B. (2003). Canonical histidine kinase activity of the transmitter domain of the ETR1 ethylene receptor from *Arabidopsis* is not required for signal transmission. PNAS 100:352–357.

Wilkinson, J.Q., Lanahan, M.B., Yen, H.C., Giovannoni, J.J. and Klee H.J. (1995). An ethylene-inducible component of signal transduction encoded by *Never-ripe*. Science 270:1807–1809.

Yang, S.F. (1987). The role of ethylene and ethylene synthesis in fruit ripening. In W Thompson, E Nothnagel, R Huffaker, eds. *Plant Senescence: Its Biochemistry and Physiology*. The American Society of Plant Physiologists, Rockville, MD, pp 156–165.

Yen, H.C., Lee, S., Tanksley, S.D., Lanahan, M.B., Klee, H.J. and Giovannoni, J.J. (1995). The tomato Never-ripe locus regulates ethylene-inducible gene expression and is linked to a homolog of the Arabidopsis ETR1 gene. Plant Physiol 107:1343–1353

Zhou D., Kalaitzis, P., Mattoo, A. and Tucker, M. (1996a). The mRNA for an *ETR1* homologue in tomato is constitutively expressed in vegetative and reproductive tissues. Plant Mol Biol 30:1331–1338.

Zhou, D., Mattoo, A. and Tucker, M. (1996b). Molecular cloning of a tomato cDNA encoding and ethylene receptor. Plant Physiol 110:1435–143

Ethylene induces sexual development through the enhanced expression of a novel *zyg1* gene in *Dictyostelium*

Amagai, A.*

Department of Developmental Biology and Neurosciences, Graduate School of Life Sciences, Tohoku University, Aoba, Sendai 980-8578, Japan. (*Corresponding author: aiamagai@mail.tains.tohoku.ac.jp)

1. Introduction

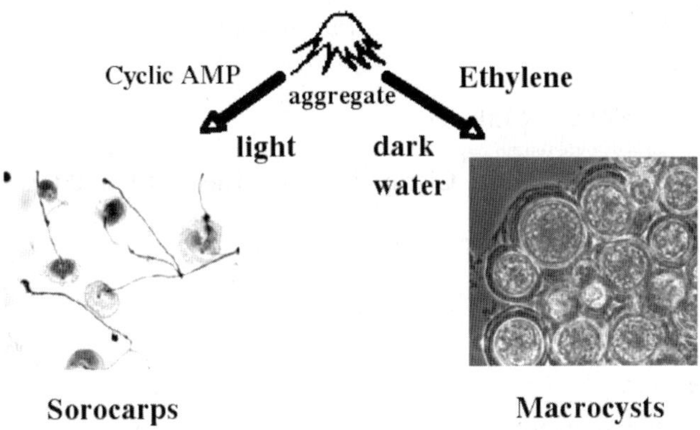

Fig. 1. Physical and chemical regulators controlling development in *Dictyostelium mucoroides-7* (Dm7).

It is generally accepted that ethylene is a potent hormone involved in regulation of many events in plant growth and development (Abeles, 1973). Ethylene is produced and released from fungi and bacteria as well as from higher plants. The cellular slime molds were also found to produce ethylene for regulation of sexual development including zygote formation, as shown in *Dictyostelium mucoroides* (Amagai, 1984 ; Amagai, 1989). *D. mucoroides-7* (Dm7) exhibits clear dimorphism in development depending upon environmental conditions such as light and water: fruiting body formation occurs as the asexual process in the light, while macrocyst formation as the sexual process either in the dark or under water (Weinkauf and Filosa, 1965). One of the most remarkable events characteristic of macrocyst

A. Ramina et al. (eds.), Advances in Plant Ethylene Research: Proceedings of the 7th International Symposium on the Plant Hormone Ethylene, 129–133.

formation is the appearance of giant cells at the aggregation stage (Filosa and Dengler, 1972). The giant cell has been known to be a true zygote, which is formed by sexual cell fusion and subsequent nuclear fusion (Amagai, 1989). The environmental conditions determine the developmental fates by controlling the production of two chemicals, cyclic AMP (cAMP) and ethylene (Amagai, 1984; Amagai and Filosa, 1984). Dm7 cells forming sorocarps in the light condition produce a higher amount of cAMP than those forming macrocysts in the dark condition (Amagai, 1984; Amagai and Filosa, 1984; Amagai. 1987) (Fig. 1). When Dm7 cells were incubated in the presence of AOA (amino-oxy-acetic acid), a potent inhibitor of ethylene biosynthesis, the production of ethylene is decreased, thus resulting in inhibition of macrocyst formation (Amagai, 1987). From these findings, we have proposed an idea that the choice of two developmental forms may be determined by the ratio of ethylene/cAMP at the aggregate stage when the developmental fate is determined. In facts, these two chemicals greatly affect zygote formation: ethylene induces zygote formation, while cAMP inhibits it (Amagai, 1989; Suzuki *et al.*, 1992) (Fig. 2).

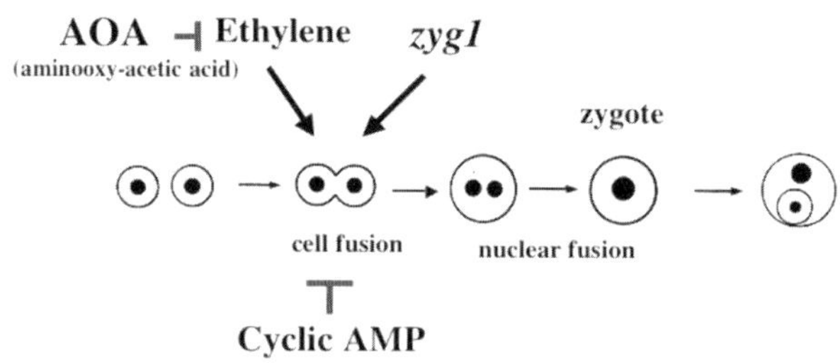

Fig. 2. Regulation of zygote formation.

As a novel gene expressed specifically during zygote formation, we have isolated *zyg1* from the cDNA library of Dm7 cells (Amagai, 2002). The expression of *zyg1* is developmentally regulated, and the overexpression of *zyg1* induces macrocyst formation through the augmented zygote formation, thus suggesting that the *zyg1* expression is tightly coupled with zygote formation (Amagai, 2002).

In *Dictyostelium*, ethylene is biosynthesized from methionine through S-adenosyl-L-methionine (SAM) and 1-aminocyclopropane-1-carboxylic acid (ACC), as in higher plants (Amagai and Maeda, 1992). In this context, cDNAs corresponding to ACC synthase and ACC oxidase (*Dd-aco*) genes were recently isolated by the Japanese *Dictyostelium* cDNA Project. To reveal the precise relationship between the amount of ethylene and the zygote formation, and also the relation of the *zyg1* expression to ethylene, we prepared two kinds of transformants that over- or under-produce ethylene by the regulated expression of ACC-oxidase homologue gene (*Dd-aco*) in Dm7 cells and analyzed their developmental behaviors.

2. Results

2.1 Relationship between the amount of ethylene and the zygote formation

2.1.1 Enhanced zygote formation by the over-expression of **Dd-aco**

Using cDNAs corresponding to ACC oxidase genes (*Dd-aco*) provided by the Japanese *Dictyostelium* cDNA Project, ACO^{OE} cells over-expressing the *Dd-aco* gene were prepared. ACO^{OE} cells exhibited strong expression of *Dd-aco* throughout development, compared to the parental Dm7. Using the two-facing culture method, ACO^{OE} cells were also shown to produce a higher amount of ethylene than Dm7 and ACO^{OE} formed macrocysts even under the conditions (in the light) favoring sorocarp formation.

2.1.2 Inhibition of zygote formation by **Dd-aco**-*knockdown*

ACO-RNAi cells under-expressing the *Dd-aco* were prepared by the RNAi method, and their decreased ethylene production was confirmed by gas chromatography. They exhibited sorocarp formation in the dark as well as in the light, and also failed to form macrocysts under water, result-ing in the formation of loose aggregates. In fact, zygote formation that is prerequisite for macrocyst formation was markedly suppressed in ACO-RNAi cells. Thus, it is evident that zygote and subsequent macrocyst formation are inhibited by the decreased production of ethylene.

2.1.3 Relationship between the amount of ethylene and the **zyg1** *expression*

Since *zyg1* is known to be involved in zygote formation, its relation to ethylene was examined by comparing *zyg1* expressions in ACO^{OE}, ACO-RNAi, and parental Dm7 cells under the submerged conditions. As a result, it was found that Dm7 and ACO^{OE} cells exhibited high levels of *zyg1* expression, coupling with zygote and macrocyst formation, while ACO-RNAi cells gave a significantly lower level of *zyg1* expression, thus resulting in their failure to form zygotes. Moreover, the *zyg1* expression was greatly inhibited by AOA in a dose-dependent manner. Taken together, the correlation among the amount of ethylene, *zyg1* expression, and developmental modes of cells is summarized in Table 1.

Table 1. The correlation among the amount of ethylene, *zyg1* expression and developmental modes.

Cells	Ethylene amount	*Zyg1* expression	Developmental modes
Dm7	High	High	Macrocysts
ACO^{OE}	High	High	Macrocysts
ACO-RNAi	Low	Low	Loose aggregates

3. Conclusions

Here we have succeeded in finely regulating the amount of ethylene production by cells, preparing transformants that over- or under-produce *Dd-aco*. The use of these transformants provided a valuable way for clarification of the relationship between ethylene production and zygote (macrocyst) formation. The level of *zyg1* expression was also found to change depending on the amount of ethylene biosynthesized in the cells. From these results, we have concluded that ethylene induces zygote (macrocyst) formation through enhanced expression of *zyg1*. This is the first demonstration of a marked function of ethylene at the cellular level. Thus, analyses using the developmental system of *Dictyostelium* offer promise for elucidation of the mechanism of sexual development coupled with the ethylene action.

References

Abeles FB (1973) Ethylene in Plant Biology. Academic Press, New York

Amagai A (1984) Induction by ethylene of macrocyst formation in the cellular slime mould *Dictyostelium mucoroides*. J. Gen. Microbiol. 130:2961–2965

Amagai A (1987) Regulation of the developmental modes in *Dictyostelium mucoroides* by cAMP and ethylene. Differentiation 36:111–115

Amagai A (1989) Induction of zygote formation by ethyl-ene during the sexual development of the cellular slime mold *Dictyostelium mucoroides*. Differentiation 41:176–183

Amagai A (2002) Involvement of a novel gene, *zyg1*, in zygote formation of *Dictyostelium mucoroides*. J. Muscle Res. Cell Motil. 23:867–874

Amagai A and Filosa MF (1984) The possible involvement of cyclic AMP and volatile substance(s) in the development of a macrocyst-forming strain of *Dictyostelium mucoroides*. Dev. Growth Differ. 26:583–589

Amagai A and Maeda Y (1992) The ethylene action in the development of cellular slime molds: an analogy to higher plants. Protoplasma 167:159–168

Filosa MF and Dengler RE (1972) Ultrastructure of macrocyst formation in the cellular slime mold *Dictyostelium mucoroides* Extensive phagocytosis of amoebae by a specialized cell. Dev. Biol. 29:1–16

Suzuki T, Amagai A, Maeda Y (1992) Cyclic AMP and Ca^{2+} as regulators of zygote formation in the cellular slime mold *Dictyostelium mucoroides*. Differentiation 49:127–132

Weinkauf AM and Filosa MF (1965) Factors involved in the formation of macrocysts by the cellular slime mold, *Dictyostelium mucoroides* Can. J. Microbiol. 11:385–387

Ethylene controls the development, germination, and growth of petunia male gametophytes in the progamic phase of fertilization

Kovaleva, L.*, Timofeeva, G., Dobrovolskaya, A. and Rakitin, V.

Institute of Plant Physiology, Russian Academy of Sciences, Botanicheskaya 35, tel 7(495)9039306 fax 7(495)9778018.
(*Corresponding author: kovaleva_l@mail.ru)

1. Introduction

In flowering plants, double fertilization involves a complex series of interactions between essentially three plants – male gametophyte, female gametophyte, and sporophyte – culminating in the fusion of sexual cells and nuclei and the formation of an embryo and endosperm. The progamic phase of fertilization that consists in delivery of sperms to ovule and includes a cascade of events triggering pollen germination on the surface of stigma and growth of pollen tubes within tissues of the style and ovary plays a crucial role.

In *Petunia hybrida*, a species with gametophytic self-incompatibility, pollination has long been known to trigger a number of physiological responses in the flower, including alteration in metabolite source-sink relations in floral organs (Linskens, 1974), ethylene synthesis (Hoekstra and Roekel, 1988; Pech *et al.*, 1987; Whitehead *et al.*, 1984). Two phases of pollination-induced ethylene production have been described in petunia, an early phase which reaches a maximum level 3 h after pollination and a late phase beginning at about 20 h after pollination (Whitehead *et al.*, 1984).

To elucidate the role of ethylene in male gametophyte development and growth, we have examined ethylene production by developing anthers of three petunia clones and by the pollen–pistil system after self-compatible (SC) and self-incompatible (SI) pollinations.

A. Ramina et al. (eds.), Advances in Plant Ethylene Research: Proceedings of the 7th International Symposium on the Plant Hormone Ethylene, 135–139.

2. Materials and Methods

2.1 Plant material and experimental design

Experiments were performed with flowers of three petunia (*Petunia hybrida* L.) clones: (1) SC (development of fertile male gametophyte; fertilization and seed set occur after self- and cross-pollination); (2) SI (development of fertile male gametophyte; fertilization and seed set occur after cross-pollination; after self-pollination gametophyte self-incompatibility (GSI) occurs in the sporophyte style tissues); (3) sterile (abortion of male gametophyte development).

Vegetatively propagated plants were grown in the greenhouse at 25°C in a soil culture (5-l plastic vessels) under ambient illumination.

On the day before flower opening, the stamens were removed from flowers and each castrated flower was isolated inside a small agril bag. On the following day, some flowers were pollinated by pollen collected from opened flowers of the same clone (SC or SI pollination), whereas other castrated flowers were pollinated with pollen from opened flowers of the different clone (cross-pollination); unpollinated pistils from flowers castrated on the day before bud opening were used as a control.

2.2 Effects of exogenous ethylene on developing fertile male gametophyte

Flowering petunia shoots were treated in sealed vessels for a day, and then the isolated gametophyte cells from the anthers were sampled over 3 days.

2.3 Ethylene measurement

Before 0.5 h to beginning of the each experiment, the flowers were excised and placed into bottles with H_2O and immediately transported to laboratory. Pistils of five to seven flowers were placed individually in 15.0-ml glass vial and hermetically sealed with polypropylene caps plugs (Septa, Red Rubber, Aldrich, USA).Vials containing no flower parts were also sealed for measuring of initial level of ethylene in air. Bottles with patterns were incubated in the dark at 26°C in a thermostat TCH-100 for no more than 0.5 h before ethylene measurement.

It was shown in preliminary experiments that within 30 min after dissection of unpollinated flowers, the rate of ethylene production in pistil parts remained unchanged. Evolution of wound ethylene was observed only upon longer incubation. The rate of ethylene production was monitored by changes in the ethylene amount in incubation vials by gas chromatography. A gas chromatograph equipped with a system for concentrating hydrocarbons allows us to analyze large air volumes, thereby many times increasing the sensitivity of the method (Rakitin and Rakitin, 1986).

Ethylene was measured in 1.0 ml of the headspace 20 min after enclosure using a Hewlett-Packard gas chromatograph model 5840 A (Palo Alto, CA) equipped with flame ionization detector and a column of activated alumina (an alumina-filled glass column), helium being used as a carrier gas.

3. Results and Discussion

3.1 The role of ethylene in male gametophyte development

The male gametophyte development takes place in such a manner that each phase of differentiation of sporogenous tissue corresponds to a specific state of anther tissues.

In fertile (SC and SI) petunia clones, disintegration of the tapetal cells started before the beginning of the male gametophyte development. Complete degeneration of the tapetum occurred before the formation of binucleate pollen. Formation of fertile male gametophyte in both (SC and SI) clones was accompanied by increased ethylene evolution by anthers at the stage of vacuolated microspores. The final stage of pollen grain maturing is characterized by greatest ACC content and ethylene evolution.

In the sterile clone, changes were observed in both tapetum and microsporocytes at the beginning of the meiosis. The greatest ethylene evolution by sterile anthers (threefold exceeded ethylene evolution by fertile anthers) at the pollen mother cell stage accompanied the degeneration of microsporocytes and tapetum.

Exogenous ethylene induced the death of male generative cells in fertile clones from meiosis to early microspore stage.

We conclude that ethylene regulates the development of male gametophyte by coordination of the proper timing of anther tissue modifications. In fertile clones, complete tapetum degeneration occurs at the stage of late microspores and coincides with active ethylene production by anthers. Pollen abortion in sterile clones is accompanied by precocious degeneration of the tapetum during meiosis. It is due to high ethylene levels in anther tissues at mother pollen cell stage.

3.2 Ethylene is a factor of pollen tube germination and growth at progamic phase of fertilization

Under self-pollination, pollen grains of both fertile (SC and SI) clones land to stigma surface, adhere, hydrate, and germinate into stigma tissues and then – inside style tissues. However, 8–10 h later, the pollen tube growth of the SI clone is blocked due to GSI. To investigate the role of ethylene in male gametophyte germination and growth, we have examined ethylene production by germinating *in vitro* pollen tubes and by pollen–pistil system after SC and SI pollinations.

In both fertile clones, male gametophyte germination *in vitro* was accompanied by ethylene peak in 2 h of cultivation.

Germination of pollen grains on the stigma were accompanied by a fivefold increase in ethylene production in the case of SC pollination and a sevenfold increase in ethylene production in the case of SI pollination (Fig. 1A). Later ethylene production by pistil tissues decreased in both cases; however, in the case of SI pollination, ethylene declined more slowly.

The dynamics of ethylene production by the petunia pollen–pistil system indicated that the stigma is the main site of ethylene synthesis. Ethylene production by the stigma increased during pollen germination and decreased during the pollen tube growth. In the case of SI pollination the level is higher.

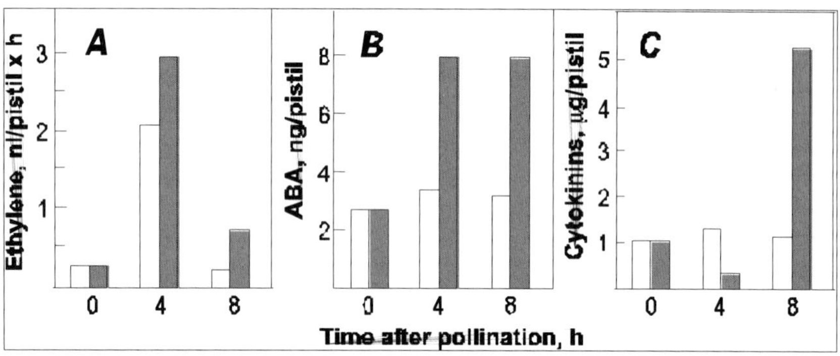

Fig. 1. Ethylene production (**A**) and ABA (**B**), and cytokinin (**C**) content in the petunia pollen–pistil system after SC (*blank bars*) and SI (*dark bars*) pollinations.

Each type of pollination (self- and cross-pollination of SC and SI clones) exhibited its time course of ethylene evolution by stigma and ACC content in it. Pollen grain germination after SC pollination was characterized by high ACC content, after SI pollination by the lowest ACC level. The greatest ethylene evolution was recorded in the case of self-pollination of SI clone. Both cases of cross-pollination occupied an intermediate position.

Thus, ethylene participates in the pollen–pistil interactions at the progamic phase of fertilization, controlling male gametophyte development and pollen tube germination and growth.

3.3 Ethylene, ABA, and cytokinins take part in the GSI?

The hormonal status of the petunia pollen–pistil system under SI pollination differed dramatically from that under SC pollination (Kovaleva and Zakharova, 2003). In the case of SI pollination, pollen germination was accompanied by ethylene evolution exceeding that in SC pollination (Fig. 1A) and by a threefold increase in ABA content (Fig. 1B), whereas growth inhibition of pollen tubes through the conducting tissue of style was accompanied by a fivefold increase in cytokinin content (Fig. 1C), and high levels of ABA (Fig. 1B) in the pollen–pistil system.

Taken together, the hormonal status of petunia pollen–pistil system indicates that ethylene signal pathway is integrated through a network of cross-talking connections with ABA and cytokinins that appear to coordinate responses in the pollen–pistil interactions after SI pollination. Ethylene, together with ABA and cytokinins, appear to be implicated in the GSI phenomenon.

References

Hoekstra FA, Roekel T (1988) Effects of previous pollination and stylar ethylene on pollen tube growth in *Petunia hybrida* styles. Plant Physiol 86:4–6.

Kovaleva L., Zakharova E (2003) Hormonal status of the pollen-pistil system at the progamic phase of fertilization after compatible and incompatible pollination in *Petunia hybrida* L. Sex Plant Reprod. 16:191–196.

Linskens HF (1974) Translocation phenomena in the *Petrunia* flowers after self- and cross-pollination. In: Linskens HF (ed) Fertilization in higher plants. North Holland Publishing Company, Amsterdam, pp 285–292.

Pech JC, Latche A, Larriguadiere C, Reid MS (1987) Control of early ethylene synthesis in pollinated petunia flowers. Plant Physiol Biochem 25:431–437.

Rakitin V, Rakitin L (1986). Gas exchange and the content of ethylene, carbon dioxide, and oxygen in plant tissues (in Russian). Fiziol Rast 33:403–413.

Whitehead CS, Halevy AH, Reid MS (1984) Roles of ethylene and ACC in pollination and wound-induced senescence of *Petunia hybrida* flowers. Physiol Plant 61:643–648.

Ethylene-regulated floral volatile synthesis in *Petunia × hybrida*

Dexter, R.J., Underwood, B.A. and Clark, D.G.*

Plant Molecular and Cellular Biology Program (PMCB), Environmental Horticulture Department, University of Florida, Gainesville, Florida. (*Corresponding author: geranium@ufl.edu)

Abstract

Ethylene is involved in the regulation of many physiological responses throughout plant development. In *Petunia* × *hybrida* "Mitchell Diploid" (MD), a rapid increase in endogenous ethylene is synthesized in the ovary coinciding with fertilization, and signaling a change in floral function from one of pollinator attraction to one of seed development. Following treatment of petunia flowers with exogenous ethylene, 76% of human olfactory panelists correctly identify a difference in floral aroma when compared to the control (air treatment). Utilizing both MD and transgenic CaMV35S::*etr1-1* (44568 – transgenic petunia line with reduced ethylene sensitivity) plants, we studied the effects of ethylene perception on floral volatile synthesis. Following a successful pollination, perception of endogenous ethylene results in decreased floral VOC emission and ultimately petal senescence in MD plants, but not 44568 plants. This perception of post-pollination endogenous ethylene also results in a subsequent decrease in mRNA accumulation of at least four genes involved in floral volatile synthesis in MD corollas, but not in 44568 corollas. These genes include *PhBSMT1* and *2*, which encode two benzoic acid/ salicylic acid carboxyl methyltransferases, *PhBPBT*, which encodes benzoyl-CoA: benzyl alcohol/phenylethanol benzoyltransferase, and *PhCFAT*, which encodes a coniferyl alcohol acyltransferase. These data indicate that ethylene is a broad regulator of the expression of volatile biosynthetic genes and subsequent volatile biosynthesis following pollination in the petunia flower.

A. Ramina et al. (eds.), Advances in Plant Ethylene Research: Proceedings of the 7th International Symposium on the Plant Hormone Ethylene, 141–146.

1. Introduction

Floral aroma is comprised of a unique blend of volatiles specific to individual angiosperms and critical to many processes including pollinator signaling. Over the past 5 years, *Petunia* × *hybrida* cv "Mitchell Diploid" has become a model system for the study of floral aroma in angiosperms (Boatright *et al.*, 2004; Negre *et al.*, 2003; Underwood *et al.*, 2005). Localized primarily to the petal limb, emission of floral volatiles in petunia is rhythmic with highest emission in the evening coinciding with peak pollinator (hawkmoth) activity (Verdonk *et al.*, 2005). GCMS analysis identified seven compounds (methylbenzoate, benzylbenzoate, phenylethyl alcohol, phenylacetaldehyde, benzaldehyde, benzyl alcohol, and isoeugenol) which primarily compose MD floral aroma (Verdonk *et al.*, 2005).

Following a successful pollination in *Petunia* × *hybrida*, ethylene is synthesized and perceived sequentially throughout the flower, ultimately resulting in petal senescence (Holden *et al.*, 2003; Jones *et al.*, 2003; Tang and Woodson, 1996; Wilkinson *et al.*, 1997). Initially, ethylene is synthesized in the stigma/style 2–4 h after pollination coinciding with pollen tube growth. This is later followed by a second period of ethylene synthesis coinciding with a successful fertilization, localized first to the stigma/style and ovary 12–24 h after pollination, and later followed by autocatalytic ethylene synthesis in the corolla from 24–36 h after pollination (Jones *et al.*, 2003; Tang and Woodson, 1996). In *Petunia* × *hybrida*, constitutive expression of the dominant mutant allele (*etr1-1*) from *Arabidopsis thaliana*, resulted in a transgenic petunia (44568) with reduced ethylene sensitivity (Wilkinson *et al.*, 1997). This line has now become a powerful tool to study the effects of ethylene perception in petunia.

2. Materials and Methods

Volatile emission and gene expression analysis were analyzed as previously described in Underwood *et al.* (2005). For human olfactory panels, MD whole flowers were collected at anthesis at 6 pm and treated with 2–3 μl/l ethylene or air for 12 h overnight. In conjunction with Dr. Charles Sims (Food and Agricultural Sciences, University of Florida) flowers were placed in individual glass jars and capped for 30 min prior to sampling. Using a triangle test format, three jars (two ethylene treated/one air treated or two air treated/one ethylene treated) were presented to a human panelist as a triangle test in which each panelist was asked to select the jar whose aroma differed from that of the other two. Statistical significance was determined by a p value less than 0.01.

3. Results and Discussion

3.1 Ethylene-dependent regulation of floral volatile emission in Petunia x hybrida

Microarray analysis of a nonredundant set of clones from three petunia cDNA libraries led to the isolation of a subset of candidate genes down-regulated following ethylene perception (Underwood, 2003). Further analysis of this subset led to the identification of several genes thought to be involved in floral volatile biosynthesis in petunia. To determine if ethylene perception also resulted in a decrease, a floral volatile biosynthesis and emission, excised MD, and 44568 petunia flowers were collected and treated with 2–3 μl/l exogenous ethylene for 24 h (Table 1). When compared to volatile levels measured in 44568, all seven primary volatiles were decreased in ethylene-treated MD flowers. Similarly, following pollination, ethylene produced in the corolla (24–36 h after pollination) coincided with decreased volatile when compared to 44568 (Table 1), demonstrating a direct correlation between ethylene perception and decreased volatile emission.

Table 1. Petunia floral volatile emission following perception of exogenous or endogenous (post-pollination) ethylene. MD and 44568 volatile emission was collected for 1 h and analyzed to determine MD volatile levels as compared to 44568.

Floral volatile	MD volatile levels 10 h following exogenous ethylene treatment	Volatile levels 36 h after pollination
Benzaldehyde	Decreased	Decreased
Benzyl alcohol	Decreased	Unchanged
Phenylacetaldehyde	Decreased	Decreased
Methylbenzoate	Decreased	Decreased
2-Phenylethanol	Decreased	Decreased
Isoeugenol	Decreased	Decreased
Benzylbenzoate	Decreased	Decreased

3.2 Human olfactory panel

To determine if humans could perceive this decrease in volatile emission, excised MD flowers were treated with ethylene or air, and presented to 60 panelists who were subsequently asked to smell each of three jars and determine which flower differed from the other two (Table 2a). Forty-six out of 60 panelists were able to identify the correct flower, indicating that a statistically significant ($p < 0.01$) number of panelists could correctly discern ethylene-treated flowers from those treated with air (Table 2b). Those panelists who correctly identified the MD flower treated with ethylene described it as less fragrant, earthy, or musty as compared to air-treated MD flowers. This observation indicates that human olfaction can discern a decrease in volatile emission as a result of ethylene perception. Commercially, this indicates that ethylene produced in response to wounding (pruning) or pollination could perceivably decrease floral aroma of cut-flowers during the shipping process resulting in a less desirable product.

Table 2a. The demographic breakdown of human olfactory panelists.

Panelist age (Years)	Under 18	18–29	30–44	45–65	Over 65	Total
Men	0	22	5	4	1	32
Women	0	25	2	1	0	28
Total	0	47	7	5	1	60

Table 2b. MD ethylene- vs. air-treated human olfactory results. Panelists were given three jars and asked to identify the flower whose aroma differed from that of the other two samples.

	Results
Correct	46
Incorrect	14
Total	60

***33 out of 60 needed to attain significance ($p<0.01$)

3.3 Transcriptional regulation of floral volatile biosynthesis genes following ethylene perception

As discussed previously, microarray analysis led to the isolation of a subset of genes thought to be involved in floral volatile biosynthesis. Using RNAi-based gene silencing in conjunction with *in vitro* enzyme activity studies, four of these genes have been identified and characterized in *Petunia* × *hybrida*. These genes include benzoic acid/salicylic acid carboxyl methyltransferase 1 and 2 (*PhBSMT1* and *2*), which catalyze the conversion of benzoic acid or salicylic acid to methylbenzoate or methyl-salicylate, respectively (Negre *et al.*, 2003; Underwood *et al.*, 2005), benzoyl-CoA: benzyl alcohol/phenylethanol benzoyltransferase (*PhBPBT*), which catalyzes the transfer of a benzyl group to benzyl alcohol or 2-phenylethanol producing benzylbenzoate and phenylethylbenzoate, respectively (Boatright *et al.*, 204), and coniferyl alcohol acyltransferase (*PhCFAT*), a general acyltransferase critical to the production of the floral volatile, isoeugenol (Dexter *et al.*, in press).

Analysis of the these four genes at the transcript level revealed three modes of regulation common to genes involved in floral volatile biosynthesis (Table 3). In all cases, transcript was primarily restricted to the petal limb, rhythmically expressed, and down-regulated following ethylene perception. This observation of conserved modes of regulation common to these genes provides a powerful tool for the identification of future genes involved in floral volatile biosynthesis.

Table 3. Transcription regulation of four floral volatile biosynthesis genes from *Petunia x hybrida*.

	Tissue of highest spatial expression	Rhythmically expressed	Regulation of transcript following ethylene perception
PhBSMT1	Petal Limb	Yes	Down
PhBSMT2	Petal Limb	Yes	Down
PhBPBT	Petal Limb	Yes	Down
PhCFAT	Petal Limb	Yes	Down

3.4 Ethylene is a key modulator of the floral transition from pollinator attraction to seed development in Petunia x hybrida

Prior to pollination, the primary function of the flower is to attract a pollinator and achieve successful fertilization. During this stage of floral development, the corolla is turgid and floral volatiles are rhythmically emitted (Verdonk *et al.*, 2005). Following pollination and a successful fertilization, ethylene is synthesized from the stigma/style and ovary subsequently resulting in autocatalytic ethylene production in the corolla (Jones *et al.*, 2003). This sequential burst in ethylene signifies a shift in floral function from one of pollinator attraction to one of ovary and seed development. As part of this transition, the biosynthesis of floral volatiles, a metabolically expensive process, is down-regulated and ultimately the petunia corolla is senesced (Underwood *et al.*, 2005). This transition efficiently reallocates valuable assets to ovary and seed development critical to reproductive success.

References

Boatright J, Negre F, Chen X, Kish CM, Wood B, Peel G, Orlova I, Gang D, Rhodes D, Dudareva N (2004) Understanding in vivo benzenoid metabolism in petunia corolla tissue. Plant Physiology 135: 1993–2011.

Dexter RJ, Qualley A, Marshall A, Ma CJ, Wood B, Dudareva N, Pichersky E, Clark DG (in press) Characterization of a petunia acetyltransferase involved in the biosynthesis of the floral volatile isoeugenol. The Plant Journal.

Holden MJ, Marty JA, Singh-Cundy A (2003) Pollination-induced ethylene promotes the early phase of pollen tube growth in *Petunia inflata*. Journal of Plant Physiology 160: 261–269.

Jones ML, Langston BJ, Johnson F (2003) Pollination-induced senescence of ethylene sensitive and insensitive petunias. In M Vendrell, H Klee, JC Pech, F Romojaro, eds, Biology and Biotechnology of the Plant Hormone Ethylene III, IOS Press, Amsterdam, The Netherlands, pp 324–326.

Negre F, Kish CM, Boatright J, Underwood B, Shibuya K, Wagner C, Clark DG, Dudareva N (2003) Regulation of methylbenzoate emission after pollination in snapdragon and petunia flowers. The Plant Cell 15: 2992–3006.

Tang XY, Woodson WR (1996) Temporal and spatial expression of 1-aminocyclopropane-1-carboxylate oxidase mRNA following pollination of immature and mature *Petunia* flowers. Plant Physiology 112: 503–511.

Underwood BA, Tieman DM, Shibuya K, Dexter RJ, Loucas HM, Simkin AJ, Sims CA, Schmelz EA, Klee HJ, Clark DG (2005) Ethylene-regulated floral volatile synthesis in petunia corollas. Plant Physiology 138: 255–266.

Underwood BA (2003) Effects of Ethylene on Floral Fragrance of Petunia x hybrida 'Mitchell Diploid'. Dissertation: University of Florida (PMCB).

Verdonk JC, Haring MA, Tunen AJ, Schurrink RC (2005) *ODORANT1* regulates fragrance biosynthesis in petunia flowers. The Plant Cell 17(5): 1612–1624.

Wilkinson JQ, Lanahan MB, Clark DG, Bleecker AB, Chang C, Meyerowitz EM, Klee HJ (1997) A dominant mutant receptor from Arabidopsis confers ethylene insensitivity in hererologous plants. Nature Biotechnology 15: 444–447.

Unravelling ethylene biosynthesis and its role during tracheary element formation in *Zinnia elegans*

Pesquet, E.* and Tuominen, H.

Umeå Plant Science Centre (UPSC), Department of Plant Physiology, Umeå University, SE-901 87 UMEÅ, Sweden. (*Corresponding author: edouard.pesquet@plantphys.umu.se)

Xylem is the plant vascular tissue responsible for raw sap conduction. It comprises two main types of cells that are derived from differentiating cambium: tracheary elements (TEs) and fibres that have conducting and mechanical role, respectively. Xylem formation is a developmental process and is under strict hormonal control involving a stew of phytohormones including auxin, cytokinin and ethylene. The relationship between ethylene and TE/xylem formation is supported by several lines of evidence: (i) ethylene is produced in the cambium of *Abies balsamea* (Little and Eklund, 1999) and ACC is actively transported in xylem sap (Else *et al.*, 1995); (ii) ACC oxidase is induced in actively growing compression wood in poplar stems submitted to physical tension and ACC accumulates on the opposing side where differentiation stops (Andersson-Gunneras *et al.*, 2003); (iii) furthermore, ethylene promotes TE formation in spruce trees (Eklund and Tiltu, 1999), in *Abies balsamea* (Little and Eklund, 1999) and, concomitantly with auxin, in pine hypocotyls (Kalev and Aloni, 1999) and lettuce (Miller *et al.*, 1984); (iv) an ACC synthase has been recently identified as being expressed during *Zinnia elegans in vitro* TE differentiation (Pesquet *et al.*, 2005). This suggests that ethylene/ACC could play an important role in signalling during TE formation. We have used the *Zinnia elegans in vitro* TE system (Fukuda and Komamine, 1980; Pesquet *et al.*, 2003) to assess the role of ethylene during xylogenesis.

In this *in vitro* system, ethylene is produced in a tri-phasic manner during TE differentiation, with a major peak prior to TE programmed cell death and lignification. Intracellular ACC continues to accumulate during TE differentiation, while ACC exports mimic the tri-phasic ethylene peaks. ACC synthase and oxidase activities mirrored their product accumulation, suggesting that ethylene production is controlled by ACC oxidase rather than by ACC synthase which is constantly active.

A. Ramina et al. (eds.), Advances in Plant Ethylene Research: Proceedings of the 7th International Symposium on the Plant Hormone Ethylene, 147–149.

Pharmacological modulation of ethylene synthesis using inhibitors of ACC synthase (AVG) and of ACC oxidase (CoCl2 and AIB) showed a dose-dependant reduction of TE differentiation. This confirms that ethylene is part of the signals necessary for TE differentiation. Providing cells with ACC and ethephon, nevertheless, limits TE differentiation. This suggests that, even though ACC/ethylene are necessary for TE differentiation, over-accumulation of these compounds limits TE differentiation. Modulation of ethylene perception, using MCP and STS, blocks TE maturation prior to programmed cell death and lignification.

Our results show that ethylene is specifically associated with TE formation and has a dual role in specific processes of TE differentiation, influencing both the efficiency of TE differentiation and the proper maturation of TEs.

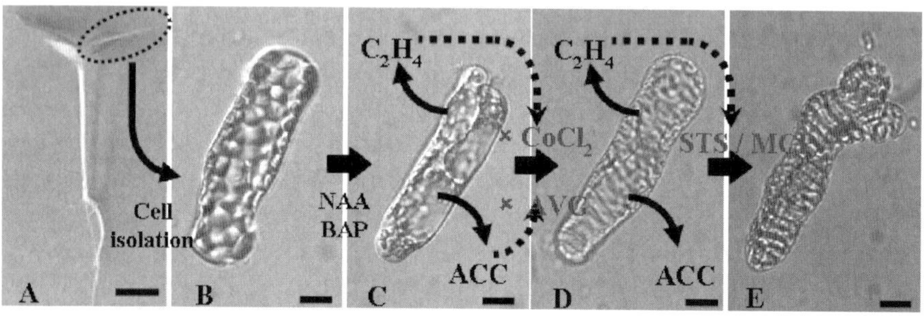

Fig. 1. (Colour figure in the Annex, p.456) **A**: Two-weeks old seedlings, first leaf used for cell isolation is circled, bar = 1 cm. **B**: Freshly isolated cell (0 h) **C**: Divided 36 h cell producing first ethylene peak. Synthesis inhibitors block differentiation progression. **D**: 60 h differentiating TE producing second ethylene peak. Perception inhibitors block TE maturation. **E**: Mature 120 h TE. Bars = 8 μm.

References

Little C.H.A. and Eklund L. (1999) Trees, 13: 173–177

Else M.A., Hall K.C., Arnold G.M., Davies W.J. and Jackson M.B. (1995) Plant Physiol., 107: 377–384

Andersson-Gunneras S., Hellgren J.M., Bjorklund S., Regan S., Moritz T. and Sundberg B. (2003) Plant J., 34: 339–349

Eklund L. and Tiltu A. (1999) J. Exp. Bot., 50: 1489–1493

Kalev N. and Aloni R. (1999). New Phytol., 142: 307–313

Miller A.R., Pengelly W.L. and Roberts L.W. (1984) Plant Physiol., 75:1165–1166

Pesquet E., Ranocha P., Legay S., Digonnet C., Barbier O., Pichon M. and Goffner D. (2005) Plant Physiol., 139: 1821–1839.

Fukuda H. and Komamine A. (1980). Plant Physiol., 65: 57–60
Pesquet E., Jauneau A., Digonnet C., Boudet A.M., Pichon M. and Goffner D. (2003) Physiol. Plant, 119: 463–468

Leaf senescence in tissue culture of *Passiflora incarnata* L.: the role of ethylene

Mensuali-Sodi, A.[1]*, Lucchesini, M.[2], Maltinti, S.[2], Serra, G.[1] and Tognoni, F.[2]

[1]Scuola Superiore Sant'Anna,.piazza Martiri della Libertà 33, Pisa, Italy; [2]Dip. Biologia delle Piante Agrarie, Univ. di Pisa, viale delle Piagge 23, Pisa, Italy. (*Corresponding author: mensuali@sssup.it)

The work was conducted with the aim to improve the quality of *in vitro* propagated shoots of the medicinal species *Passiflora incarnata* L.

Plantlets belonging to this genus have difficulty growing under *in vitro* conditions with reduced gas exchange. As a climacteric fruit crop, *Passiflora* spp. shows a high rate of ethylene production which could limit *in vitro* growth and development of plantlets (Faria and Segura, 1997). In our previous work, a method for *in vitro* plantlet multiplication of *P. incarnata* L. was developed (Lucchesini et. al, 2003) but during active shoot proliferation, leaf senescence symptoms occurred impairing the successful outcome of the cultivation.

P. incarnata shoots were cultured on MS macro and microelements with Gamborg's B5 vitamins, reduced glutathione 300 mg L^{-1}, MES 500 mg L^{-1}, sucrose 20 g L^{-1}, kinetin 2 mg L^{-1}, agar 7 g L^{-1}. Three different culture vessels were used: Ury vials (PBI, Milan, Italy) 30 mL; Red screw cap (PBI, Milan, Italy) disposable 200 mL vessels; Magenta GA-7 (Sigma Chemical Co., St. Louis, USA). To elucidate the role of ethylene in leaf senescence ACC 10 μM and AVG 10 μM were added to the basal medium in Ury vials by microfiltration and SmartFreshSM (AgroFresh Inc. c/o Rohm & Haas Italia srl) powder was used to provides the desired concentration of 1-MCP (500 ppm) in the internal headspace of the culture vessels. Chlorophyll a and b were detected spectrophotometrically on fresh leafy shoots at the end of the *in vitro* cultures. Ethylene was determined by gas chromatography.

Leaf senescence after 2, 3 and 4 weeks of culture (Fig. 1A) was principally correlated with the characteristics of the vessels employed. Accumulation of abiotic ethylene (Fig. 1B) was monitored in the different vessels. ACC added to the medium caused deleterious effects on the shoot development (Fig. 1C) and pigment content (Fig. 1D). 1-MCP treatments

A. Ramina et al. (eds.), Advances in Plant Ethylene Research: Proceedings of the 7th International Symposium on the Plant Hormone Ethylene, 151–152.

given every week from the beginning of the culture did not affect shoots quality and development in comparison with control. A marked renewal of the shoots was observed when 1-MCP was provided for the second half of the culture period (Fig. 1D).

The results obtained demonstrate the sensitivity of *P. incarnata* to ethylene accumulation and suggest the possibility to employ 1-MCP to inhibit deleterious ethylene effects in *in vitro* culture.

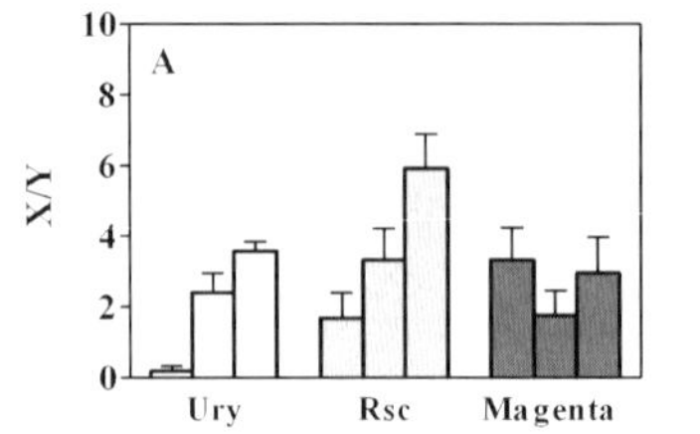

Leaf senescence during the culture period (weeks) in the different type of vessels: Ury Red screw cap (Rsc) and Magenta. x/y= yellow+ abscissed leaves/ green leaves.

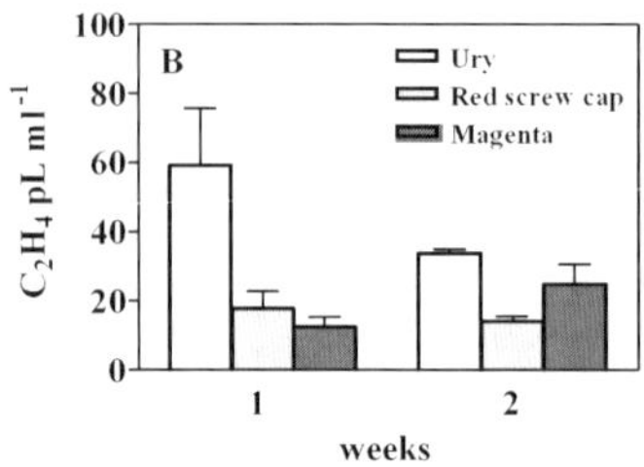

Ethylene accumulation during the first and second week in different culture vessels with culture medium but without plants

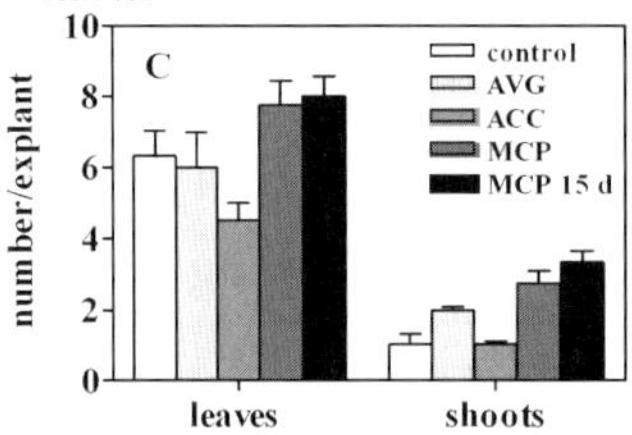

Leaf and shoot number per explant in *P. incarnata* multiplication phase treated with ACC and AVG or 1-MCP.

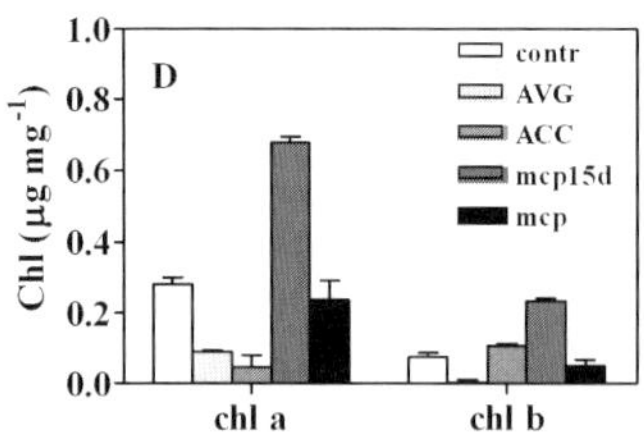

Pigment concentration per mg of shoot fresh mass at the end of the culture period of *P. incarnata* subjected to AVG, ACC, 1-MCP treatments and 1-MCP given after the second week of culture.

References

Faria J.L.C., Segura J., 1997. In vitro control of adventitious bud differentiation by inorganic medium components and silver thiosulfate in explants of *Passiflora edulis* f. *flavicarpa*. In vitro Cell Dev. Boil.-Plant 33:209–212.

Lucchesini M., Mingozzi M., Mensuali Sodi A., 2003. Callus formation and shoot regeneration of *Passiflora incarnata* L. from seed explants, Agricoltura Mediterranea 133:72–80.

Synergistic effect of kinetin and benzyl adenine improves the regeneration of cotyledon explants of bottle gourd (*Lagenaria siceraria*) on ethylene production

Shyamali, S.* and Kazumi, H.

Laboratory of Plant Genetics and Breeding, Graduate School of Bioagricultural Sciences, Nagoya University, Furo-cho, Chikusa-ku, Nagoya 464-8601 Japan. *Corresponding author: Shyamali Saha (*Corresponding author: shyamali2000@gmail.com)

In this study, we show the capacity of de novo shoot organogenesis from cotyledon explants of bottle gourd in vitro in relation to ethylene, ethylene inhibitor, phytohormones and growth supplements. The physiological relevance of the hormonal interaction with respect to regulation of ethylene initiation and effect on regeneration is discussed. In this communication, we report the stimulatory effect of N6-benzyladenine (BA) and kinetin on adventitious shoot organogenesis leading to high frequency plant regeneration from cotyledon explants of bottle gourd on ethylene production. In recent years, there has been increasing evidence that the occurrence of morphogenesis in cultured plant cells may be associated with ethylene. Cytokinins are also known to promote ethylene production several-fold in many plants partially through the increase in ACC synthase activity (Abeles *et al.*, 1992). BA, a synthetic cytokinin, synergistically enhances ethylene production in the presence of IAA in mungbean hypocotyls (Yoshii and Imaseki, 1982).

Table 1. Effect of different plant growth regulators on shoot regeneration.

BA	kinetin	Bud proliferation	Regeneration %	Shoot/explants	Shoot length (cm)
2	0	3.5	25.33bc	3.80a	2.85c
3	0	3.5	20.00cd	3.07b	2.34d
0	1	2.0	18.67d	2.66bc	3.70a
0	2	2.0	24.00bcd	2.80bc	3.36b
2	1	3.8	80.60a	4.06a	1.18e
2	2	3.8	33.33b	1.01d	0.70f

Decoated seeds were sterilized by soaking in 70% ethanol for 1 min followed by 45 min in 20% sodium hypochlorite (1% a.i.) containing

A. Ramina et al. (eds.), Advances in Plant Ethylene Research: Proceedings of the 7th International Symposium on the Plant Hormone Ethylene, 153–155.

0.2% Tween-20, rinsed 4 times and then sterilized seeds were placed on germination medium containing Murashige and Skoog (MS) basal salts and vitamina with 2% (w/v) sucrose. The pH of the medium was adjusted to 5.8 before the addition of the 0.8% agar (INA AGAR BA-30). After five days proximal parts of the cotyledons were isolated from seedlings and cultured on basal medium supplemented with different levels of BA and kinetin alone or in combination. Data on the bud proliferation state, percentage of regeneration, number of shoots per explants and shoot length were statistically tested by analysis of variance (ANOVA). Ethylene was quantified by comparison with an ethylene standard.

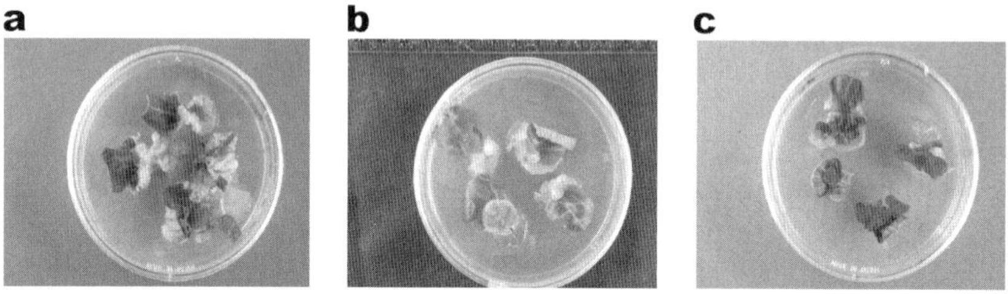

Fig. 1. (Color figure in the Annex, p.457) (**a**) Bud state showed a higher brownish expression on MS medium containing 2 mg/l BA. (**b**) No brownish tissue was observed in 2 mg/l kinetin and (**c**) Bud showed a moderate brownish tissue on MS medium containing 2 mg/l BA with 1 mg/l kinetin combination.

Bud proliferation, number of shoots per explant, shoot regeneration and shoot elongation varied depending upon the kind and concentration of cytokinins, which are shown in Table 1. Combined cytokinins (BA and kinetin) produced more bud (3.8) than when cytokinins were used individually. Kinetin was less effective as compared to BA in terms of bud proliferation as well as shoot bud induction but positively influenced shoot elongation (>3.36 cm). Better shoot bud differentiation on combined BA and kinetin has been reported in *Feronia limonia* L. hypocotyle explants (Vyas *et al.*, 2005), cotyledonary explants (Hossain *et al.*, 1994) and *Vigna radiata* L. (Gulati and Jaiwal, 1990). (Fig. 1). BA showed a more brownish phenotypic expression (Fig. 1a) than the BA and kinetin combination (Fig. 1c). No browning of tissue was observed when cotyledon explants were cultured on the kinetin containing medium (Fig. 1b). A positive relation was found between the tissue browning and ethylene production. A shoot generation capacity of explants and stimulation of ethylene biosynthesis may vary depending on the growth regulator used (Kumar *et al.*, 1987). From these data it can be concluded that the combination of kinetin and BA in the growth medium induces less ethylene emissions than BA and decreased brownish content, suggesting cell differentiation and ensured higher regeneration.

References

Abeles FB, Morgan PW and Salveit ME (1992) Ethylene in Plant Biology. 2nd Edn. Academic Press, San Diego, Canada.

Gulati A and Jaiwal PK (1990) Culture conditions beffecting plant regeneration from cotyledons of *Vigna radiata*(L.) Wilczek. Plant Cell Tissue and Organ Cult 23:1–7.

Hossain M, Biswas BK, Karim MR, Rahman S, Islam R and Joarder OI (1994) In vitro organogenesis of elephant apple (*Feronia limonia*). Plant Cell Tissue Organ Cult 39:265–268.

Kumar PP, Reid DM and Thorpe TA (1987) The role of ethylene and carbon di oxide in differentiation of shoot buds in excised cotyledons of Pinnus radiata in vitro. Physiol. plant 69:244–252.

Vyas S, Joshi N, Tak K and Purohit SD (2005) In vitro adventitious shoot bud differentiation and plantlet regeneration in *Feronia limonia* L.(Swingle). In Vitro Cell. Dev. Biol. plant 41:296–302.

Yoshii H and Imaseki H (1982) Regulation of auxin- induced ethylene biosynthesis.Repression of inductive formation of 1-aminocyclopropane-1-carboxylate synthase by ethylene. Plant Cell Physiol 23:639–649.

Role of the *Cucurbita pepo* ethylene receptor cpers1 in sex determination

Peñaranda, A.[1], Payán, C.[1], Garrido, D.[3], Jamilena, M.[2] and Gómez, P.[1]

[1]Mejora y Biotecnología de cultivos, IFAPA, CIFA Almería. España.
[2]Dept. Biología Aplicada. Genética. Universidad de Almería. España.
[3]Dept. Fisiología Vegetal. Universidad de Granada. España.
(*Corresponding authors: cisneros@ual.es, mjamille@ual.es)

Species of the monoecious family *Cucurbitaceae* change sex determination of flowers when levels of ethylene change. Thus, the exogenous application of ethylene releasing agents such as ethephon inhibits the production of staminate flowers and increases the number of pistillate flowers in squash, being therefore used in the commercial production of hybrid seed (Rudich *et al.*, 1970). On the other hand, the application of AVG (aminoethoxyvinylglycine), an inhibitor of ethylene biosynthesis, is able to induce a reduction of femaleness. We have increased or reduced the level of ethylene in floral buds by treating the apical meristems of plants with ethephon or AVG, and also we use $AgNO_3$ in order to inhibit ethylene perception. Our results show changes not only in sex determination of floral buds along the main axis of the plant, but also changes in the morphology of stamens in female flowers. When plants have been sprayed with AVG and $AgNO_3$ (not ethephon) female flowers show diverse degree of stamen development, which are generally arrested in a very early stage.

Linked with these results, and in order to assess the functional role of ethylene perception in *Cucurbita pepo*, we have cloned and characterized an ethylene receptor of this specie which is highly homologous to *ERS1* of *Arabidopsis thaliana*. The gene was cloned from a zucchini cultivar by using degenerate primers derived from conserved regions of ethylene receptor genes in different species. The structure of the isolated gene is similar to the *ERS1* gene of Arabidopsis (Fig. 1), showing the same number and location of introns. The encoded protein of 640 amino acids, has three characteristic domains of ethylene receptors, except the receiver domain, and has an overall 73% identity withArabidopsis *ERS1* (Hua *et al.*, 1995), as well as 91% and 90% identity with *ERS1* homologous in *Cucumis melo* and *Cucumis sativus* (Yamasaki *et al.*, 2000) respectively.

A. Ramina et al. (eds.), Advances in Plant Ethylene Research: Proceedings of the 7th International Symposium on the Plant Hormone Ethylene, 157–158.

Therefore, we have named this protein CpERS1. The homology of conserved domains is detailed in Fig. 1.

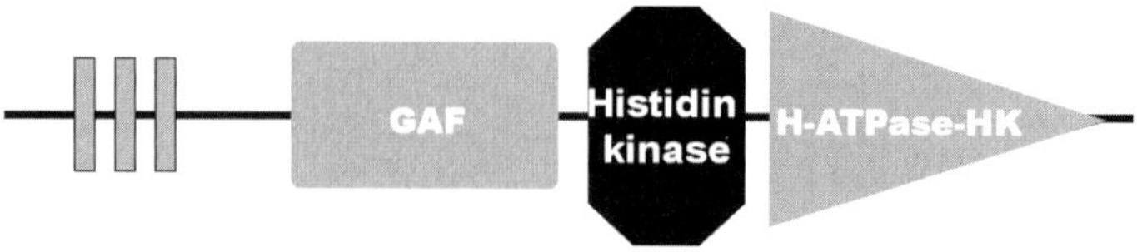

C.Sativus	84	95,52	74,83
C.Melo	88,3	97,01	72,54
A. thaliana	50,7	85,91	45,35
L. esculentum	45,45	88,57	42,24

Fig. 1. Distribution of the conserved domains in *Cp-ERS*. Under every conserved domains is found the homology with the respective conserved domain of *ERS-1* homolog of *C. sativus*, *C. melo*, *A. thaliana*, and *L. esculentum*.

Analysis of the expression of *CpERS1* and of other ethylene receptors including *CpETR1* (Gómez *et al*., 2004) have been performed using RT-PCR and *in situ* hybridization, confirming their expression in floral organs, including stamens.

Acknowledgements

This work was supported by project AGL2005-0667700-CO2 (Ministerio de Educación y Ciencia, Spain) and C03-180 awarded by the Consejería de Agricultura y Pesca de la Junta de Andalucia (Andalucia, Spain).

References

Gómez P, Peñaranda A, Garrido D, García S, Rosales R and Jamilena M. 2004. Role of ethylene in the control of flower abscission in zucchini squash (*cucurbita pepo*). APP 20: 74–75

Hua J, Chang C, Sun Q and Meyerowitz EM. 1995. Ethylene insensivity conferred by Arabidopsis ERS gene. Science 269: 1712–1714

Rudich J, Kedar N and Halevy A.H. 1970. Changed sex expression and possibilities for F1 hybrid seed production in some cucurbits by application of ethrel and Alar (B-995). Euphytica 19: 47–53.

Yamasaki S, Fujii N and Takahashi H. 2000. The ethylene regulated expression of *CS-ETR2* and *CS-ERS* genes in cucumber plants and their possible involvement with sex expression of flowers. Plant Cell Physiology 41: 608–616.

Populus genomics as a tool to unravel ethylene-dependent wood formation

Vahala, J.[1], Love, J.[2], Björklund, S.[2], Tuominen, H.[3], Sundberg, B.[2] and Kangasjärvi, J.[1]*

[1]Plant Biology, Department of Biological and Environmental Sciences, POB65 (Viikinkaari 1), 00014 University of Helsinki, Finland; [2]Umeå Plant Science Centre, Department of Forest Genetics and Plant Physiology, SLU, 901 83 Umeå, Sweden; [3]Umeå Plant Science Centre, Department of Plant Physiology, Umeå University, 901 83 Umeå, Sweden (*Corresponding author: jaakko.kangasjarvi@helsinki.fi)

Plant hormones, including the gaseous plant hormone ethylene, are important regulators of wood formation both due to environmental cues and in determining growth patterns. However, little is known about the role of ethylene in these processes at molecular level. Thus, sequencing of the black cottonwood (*Populus trichocarpa*) genome facilities a great tool for such research.

Populus, Arabidopsis, and rice genomes contain almost identical number of genes for the three enzymes of ethylene biosynthesis, while the number of genes for proteins involved in ethylene perception and signaling is higher. For example, *Populus* has seven predicted genes for ethylene receptor proteins, while Arabidopsis and rice genomes have five. The CONSTITUTIVE TRIPLE RESPONSE (CTR) protein that acts just downstream of the receptor is encoded in *Populus* by four genes while in *Arabidopsis* by only one, and in rice by two genes. Especially the number of APETALA2/ETHYLENE-RESPONSIVE FACTOR (AP2/ERF) domain proteins is significantly higher in *Populus* (173 genes) than in Arabidopsis (122 genes) and rice (139 genes). See the table for more details.

The increased number of AP2/ERF domain proteins, which directly regulate ethylene-responsive genes, is most likely a result of the recent genome duplication in *Populus*, but may also indicate the existence of tree-specific AP2/ERF transcription factors. It seems that the highly conserved regulation of ethylene biosynthesis has not allowed more variation in the number of biosynthesis enzymes. However, increased variation in the regulation of target genes by the AP2/ERF domain proteins and other

A. Ramina et al. (eds.), Advances in Plant Ethylene Research: Proceedings of the 7th International Symposium on the Plant Hormone Ethylene, 159–160.

certain gene family members in *Populus* may have been utilized in ethylene-dependent processes due to more complex aerial growth patterns than in annual plants, including branching patterns and extensive secondary growth, as well as the ability of trees to control the process of dormancy (Ruonala *et al.* 2006).

Ethylene treatment of *Populus* stem with a flow-through chamber system shows striking stimulation of cambial layer followed by extensive xylem formation, most likely due to stimulated AP2/ERF domain proteins that in turn activate wood forming enzymes. Therefore, the ultimate goal is to detect relevant AP2/ERF domain and then apply gene technology to stimulate wood formation.

Table 1. Number of ethylene pathway genes in *Populus*, *Arabidopsis*, and rice.

	Populus	Arabidopsis	Rice
Biosynthesis			
SAM synthetases	6	5	5
ACC synthases	8	9	6
ACC oxidases	7	6	9
Perception			
ERS/ETR/EIN4 receptors	7	5	5
CTR	4	1	2
Signal transduction			
EIN2	2	1	4
Gene induction			
Primary transcription factors			
EIN3/EIN3-like (EIL)	6	6	7
EBF (EIN3/EIL degradation)	4	2	2
Secondary transcription factors			
AP2/ERF domain proteins	173	122	139

References

Ruonala R, Rinne PLH, Baghour M, Moritz T, Tuominen H, Kangasjärvi J (2006) Transitions in the functioning of the shoot apical meristem in birch (*Betula pendula*) involve ethylene. Plant J 46:628–640

Control of vegetative growth of 'Verdejo' grapevines with ethephon

González, R., Villalba, P., González, M.R., Martín, P.*

Dpto. de Producción Vegetal y Recursos Forestales. ETS de Ingenierías Agrarias Universidad de Valladolid. Avda. de Madrid, 57, 34004 Palencia, Spain. Tel: +34 979108459. Fax: +34 979108301. (*Corresponding author: pmartin@pvs.uva.es)

Some vineyards from 'Rueda' *Appellation d`Origine* area (Spain) tends to develop an excessive vegetative growth which often leads to massive shading. The resulting shade favours long growth periods, early fruit ripening and fruit size and can reduce flower differentiation for the following year. Grape composition might be negatively affected by shading as a result of an unfavourable light microclimate in leaves and clusters (Smart, 1987). Many studies (Szyewicz *et al.*, 1984; Gallegos *et al.*, 2003) show the chemical control of excess growth by means of ethephon (2-chloroethyl phosphonic acid) applications can improve the light penetration to canopies and grape ripening.

The aim of this work was to evaluate the decrease of lateral shoot growth induced with ethephon, and its effect on grape composition and agronomic performance of 'Verdejo' vineyards.

1. Material and Methods

The field experiment was conducted in 2004, on a 7-year-old 'Verdejo' vineyard located in 'Rueda' *Appellation d`Origine* area. Vines are grafted on 110-Richter rootstocks and are pruned to a double Guyot system. The vine spacing is 3.0 × 1.5 m. Doses of 0 (control), 400 and 800 $mg{\cdot}L^{-1}$ of ethephon were studied in a randomised complete block design with four replications. The product was sprayed after fruit set over the upper third of foliage. Photosynthetic activity, chlorophyll content, leaf area index, vigour, yield components and must composition parameters were measured.

A. Ramina et al. (eds.), Advances in Plant Ethylene Research: Proceedings of the 7th International Symposium on the Plant Hormone Ethylene, 161–163.

2. Results and Discussion

No significant differences among experimental treatments were detected on total pruning weight (Table 1), contrary to the results observed by Gallegos *et al.* (2003) in Tempranillo grapevines. Compared with control vines, the ethephon treatments significantly reduced the sprouting of lateral buds, the growth of secondary shoot already developed (Table 1) and the percentage of total leaf area index developed after spraying (46% for 400 mg·L^{-1} and 59% for 800 mg·L^{-1} dose).

Table 1. Mean values of pruning parameters, according to the treatments applied.

Ethephon dose [mg.L^{-1}]	Lateral shoot per cane	Total pruning weight per vine [g]	Lateral shoot per total pruning weight [%]
0	5.20 a	1.527 a	15.0 a
400	3.75 b	1.597 a	12.9 ab
800	2.75 b	1.530 a	11.1 b

The values followed by distinct letters are significantly different at $p < 0.05$

These effects could improve the sunlight exposure of leaves and clusters. However, ethephon applications also reduced the foliar chlorophyll content and photosynthesis rate measured at veraison (Table 2). Mean cluster weight in treated plants was reduced by 6–20%, depending on dose, in relation to the controls, but there were no differences among treatments for yield (about 7.5 kg/vine).

Table 2. Mean values of photosynthetic activity and foliar chlorophyll content at veraison.

Ethephon dose [mg.L^{-1}]	Photosynthesis rate [μmol·m^{2}·s^{-1}]	Chlorophyll content [μg·cm^{2}]
0	15.30 a	46.86 a
400	14.01 ab	37.54 b
800	13.10 b	36.88 b

The values followed by distinct letters are significantly different at $p < 0.05$

The must composition (Table 3) was not modified by ethephon treatments. Contradictory results had been focused in effect of ethephon on total solids content, pH, and total acidity of must, mainly ascribing the differences to cultivars, timing, concentration and applications methods (Szyjewicz *et al.*, 1984). The tartaric acid content in grapes was lower in treated plants with 400 mg·L^{-1} than controls (7.6 until 6.7 mg·L^{-1}).

Table 3. Mean values of must composition parameters.

Ethephon dose [mg.L^{-1}]	Total soluble solids [%]	pH	Total acidity [g·L^{-1} tartaric acid]
0	11.65	3.68	5.16
400	11.20	3.60	5.19
800	11.60	3.62	5.17

References

Gallegos JI, De la Rosa M T, Becoechea A, González M R, Martín P (2003) Efectos de las aplicaciones de etefón en cuajado y envero sobre el comportamiento agronómico de la variedad Tempranillo. XIII Jornadas del Grupo de Estudio de los Sistemas de Conducción de la Vid. Montevideo, Uruguay

Smart RE (1987) Influence of light on composition and quality of grapes. Acta Hortic 206: 37–43

Szyjewicz E, Rosner N, Kliewer WM (1984) Ethephon (2-chloroethyl phosphonic acid, Ethrel, CEPA) in viticulture–a review. Am J Enol Vitic 35: 117–123

4. FRUIT DEVELOPMENT, RIPENING AND QUALITY

1-MCP effects on ethylene emission and fruit quality traits of peaches and nectarines

Ziosi, V.[1], Bregoli, A.M.[1], Fiori, G.[1], Noferini, M.[1] and Costa, G.[1]*

[1]Dipartimento di Colture Arboree, Università di Bologna, via Fanin 46, 40127 Bologna, Italy. (*Corresponding author: guglielmo.costa@unibo.it)

Abstract

1-methylcyclopropene (1-MCP), an ethylene antagonist, is an effective device to improve the shelf life and quality of fruits, vegetables and ornamentals. It also represents an useful tool in studies aimed at clarifying the ethylene physiology in plants. Peaches and nectarines are climacteric fruits which exhibit a sharp rise in ethylene synthesis at the onset of ripening associated with changes in colour, texture, aroma and other biochemical features. Therefore, treatments with ethylene antagonist such as 1-MCP has been performed for extending post-harvest storage and shelf life of peach fruit. Fruit developmental stage must be considered when applying 1-MCP, as its effects vary with fruit maturity. In fact, the efficacy of 1-MCP treatment decreases with advanced fruit development. In order to study the effects of 1-MCP on peaches and nectarines at different ripening stages, the fruit was divided into homogeneous classes by using a non-destructive technology (NIRs = near-infrared spectroscopy). This technology allowed to correlate the difference of absorbance at two specific wavelengths (DA index) to ethylene emission levels in peach fruit. Immediately after harvest, three groups of fruit were set, with different AD index intervals: the first one included 'pre-climacteric' fruit, characterised by very low or undetectable ethylene emission levels, the second one included 'early-climacteric' fruit (low ethylene emission levels) and the third one 'late-climacteric' fruit (high ethylene emission levels). The fruit of the three classes was treated with 1-MCP (1 $\mu l\ l^{-1}$) for 12 h at 25°C. Treated and control fruit was then kept at 25°C for about 1 week and the effects on ethylene emission and fruit quality traits (flesh firmness, total soluble solids content, titratable acidity) were evaluated.

Keywords: Ethylene, 1-methylcyclopropene, NIRs, peach, ripening

A. Ramina et al. (eds.), Advances in Plant Ethylene Research: Proceedings of the 7th International Symposium on the Plant Hormone Ethylene, 167–174.

1. Introduction

Melting flesh peaches and nectarines are climacteric fruits which exhibit a sharp rise in ethylene production at the onset of ripening and undergo rapid flesh softening after harvest, leading to losses in the marketing chain. Therefore, they are often picked at an early stage of ripening to better withstand handling, and they never reach full flavour and aroma. Thus, a treatment to inhibit rapid softening after harvest would allow fruit to be picked at a tree-ripe stage (Bregoli *et al.*, 2005).

Ripening is a complex and genetically programmed process leading to biochemical, physiological and structural changes which affect fruit colour, texture, flavour and aroma. Ethylene plays a key role in climacteric fruit ripening by coordinating the expression of ripening-related genes (Giovannoni, 2004).

Since the discovery of 1-methylcyclopropene (1-MCP) as an ethylene antagonist, a great number of studies have examined its mechanism of action, and its effect on extending the shelf life and quality of fruit. 1-MCP also represents a powerful tool in studies aimed at clarifying the role of ethylene in the ripening process. The efficacy of 1-MCP application in delaying fruit ripening depends upon several factors, such as chemical concentration, temperature, time of exposure, specie, and cultivar. Moreover, fruit developmental stage must be carefully considered when applying 1-MCP, as its effects change widely with fruit maturity (Blankenship and Dole, 2003). As a consequence, 1-MCP has often a limited commercial potential, since in any commercial consignment there is mixture of fruit maturity. Therefore, in order to improve 1-MCP effectiveness, the chemical should be applied on fruit at the same ripening stage.

In recent years, extensive research has been focussed on the development of non-destructive techniques, such as near-infrared spectroscopy (NIRs), for grouping the fruits based on their maturity. NIRs technique allows to determine several fruit quality traits in a non-destructive manner, thus offering the opportunity to extend measurements on a high number of fruit, and to repeat the analysis on the same samples. Moreover, NIRs technology is able to evaluate several fruit quality traits with the same measure, providing information on the overall quality of the fruit and its ripening stage (Costa *et al.*, 2002).

The aim of the present work was to investigate the effects of 1-MCP application on peach fruit at different ripening stages. With this purpose, peaches and nectarines were grouped into homogeneous classes of ripening by using the NIRs technology, and the effects of 1-MCP application on ethylene production and fruit quality traits were evaluated.

2. Materials and Methods

2.1 Plant material and NIRs equipment

Trials were carried out at the experimental farm of the University of Bologna on an early season nectarine (*Prunus persica* L. Batsch, cv 'Laura') and a late-season peach (*Prunus persica* L. Batsch, cv 'Fayette'). 'Laura' nectarines were collected from 8-year-old trees grafted on seedling rootstock and trained to open vase. 'Fayette' peach fruits were harvested from 9-year-old trees grafted on seedling rootstock and trained to Y-shape.

At harvest, nectarines and peach fruits were divided according to their ripening stage, as determined by using a NIRs device. The NIRs equipment used for the experiments was developed by the Department of Colture Arboree by using a spectrometer operating at a wavelength range between 650 and 1200 nm (Costa *et al.*, 2003). The NIRs instrument allowed to group the fruits according to a new maturity index, measured on the basis of the difference in absorbance (DA) at two specific wavelengths, whose values correlate with ethylene emission and quality traits practically used to define fruit ripening stage (flesh firmness, soluble solids content, titratable acidity). The NIRs instrument and the DA index were patented by the University of Bologna (patent identification n° MO 2005000211). 'Laura' nectarines (harvested 103 dAFB) were divided in two classes defined by ranges of AD index between 0.8 and 0.6 (class 1) and 0.6 and 0.4 (class 2). Analogously, 'Fayette' fruits (harvested 130 dAFB) were graded in two classes on the basis of the DA index (class 1: DA index between 1.5 and 1.1; class 2: DA index between 1.1 and 0.8).

2.2 1-MCP treatments

Fifty nectarines and 50 peach fruits per class were placed in two sealed 30-l plastic jars and incubated with 1-MCP (final concentration 1 $\mu l\ l^{-1}$ equivalent to 1 ppm) for 12 h at 25°C as described by Bregoli *et al*. (2005). The same number of fruits per class was kept in two sealed 30-l jars for 12 h at 25°C without 1-MCP (controls). Treated and control fruits of each class were then transferred in a growth chamber at 25°C for 60 h. At harvest, at the end of treatment (12 h), and at each following sampling time (36 and 60 h), ethylene production, flesh firmness (FF), soluble solids content (SSC), and titratable acidity (TA) were determined on ten control and ten treated fruits, as described by Torrigiani *et al*. (2004).

2.3 Statistical analysis

Data on ethylene production and fruit quality traits represent the means ± standard error and were analysed by using the Student's *t*-test (n = 10).

3. Results

3.1 Fruit grouping on the basis of the DA index

At harvest, DA index allowed to group both 'Laura' nectarines and 'Fayette' peaches in two homogeneous classes of ripening characterised by different ethylene emission, FF, and TA levels. On the contrary, no differences were observed in SSC (Tables 1, 2). Reported values indicated that class-1 'Laura' nectarines were at the pre-climacteric stage, while class-2 ones were at the onset of ethylene climacteric (Table 1). Analogously, 'Fayette' peaches were at the onset (class 1), or at a more advanced stage (class 2) of ethylene climacteric (Table 2).

Table 1. DA index, ethylene production and fruit quality traits of class-1 and class-2 'Laura' nectarines at harvest.

Class	**DA** index	**Ethylene** ($nl\ l^{-1}\ h^{-1}\ g^{-1}$ FW)	**SSC** (°Brix)	**FF** (N)	**TA** ($g\ l^{-1}$ malic acid)
1	0.8–0.6	0.1 ± 0.1	10.1 ± 0.2	49.0 ± 1.0	13.0 ± 0.6
2	0.6–0.4	0.7 ± 0.5	10.3 ± 0.2	37.0 ± 1.0	11.0 ± 0.5

Table 2. DA index, ethylene production and fruit quality traits of class-1 and class-2 'Fayette' peaches at harvest.

Class	**DA** index	**Ethylene** ($nl\ l^{-1}\ h^{-1}\ g^{-1}$FW)	**SSC** (°Brix)	**FF** (N)	**TA** ($g\ l^{-1}$ malic acid)
1	1.5–1.1	0.5 ± 0.1	12.7 ± 0.2	34.0 ± 1.0	10.4 ± 0.1
2	1.1–0.6	1.6 ± 0.6	12.6 ± 0.2	26.0 ± 2.0	6.6 ± 0.7

3.2 1-MCP effects on ethylene production and fruit quality traits

3.2.1 'Laura' nectarines

In control fruit of both classes, ethylene production increased from harvest, reaching the climacteric peak at 36 h (class 1) or 12 h (class 2) (Table 3). In class-1 fruits, 1-MCP application significantly inhibited ethylene production at 12 h (about fivefold) and 36 h (about threefold), while no difference was observed at 60 h. On the contrary, in class-2 nectarines, ethylene emission was dramatically stimulated by 1-MCP application at each sampling time, and especially at 36 h (about sixfold).

In class-1 control fruits, FF did not change from harvest (49.0 ± 1.0 N) up to 12 h, and then gradually decreased during the following trial period. Otherwise, FF of class-2 control nectarines dropped from 37.0 ± 1.0 N to 5.0 ± 1.0 N within 36 h after harvest (Table 3). In both classes, no changes in SSC were recorded during the considered span of time, while TA gradually decreased from harvest up to 60 h. In both cases, flesh softening was significantly delayed by 1-MCP application: in fact, at 36 h and 60 h, 1-MCP-treated fruits were firmer (about 1.5–2-fold) than controls. No differences in SSC and TA were observed between untreated and 1-MCP-treated fruits of both classes (Table 3).

Table 3. Ethylene production and fruit quality traits in control (C) and treated (T) class-1 and class-2 'Laura' nectarines. Asterisks (*) indicate significant differences, as compared with relative controls, at $P<0.05$ (*).

		Class 1			Class 2		
		12 h	36 h	60 h	12 h	36 h	60 h
Ethylene	C	1.0 ± 0.3	1.5 ±0.2	1.1±0.2	2.3±0.5	2.0±0.3	2.5±0.5
	T	0.2 ±0.04*	0.5 ±0.2*	1.1±0.3	4.7±0.6*	12.2±1.2*	7.0±0.9*
FF	C	40.0 ±3.0	28.0±2.0	14.0±4.0	25.0±2.0	5.0±1.0	5.0±0.3
	T	43.0 ±1.0	38.0 ± 3.0*	27.0±4.0*	28.0±1.0	13.0±2.0*	9.0±1.0*
SSC	C	10.5 ±0.5	10.9 ± 0.2	11.1±0.2	10.4±0.4	11.0±0.4	11.6±0.7
	T	10.2 ±0.5	11.0 ± 0.3	11.2±0.3	10.5±0.6	11.1±0.3	11.6±0.4
TA	C	12.6 ±0.6	12.3 ± 0.4	11.8±0.4	10.5±0.5	10.5±0.6	9.0±0.4
	T	12.8 ±0.5	12.5 ± 0.5	12±0.4	10.8±0.5	10.8±0.6	10.6±0.6

3.2.2 'Fayette' peaches

In control fruits of both classes, ethylene production peaked at 36 h, although class-2 fruits reached the highest levels (Table 4). In both cases, 1-MCP application led to a slight, even though statistically significant, stimulation of fruit ethylene production at 36 h (classes 1 and 2) and 60 h (class 1).

Class-1 control fruits gradually softened from harvest (34.0 ± 1.0 N) up to 60 h (6.0 ± 1.0 N), while FF of class-2 control nectarines dropped from 26.0 ± 2.0 N to 7.0 ± 2.0 N within 36 h after harvest (Table 4). Analogously, TA of untreated class-1 peach fruits gradually decreased (from 10.4 ± 0.1 to 5.6 ± 0.3 g l^{-1} malic acid) during the considered span of time, while that of class-2 fruits reached the lowest values (5.5 ± 0.3 g l^{-1} malic acid) within 36 h. In both classes, SSC remained constant throughout the trial period (Table 4). In fruit of both classes, 1-MCP application slowed down flesh softening and prevented loss in acidity. In fact, at 36 h (class 2) and 60 h (classes 1 and 2), FF and TA of 1-MCP-treated fruits were higher than relative controls. No differences in SSC between controls and 1-MCP-treated fruits were observed (Table 4).

Table 4. Ethylene production and fruit quality traits in control (C) and treated (T) class-1 and class-2 'Fayette' peaches. Asterisks (*) indicate significant differences, as compared with relative controls, at P <0.05 (*).

		Class 1			Class 2		
		12 h	**36 h**	**60 h**	**12 h**	**36 h**	**60 h**
Ethylene	C	0.5±0.1	3.3±0.6	3.1±0.5	2.6±0.8	7.9±1.3	6.3±1.2
	T	0.6±0.1	4.9±0.8*	4.8±0.9*	1.1±0.1	10.8±0.8*	6.8±0.7
FF	C	29.0±7.0	26.0±3.0	6.0±1.0	17.0±2.0	7.0±2.0	5.0±0.1
	T	28.0±2.0	27.0±2.0	16.0±2.0*	23.0±3.0	13.0±2.0*	10.0±2.0*
SSC	C	12.5±0.2	12.6±0.2	12.7±0.3	12.8±0.2	12.8±0.2	12.7±0.2
	T	12.3±0.3	12.9±0.1	12.7±0.1	12.4±0.6	13.2±0.2	12.9±0.2
TA	C	9.1±0.4	8.5±0.5	5.6±0.3	6.0±0.5	5.5±0.3	5.7±0.2
	T	9.0±0.6	8.3±0.4	6.8±0.4*	6.8±0.2	6.7±0.3*	6.7±0.4*

4. Discussion

Present results showed that the NIRs technique can be an effective tool for grouping peach fruit into homogeneous classes of ripening. In fact, peaches and nectarines at various developmental stages, as defined on the basis of the DA index, showed a different evolution of the ripening syndrome and differentially responded to 1-MCP, whose effects are dependent on fruit maturity (Blankenship and Dole, 2003).

1-MCP treatment inhibited ethylene production when fruits were at the pre-climacteric stage, and the hormone was absent or barely detectable. On the contrary, ethylene production was slightly ('Fayette') or strongly ('Laura') enhanced when this chemical was applied after the ethylene climacteric was started, as observed in banana (Golding *et al.*, 1998). In peach fruit, contrasting results about the effects of 1-MCP on ethylene emission have been reported. In fact, when the chemical was applied to 'Hakuho' and 'Maria Marta' peaches, ethylene production was slightly reduced (Mathooko *et al.*, 2001) or enhanced (Rasori *et al.*, 2001), respectively. Present data suggested that, besides some differences in how various peach cultivars respond to 1-MCP, this discrepancy could be due to the different fruit ripening stage.

1-MCP-treated nectarines at pre-climacteric stage also showed higher FF as compared to controls, while SSC and TA were not affected. These results confirmed that, by interfering with the climacteric peak, peach fruit ripening is affected, as reported in other cultivars (Mathooko *et al.*, 2001; Bregoli *et al.*, 2005). However, in climacteric fruits, 1-MCP delayed ripening by slowing down flesh softening and counteracting loss in acidity, although ethylene production was stimulated. While SSC and TA may not be clearly under the control of ethylene, this hormone regulates gene expression and activity of several cell-wall-related enzymes (Giovannoni, 2004). However, flesh softening was almost always slowed down by 1-MCP, even when ethylene production was not affected (Blankenship and Dole, 2003) or stimulated (Rasori *et al.*, 2002). In banana, 1-MCP delayed colour development and volatile emission despite an ethylene overproduction (Golding *et al.*, 1998), suggesting that most of the biochemical changes associated with ripening are dependent upon functioning ethylene receptors.

In conclusion, present data showed that, in peaches and nectarines, 1-MCP effect on ethylene production was related to the cultivar and the fruit ripening stage. Nevertheless, 1-MCP application delayed ripening independently of its effects on ethylene climacteric, as also reported in other species (Blankenship and Dole, 2003). To unravel the relationships between ethylene and fruit quality traits evolution, in response to 1-MCP application, in pre-climacteric and climacteric fruits, further biochemical and molecular studies are needed.

References

Blankenship SM, Dole JM (2003) 1-methylcyclopropene: a review. Postharvest Biol. Technol., 28: 1–25

Bregoli AM, Ziosi V, Biondi S, Rasori A, Ciccioni M, Costa G, Torrigiani P (2005) Postharvest 1-mehylcyclopropene application in ripening control of 'Stark Red Gold' nectarines: temperature-dependent effects on ethylene production and biosynthetic gene expression, fruit quality, and polyamine levels. Postharvest Biol. Technol., 37: 111–121

Costa G, Noferini M, Fiori G, Miserocchi O, Bregoli AM (2002) NIRs evaluation of peach and nectarine fruit quality in pre- and post-harvest conditions. Acta Hortic., 592: 593–599

Costa G, Noferini M, Montefiori M, Brigati S (2003) Non-destructive assessment methods of kiwifruit quality. Acta Hortic., 610: 179–190

Giovannoni JJ (2004) Genetic regulation of fruit development and ripening. Plant Cell, 16: 170–180

Golding JB, Sheaer D., Wyllie SG., McGlasson WB (1998) Application of 1-MCP and propylene to identify ethylene-dependent ripening processes in mature banana fruit. Postharvest Biol. Technol., 14: 87–98

Mathooko FM, Tsunashima Y, Owino WZO, Kubo Y, Inaba A (2001) Regulations of genes encoding ethylene biosynthetic enzymes in peach

(*Prunus persica* L.) fruit by carbon dioxide and 1-methylcyclopropene. Postharvest Biol. Technol., 21: 265–281

Rasori A, Ruperti B, Bonghi C, Tonutti P, Ramina A (2002) Characterization of two putative ethylene receptor genes expressed during peach fruit development and abscission. J. Exp. Bot., 53: 2333–2339

Torrigiani P, Bregoli AM, Ziosi V, Scaramagli S, Ciriaci T, Rasori A, Biondi S, Costa G (2004) Pre-harvest polyamine and aminoethoxyvinylglycine (AVG) applications modulate fruit ripening in Stark Red Gold nectarines (*Prunus persica* L. Batsch). Postharvest Biol. Technol., 33: 293–308

University of Bologna patent n° MO 2005000211. "Metodo ed apparato per determinare la qualità di prodotti ortofrutticoli"

Flesh softening in melting flesh, non-melting flesh and stony hard peaches: endopolygalacturonase expression and phosphorylation of soluble polypeptides in relation to ethylene production

Ghiani, A.*, Negrini, N., Morgutti, S., Nocito, F.F., Spinardi, A.M., Ortugno, C., Mignani, I., Bassi, D. and Cocucci, M.

Dipartimento di Produzione Vegetale, University of Milan, via Celoria 2, 20133 Milan, Italy, Phone: +39-03714662474; Fax: +39-0250316521 (*Corresponding author: alessandra.ghiani@unimi.it)

1. Introduction

Ethylene plays a role in many developmental processes among which fruit ripening has practical importance to the human diet. Peach is a climacteric fruit whose ripening, characterized by increased ethylene production, is accompanied by dramatic changes in colour, flavour, aroma and flesh texture. The latter is an important quality parameter affecting fruit shelf-life. Peaches can be classified as non-melting flesh (NMF, preferred for canning), melting flesh (MF, appreciated for the fresh market, Brovelli *et al.*, 1998) and stony hard (HD, maintaining high flesh firmness also when ripe and with a long shelf-life, Goffreda, 1992). Slow ripening mutants (SR) with altered ripening phenotype (impaired fruit development and ethylene production) also there exist, representing an interesting model for studies on ripening (Brecht *et al.*, 1984). The differences in the peach fruit softening pattern depend, among other factors, on presence and activity of cell wall hydrolytic enzymes, with particular regard to endopolygalacturonase (endoPG). In peach, the *melting flesh* (*M*) locus encodes endoPG and has been proposed to control fruit firmness. Our previous data showed that softening of MF fruits was characterized by increased levels of endoPG which was also present, although in very limited amounts, in NMF fruits. The differences in endoPG levels seemed due to different transcription of an *endoPG* gene (Morgutti *et al.*, 2005; Morgutti *et al.*, 2006). These peach fruit phenotypes are characterized also by different ethylene production. MF and NMF fruits do produce ethylene, even if in different amounts, while HD fruits do not produce ethylene, even if they can soften when treated with the hormone. Genetic

A. Ramina et al. (eds.), Advances in Plant Ethylene Research: Proceedings of the 7th International Symposium on the Plant Hormone Ethylene, 175–180.

analysis indicated that *stony hard* (*hd*) is a recessive locus, different from *M/m,* and is believed to result from a mutation in ethylene production (Haji *et al.*, 2005). Different expression of the MF/NMF/HD traits may result from the ability of the tissues to respond to ethylene across a broad concentration range due to some form of signal modulation. Kinase-mediated protein phosphorylation is a common means of signal modulation (Chen *et al.*, 2005). Our previous results indicated that the Ca^{2+}-dependent phosphorylation of a soluble polypeptide (Mr approx. 52 kDa) decreased with ripening in NMF and remained constant in MF fruits. In SR phenotypes, which neither softened nor produced endoPG phosphorylation of this polypeptide was not detectable, suggesting that phosphorylation of the 52 kDa polypeptide may be involved in the modulation of ethylene signalling and flesh softening (Morgutti *et al.*, 2005).

This study has dealt with the evaluation of ethylene evolution, protein phosphorylation patterns and endoPG expression in fruits from a HD and a SR phenotype, whose fruits do not produce ethylene and remain small, green and very firm. The results have been compared with those obtained in NMF and MF fruits and discussed in relation to possible differences in ethylene perception.

2. Materials and Methods

Peach (*Prunus persica* L. Batsch.) fruits of NMF 'Oro A', MF 'Bolero', HD 'Ghiaccio' and of one nectarine (*Prunus persica* L. Batsch. var. *nectarina*) SR phenotype ('BO 95021043') were used. Fruits were harvested at one time (beginning of physiological ripening) and divided into classes, representing distinct maturity categories, based on epicarp ground colour. Ethylene evolution from fruits was determined with a Dani 3800 gas chromatograph immediately after harvest or after ethylene treatment (SR phenotype, 100 ppm for 5 days). Mesocarp samples from single fruits of known firmness (Effegi hand penetrometer) were frozen in liquid N_2, stored at –80°C, and used for *in vitro* analyses. EndoPGs were extracted from the cell walls and probed by Western analysis as previously described (Morgutti *et al.*, 2006). Soluble protein extraction, *in vitro* protein phosphorylation, SDS-PAGE and phosphorylation pattern detection were conducted as previously described (Negrini *et al.*, 2000; Morgutti *et al.*, 2001). Northern analysis was performed by hybridizing the RNA extracted from fruits with ^{32}P[dATP]-labelled cDNA of Bolero *endoPG* (Gene Bank DQ340809) as a probe (Morgutti *et al.*, 2006).

3. Results and Discussion

Peach fruits with different phenotypes were characterized for ethylene evolution. The softest NMF 'Oro A' fruits, that at ripening showed high flesh firmness (34 N), evolved about twice the ethylene than the softest (12 N) fruits of MF 'Bolero' (approx. 100 µl kg FW^{-1} h^{-} in 'Oro A' as compared to approx. 50 µl kg FW^{-1} h^{-1} in 'Bolero'). This result, consistent with data reported in the literature for NMF and MF fruits (Brovelli *et al.*, 1999), seems to suggest that softening could possibly depend on a different sensitivity of the tissue to, rather than on the levels of, ethylene. Neither in HD 'Ghiaccio' fruits, with limited softening (35 N), nor in air- (control) or ethylene-treated fruits of the SR phenotype 'BO 95021043' was the evolution of ethylene detectable.

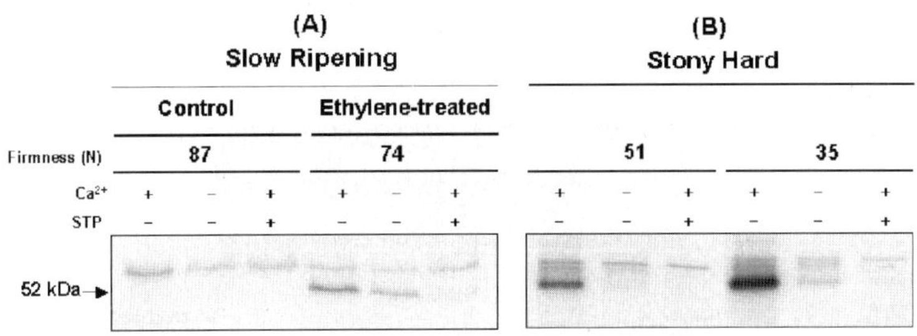

Fig. 1. Phosphorylation patterns of soluble polypeptides from the mesocarp of: **(A)** SR ('BO 95021043', treated or not with ethylene) and **(B)** Stony hard ('Ghiaccio') peach fruits, at maximum and minimum values of flesh firmness. (A): Control: 5 d in air; Ethylene-treated: 5 d in 100 ppm C_2H_4. Ca^{2+}: 0.25 mM $CaCl_2$; STP: 15 µM STP. Absence of Ca^{2+} was obtained with 1 mM EGTA. Ten µg of protein per lane. One experiment representative of three.

A soluble polypeptide (Mr approx. 52 kDa), phosphorylated on Ser/Thr residues in a Ca^{2+}-dependent manner, was present in both MF and NMF fruits. The phosphorylation intensity decreased in NMF fruits during the limited softening, while it remained high in the softer MF fruits (Morgutti *et al.*, 2005). The present work shows that in SR 'BO 95021043' fruits treatment with ethylene induced the ^{32}P-labelling of the 52 kDa soluble polypeptide, which was almost absent in the control fruits (Fig. 1A). In HD 'Ghiaccio' fruits the phosphorylation of the 52 kDa soluble polypeptide occurred also in the absence of ethylene emission and increased with the limited flesh softening (Fig. 1B). Preliminary characterization of soluble protein kinase activity/ies suggested that the phosphorylation of the 52 kDa soluble polypeptide could be due to calmodulin-independent CDPK/s (Ca^{2+}-dependent protein kinase/s) which can use an Histone type III S as a preferential substrate (data not shown).

In the control, SR 'BO 95021043' fruits lack of ethylene evolution was accompanied by the absence of both *endoPG* transcripts accumulation (Fig. 2A) and endoPG protein (Fig. 2C). Ethylene treatment triggered the accumulation of the *endoPG* transcripts (Fig. 2A) even if this was paralleled by neither fruit softening (with flesh firmness remaining high at 74 N) nor endoPG accumulation in the mesocarp cell walls (Fig. 2C).

According to the high flesh firmness, in HD fruits only very low levels of *endoPG* transcripts were detected, not accompanied by accumulation of the endoPG protein in the mesocarp cell walls (Figs. 2A and C).

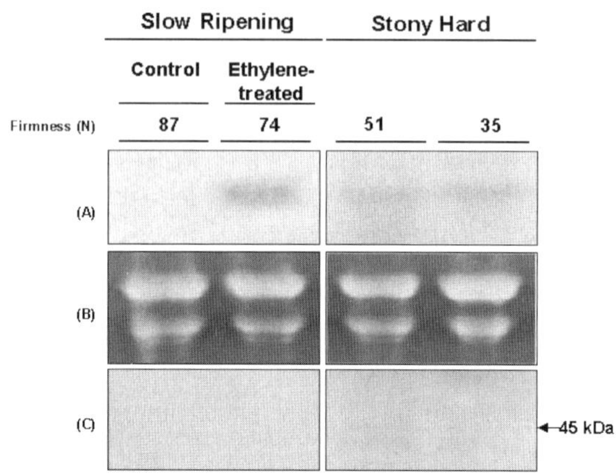

Fig. 2. Expression analysis of an *endoPG* gene in the mesocarp of SR ('BO 95021043', treated or not with ethylene) and HD ('Ghiaccio') peach fruits, at maximum and minimum values of flesh firmness. **(A)** Northern blot, 20 mg of RNA per lane; **(B)** total RNA, quantification image of the ethidium bromide-stained RNA gel; **(C)** Western blot, 15 mg of protein per lane. Control: 5 d in air; Ethylene-treated: 5 d in 100 ppm C_2H_4. One experiment representative of three.

In several species expression of the *PG* gene is related to climacteric (Lester *et al.*, 1994; Hadfield *et al.*, 1998), though in tomato the threshold concentration of ethylene required to induce *PG* mRNA accumulation is exceeded even in transgenic antisense *ACC synthase* fruits (Sitrit and Bennett, 1998). In kiwifruit appearance of PG isoforms is independent of ethylene production (Wang *et al.*, 2000). In NMF 'Oro A' fruits, higher ethylene production was accompanied by very low accumulation of the *endoPG* transcripts and endoPG protein, which in contrast were present at high levels in MF 'Bolero' fruits. In NMF 'Oro A' and MF 'Bolero' fruits *endoPG* expression, as well as endoPG production and consequent flesh softening, seemed related to the phosphorylation of the 52 kDa soluble polypeptide, probably due to the activity of a CDPK (data not shown).

In this study, a direct relationship between *endoPG* expression and ethylene was apparent in SR ('BO 95021043') and HD ('Ghiaccio') fruits. In the SR fruits the lack of ethylene biosynthesis occurs starting from the earliest stage of ripening and seems to affect the overall fruit developmental process. In these fruits, the effect of exogenous ethylene on the phosphorylation of the 52 kDa soluble polypeptide and induction of *endoPG* transcription could suggest a parallelism between protein kinase activity and ethylene-sensitive events involved in softening. In HD 'Ghiaccio' fruits, lack of ethylene production is consistent with the evidence in HD fruits of 'Yumyeong' (from which 'Ghiaccio' is derived) the expression of *ACS1* during ripening is suppressed (Tatsuki *et al.*, 2006). Nevertheless, consistently with previous data showing presence of polymorphisms in an *endoPG* gene of NMF 'Oro A' fruits that could be possibly responsible for diminished endoPG accumulation in mesocarp cell walls (Morgutti *et al.*, 2006), it can not be excluded that also in HD fruits the presence of possible mutations in the *endoPG* gene might affect its expression. Different presence and/or activity of cell wall enzymes, other from endoPG, could also be involved in the limited softening of these fruits, possibly triggered by low or undetectable levels of ethylene or by other signals (Hayama *et al.*, 2006).

Work is in progress to further characterize the differentially phosphorylated 52 kDa polypeptide and better understand its role in ethylene signalling during fruit ripening.

References

Brecht JK, Kader AA, Ramming DW (1984) Description and postharvest physiology of some slow-ripening nectarine genotypes. J Am Soc Hortic Sci 109:596–600

Brovelli EA, Brecht JK, Sherman WB, Sims CA (1998) Potential maturity indices and developmental aspects of melting- and non melting flesh peach genotypes for the fresh market. J Am Soc Hortic Sci 123:438–444

Brovelli EA, Brecht JK, Sherman WB, Sims CA (1999) Nonmelting-flesh trait in peaches is not related to low ethylene production rates. HortScience 34:313–315

Chen YF, Etheridge N, Schaller GE (2005) Ethylene signal transduction. Ann Bot 95:901–915

Goffreda JC (1992) Stony hard gene of peach alter ethylene biosynthesis, respiration, and other ripening-related characteristics. HortScience 27:122.

Hadfield KA, Rose JKC, Yaver DS, Berka RM, Bennett AB (1998) Polygalacturonase gene expression in ripe melon fruit supports a role for polygalacturonase in ripening-associated pectin disassembly. Plant Physiol 117:363–373

Haji T, Yaegaki H, Yamaguchi M (2005). Inheritance and expression of fruit texture melting, non-melting and stony hard in peach. Scientia Hortic 105:241–248

Hayama H, Tatsuki M, Ito A, Kashimura Y (2006) Ethylene and fruit softening in the stony hard mutation in peach. Postharvest Biol Technol 41:16–21

Lester DR, Speirs G, Orr G, Brady CJ (1994) Peach (*Prunus persica*) *endo-PG* cDNA isolation and mRNA analysis in melting and non-melting peach cultivars. Plant Physiol 105:225–231

Morgutti S, Negrini N, Cocucci M (2001) Caratterizzazione preliminare di attività protein chinasiche nella frazione solubile di frutti di pesco a diversi stadi di maturazione. Proc. XIX Convegno Nazionale SICA. Reggio Calabria, Italy, 25–28 September. pp 299–306

Morgutti S, Negrini N, Mignani I, Bassi D, Cocucci M (2005) Flesh softening and phosphorylation of soluble polypeptides in relation to ethylene production in *Prunus persica* fruits with different ripening patterns. Acta Hortic 682:155–162

Morgutti S, Negrini N, Nocito FF, Ghiani A, Bassi D, Cocucci M (2006) Changes in endopolygalacturonase levels and characterization of a putative *endo-PG* gene during fruit softening in peach genotypes with nonmelting and melting flesh fruit phenotypes. New Phytol 171:315–328

Negrini N, Morgutti S, Rivetta A, Lucchini G, Mignani I, Cocucci M (2000) Variazioni del trasporto del Ca^{2+} e del pattern di fosforilazione di polipeptidi in frutti di pesco (*Prunus persica* L. Batsch) a diversi stadi di maturazione. Proc. XVIII Convegno Nazionale SICA. Catania, Italy, 20–22 September. pp 332–339

Sitrit Y, Bennett AB (1998) Regulation of tomato fruit polygalacturonase mRNA accumulation by ethylene: a reexamination. Plant Physiol 116:1145-1150

Tatsuki M, Haji T, and Yamaguchi M (2006) The involvement of 1-aminocyclopropane-1-carboxylic acid synthase isogene, *Pp-ACS1*, in peach fruit softening. J Exp Bot 57:1281–1289

Wang ZY, MacRae EA, Wright MA, Bolitho KM, Ross GS, Atkinson RG (2000) Polygalacturonase gene expression in kiwifruit: relationship to fruit softening and ethylene production. Plant Mol Biol 42:317–328

Ethylene and volatile accumulation in citrus fruit

Shi, J.X.[1,a], Porat, R.[2], Goldschmidt, E.E.[1], Chen, S.B.[3], Gollop, N.[3], Ravid U.[4] and Goren R.[1]*

[1]The Kennedy-Leigh Centre for Horticultural Research, The Robert H. Smith Institute of Plant Sciences and Genetics in Agriculture, The Hebrew University of Jerusalem, P.O. Box 12, Rehovot 76100, Israel. [2]Department of Postharvest Science of Fresh Produce, ARO, the Volcani Center, P.O. Box 6, Bet-Dagan 50250, Isreal. [3]Department of Food Science, ARO, the Volcani Center, P.O. Box 6, Bet-Dagan 50250, Isreal. [4]Department of Aromatic and Medicinal Plants, ARO, Newe Ya'ar Research Center, P.O. Box 1021, Ramat Yishay 30095, Isreal. [a]Permanent Address: Shanxi Academy of Agricultural Sciences, Taiyuan 030031, Shanxi, China. (*Corresponding author: rgoren@agri.huji.ac.il)

Abstract

Physiological and molecular evidence supported the autocatalytic ethylene production in young "Star Ruby" grapefruit and "Murcott" mandarins, and the autoinhibitory ethylene production in mature fruit. Ethylene upregulated expression of pyruvate decarboxylase (*PDC*) and alcohol dehydrogenase (*ADH*) in both fruitlets and mature "Star Ruby", and increased acetaldehyde (AA) and ethanol production, while ethylene action inhibitors counteracted the effect. N_2 not only changed the expression of genes for ethylene biosynthesis but also genes for ethylene perception in both young and mature fruits. N_2 strongly increased expressions of *PDC* and *ADH* and production of AA and ethanol in young and mature fruits. Results suggest that ethanol fermentation in citrus fruit was, at least to some extent, ethylene-dependent, and that N_2 showed much stronger effect. SPME-GC-MS results indicated that ethylene and, to a lesser extent, N_2, played a role in controlling the production of various volatiles other than AA and ethanol in the juice of citrus fruit. As revealed by proteome analysis, the different effects of ethylene and N_2 on volatile buildup in citrus fruit may result from their different effects on the proteome profiles.

A. Ramina et al. (eds.), Advances in Plant Ethylene Research: Proceedings of the 7th International Symposium on the Plant Hormone Ethylene, 181–187.

1. Introduction

In mature citrus fruit, it has been reported that for both ethylene (Porat *et al.*, 1999) and anoxia (Shi *et al.*, 2005) enhanced ethanol fermentation; the underlying molecular mechanisms are not fully understood. Thus, effects of ethylene and anoxia on ethanol fermentation and the buildup of other volatiles were investigated in the present study, as well as their effects on ethylene biosynthesis and perception, at both physiological and molecular levels. Full details will be published elsewhere.

2. Materials and Methods

"Star Ruby" grapefruit (*Citrus paradisi* Macf.) and "Murcott" mandarin (*Citrus reticulata* Blanco) were picked at stage I and stage III of development (Katz *et al.*, 2004), respectively. Young fruitlets were sealed in 3.2 l glass jars with air, 10 μl l^{-1}ethylene alone or with 0.11 mM 2,5-norbornadiene (NBD) as before (Goldschmidt *et al.*, 1993), and N_2 treatment was applied as before (Shi *et al.*, 2005). Mature fruit were pretreated with air or 300 nl l^{-1} 1-methylcyclopropene (1-MCP) in 30 l plastic containers for 24 h, then ventilated, and sealed again in air or N_2 for another 24 h with or without 10 μl l^{-1} ethylene. After treatment, flavedo tissues were frozen in liquid N_2 and stored at −80°C until analyzed and physiological parameters determined immediately as before (Shi *et al.*, 2005).

For gene cloning, degenerated primers for pyruvate decarboxylase (*PDC*) and alcohol dehydrogenase (*ADH*) were designed based on the conserved domains of homologous genes in GenBank. PCRs were performed using citrus cDNA libraries as templates, and generated fragments with expected length were cloned and sequenced. *PDC* fragments were obtained from one citrus cDNA library (Jacob-Wilk *et al.*, 1997) and *ADH* from another citrus cDNA library (Kapri *et al.*, 2000). Isolated genes were submitted to GenBank under following accession number: *PDC*-DQ083540; *ADH*-DQ083539. *CsACS1*, *CsACS2*, and *CsACO1* were cloned by Katz *et al.* (2004) and *CsERS1* by Li *et al.* (2000). Northern blot analysis was done as previously (Li *et al.*, 2003) with the exception that the blots were hybridized in ULTRAhyb hybridization buffer at 42°C.

For volatile and proteome profile analysis, mature fruit were sealed with air, N_2 or 10 μl l^{-1} ethylene for 24 h at 20°C. Flavedo and juice vesicle were frozen immediately in liquid N_2 and stored at −80°C until use, while combined juice samples were extracted, and a 10 ml aliquot was poured into a

20 ml headspace vial containing 3.6 g NaCl. After addition of 2,6-dimethyl-5-hepten-2-ol as an internal standard, the vial was sealed immediately and kept at 4°C until use. The volatiles were analyzed and identified according to Holland *et al.* (2005). Protein extraction and two-dimensional gel electrophoresis were carried out as described by Chen *et al.* (2006) with the exception that 13 cm IPG strips (pH 4–7) were rehydrated overnight with 250 µl of 9 M urea IPG re-swelling buffer containing 30 µg protein. Gel matching for protein quantification was performed with Z3 software (Compugen Inc., Israel) and spot pairs were confirmed visually. Coomassie stained spots were digested by trypsin and analyzed by LC-MS/MS on DECA/LCQ and identified by Pep-Miner and Sequest software against nr database of plants.

3. Results and Discussion

3.1 Ethylene and anoxia on ethylene production and perception

Exogenous ethylene autocatalyzed ethylene production in young fruitlets while ethylene plus NBD counteracted the effect of ethylene (Fig. 1). Ethylene autoinhibited ethylene production in mature fruit and ethylene plus 1-MCP counteracted the autoinhibition (Fig. 2). The autocatalytic ethylene production of young fruitlets correlated to the up-regulated expression of *CsACO1* (Fig. 1) and autoinhibiting ethylene production of mature fruit correlated to the down-regulated expression of *CsACS1*, *CsACO1*, and *CsERS1* (Fig. 2). N_2 treatment for 6 h slightly decreased ethylene evolution in both fruitlets, reduced *CsERS1* expression but increased *CsACS1* expression (Fig. 1). N_2 treatment for 24 h increased ethylene evolution in both mature fruits (Fig. 2). Under N_2 atmosphere, ethylene increased ethylene evolution and retained expression of *CsACS2, CsACO1*, and *CsERS1*, in both fruits.

CsACS1 seems to be under stress-regulated since N_2 treatment increased its expression in young and mature "Murcott" (Figs. 1, 2). *CsACS2* expression was detected only in mature fruit. The expression of *CsACO1* preceded *CsACS1* and correlated well with ethylene evolution in young and mature "Murcott" fruit. Expression of *CsACO1* differed in these two cultivars in response to N_2 or ethylene. Anoxia (24 h) increased expressions of *CsACS1* and/or *CsACS2* and retained *CsACO1*expression, thus contributed to the increased ethylene evolution in mature "Murcott' as observed previously (Shi *et al.*, 2005). Ethylene induced the expression of *CsERS1* in young fruitlets but reduced it in mature fruits of both cultivars. N_2 reduced the expression of *CsERS1* in both cultivars irrespective of maturity, exerting its effect at ethylene perception level too. Under N_2 atmosphere, ethylene retained *CsERS1* expression in both mature fruits, thus reduced ethylene sensitivity of the N_2 stressed tissues. In the presence of an anaerobic stress, blocking ethylene binding prevented regulatory control of the ethylene biosynthetic pathway that resulted in an uninhibited expression of the *ACS* stress-related genes thus increased ethylene production.

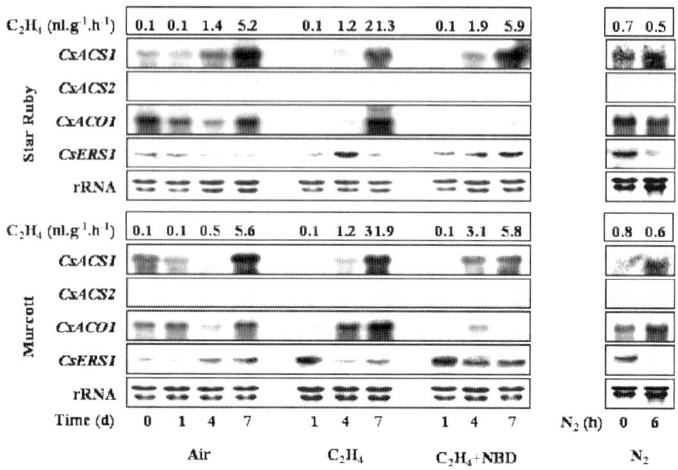

Fig. 1. Ethylene evolution and gene expression in young "Star Ruby" and "Murcott" fruitlets treated with air (Air), 10 μl l^{-1} ethylene (C_2H_4), 10 μl l^{-1} ethylene plus 0.11 mM 2,5-norbornadiene (C_2H_4 + NBD), and N_2 atmosphere (N_2), in the dark at 20°C.

3.2 Effect of ethylene and anoxia on ethanol fermentation

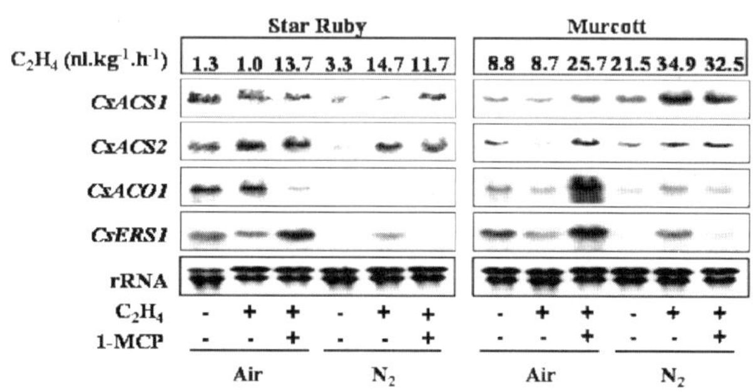

Fig. 2. Ethylene evolution and gene expression in mature "Star Ruby" and "Murcott" fruits pretreated with air or 300 nl l^{-1}1-MCP for 24 h and then stored in air (Air) or N_2 atmosphere (N_2) for another 24 h with or without 10 μl l^{-1} ethylene.

Ethylene increased AA and ethanol evolution in both young fruitlets and in mature "Star Ruby", and increased the expressions of *PDC* and *ADH*. Ethylene action inhibitors (NBD or 1-MCP) reduced the effect of ethylene on the expression of ethanol fermentative genes and the production of AA and ethanol, indicating that ethanol fermentation in citrus was, at least to some extent, ethylene dependent. Ethylene was reported to be involved in the regulation of *ADH* expression in ripening grapes (Tesniere *et al.*, 2004) and mature green tomato (Chen and Chase, 1993). N_2 treatment greatly increased AA and ethanol evolution, as well as *PDC* and *ADH* expressions in young and mature fruit of both cultivars. Under N_2 atmosphere, ethylene alone or with 1-MCP had little effect on the anaerobic-induced accumulation of AA and ethanol in mature fruits as well as corresponding expression of *PDC* and *ADH*, suggesting that ethylene does not seem to be the primary regulator of ethanol fermentation.

3.3 Effect of ethylene on volatile and proteomic profiles

Juice volatile profiles of the two cultivars differed not only in components but also in their compositions (Table.1). "Star Ruby" contained nine terpenes, one alcohol, two ketones, three esters, and two oxides, with limonene (39.48%) and trans-caryophyllene (41.66%) the two most prominent. "Murcott" had three terpenes, four alcohols, one ketone, three esters, and four aldehydes, with limonene (83.56%) the most prominent. Ethylene and N_2 treatment did not change components but changed their compositions in both cultivars. Both treatments increased total volatiles in "Star Ruby", but decreased it in "Murcott", and ethylene seemed to be more effective than N_2. It is surprising that ethylene, which enhances maturation and volatile evolution in numerous fruits, decreased total volatiles (mostly limonene) of "Murcott". In apple, ethylene affects ester biosynthesis by regulating alcohol acyltransferase (Defilippi *et al.*, 2005).

Table 1. Effect of treatment with ethylene and N_2 atmosphere on the changes in the juice head space volatile profiles of "Star Ruby" and "Murcott".

Star Ruby	Alcohols	Esters	Ketones	Oxides	Terpenes	Total
Air	59/0.89[a]	195/2.95	183/2.78	368/5.57	5792/87.81	6597[b]
C_2H_4	54/0.57	165/1.75	187/1.97	431/4.56	8613/91.15	9450
N_2	76/1.01	318/4.22	178/2.36	394/5.23	6570/87.19	7535
Murcott	Alcohols	Aldehydes	Esters	Ketones	Terpenes	Total
Air	129/3.64	145/4.08	69/1.94	4/0.12	3201/90.22	3548
C_2H_4	97/12.25	79/9.98	101/12.70	2.5/0.32	513/64.75	792
N_2	156/6.72	101/4.34	109/4.69	4/0.17	1955/84.08	2325

[a]Concentration (ml l^{-1})/Composition (%). [b]Concentration (ml l^{-1})

More than 1500 protein spots were detected by digital image analysis, and at least 400 spots gave reproducible staining patterns for all samples as judged by eye and by spot intensity ranking using Z3 software. Less spots were detected in ethylene-treated samples as compared with controls.

The identified proteins are: heat shock protein (HSP), ATP synthase β subunit, cysteine-type endopeptidase, enolase, osmotin, intracellular pathogenesis related protein (PR), glutathione S-transferase (GST), lipoxygenase (LOX), dihydrolipoamide dehydrogenase precursor (DLDH), superoxide dismutase (SOD), HSP 19 class II, alcohol dehydrogenase (ADH), putative replication factor C 36 kDa subunit (RFc), protein disulfide isomerase (PDI), putative non-LTR retroelement (RE), putative endoplasmic reticulum membrane fusion protein (ERMF), and an unknown protein. Ethylene reduced but N_2 increased the expression of some defense proteins (such as ADH, DLDH, enolase, GST, PR, SOD, and HSP 19 class II) and other proteins related to protein synthesis (such as RFc, PDI, RE, and ERMF). Ethylene accelerated senescence, thus, reduced protein synthesis, resulting in fewer detectable spots in the gel as compared with control. The observed different effect of ethylene and N_2 on the proteome profiles corresponds to their differently effects on citrus fruit volatiles.

4. Conclusions

Taken together, results suggest that ethanol fermentation in citrus fruit is, at least to some extent, ethylene dependent. Ethylene also plays a significant role in controlling the production of various groups of volatiles other than AA and ethanol in the juice of citrus fruit. Those changes may be associated with ethylene-induced senescence as revealed by proteome analysis. Compared with ethylene, N_2 had stronger effect on the ethanol fermentation but weaker effect on volatiles other than AA and ethanol. Ethylene and anoxia regulate ethylene signal by modulating the level of ethylene and its perception in citrus at the transcriptional level. Other regulatory factors can not be excluded, however.

References

Chen AR, Chase TJ (1993) Alcohol dehydrogenase 2 and pyruvate decarboxylase induction in ripening and hypoxic tomato fruit. Plant Physiol Biochem 3: 875–885.

Chen SB, Glazer I, Gollop N, Cash P, Argo E, Innes A, Stewart E, Davidson I, Wilson MJ (2006) Proteomic analysis of the entomopathogenic nematode

Steinernema feltiae IS-6IJs under evaporative and osmotic stresses. Mol Biocheml Parasitol 145: 195–204.

Defilippi BG, Kader AA, Dandekar AM (2005) Apple aroma: alcohol acyltransferase, a rate limiting step for ester biosynthesis, is regulated by ethylene. Plant Sci 168: 1199–1210.

Goldschmidt EE, Huberman M, Goren R (1993) Probing the role of endogenous ethylene in the degreening of citrus-fruit with ethylene Antagonists. Plant Growth Regul 12: 325–329.

Holland D, Larkov O, Bar-Ya'Akov I, Bar E, Zax A, Brandeis E, Ravid U, Lewinsohn E (2005) Developmental and varietal differences in volatile ester formation and Acetyl-CoA: Alcohol acetyl transferase activities in apple (Malus domestica Borkh.) fruit. J Agri Food Chem 53: 7198–7203.

Jacob-Wilk D, Goldschmidt EE, Riov J, Sadka A, Holland D (1997) Induction of a Citrus gene highly homologous to plant and yeast thi genes involved in thiamine biosynthesis during natural and ethylene-induced fruit maturation. Plant Mol Biol 35: 661–666.

Kapri R, Dahan E, Sadka A, Zehavi U, Goren R (2000) Cloning and characterization of PPi-phosphofructokinase from citrus fruit. Acta Hortic 535: 113–118.

Katz E, Lagunes PM, Riov J, Weiss D, Goldschmidt EE (2004) Molecular and physiological evidence suggests the existence of a system II-like pathway of ethylene production in non-climacteric Citrus fruit. Planta 219: 243–252.

Li CY, Zhong GY, Goren R, Jacob-Wilk D, Holland D (2000) Cloning of a full-length cDNA encoding an ethylene receptor ERS homologue from Citrus. Acta Hortic 535: 119–125.

Li CY, Shi JX, Weiss D, Goldschmidt EE (2003) Sugars regulate sucrose transporter gene expression in citrus. Biochem Biophys Res Commun 306: 402–407.

Porat R, Weiss B, Cohen L, Daus A, Goren R, Droby S (1999) Effects of ethylene and 1-methylcyclopropene on the postharvest qualities of 'Shamouti' oranges. Postharvest Biol Technol 15: 155–163.

Shi JX, Porat R, Goren R, Goldschmidt EE (2005) Physiological responses of 'Murcott' mandarins and 'Star Ruby' grapefruit to anaerobic stress conditions and their relation to fruit taste, quality and emission of off-flavor volatiles. Postharvest Biol Technol 38: 99–105.

Tesniere C, Pradal M, El-Kereamy A, Torregrosa L, Chatelet P, Roustan JP, Chervin C (2004) Involvement of ethylene signalling in a non-climacteric fruit: new elements regarding the regulation of ADH expression in grapevine. J Exp Bot 55: 2235–2240.

The role of ethylene in the expression of genes involved in the biosynthesis of aroma volatiles in melon

Lucchetta, L., Manríquez, D., El-Sharkawy, I., Flores, F.B., Latché, A. and Pech, J.C.*

UMR 990 INRA/INPT-ENSAT "Génomique et Biotechnologie des Fruits", Av. de l'Agrobiopole, BP 32607, F-31326 Castanet-Tolosan Cedex, France. (*Corresponding author: pech@ensat.fr)

Abstract

Aroma volatiles are important components of the sensory quality of fruit. In recent decades, the long shelf-life character has been introduced into commercial varieties of fruit such as tomatoes or melons through the use, in breeding programmes, of mutants impaired for ethylene perception. Extension of shelf life has resulted in a strong reduction of aroma volatiles production indicating a negative correlation between aroma production and ethylene perception or synthesis. Volatile esters are compounds that contribute to specific characteristics of the flavour of many fruit. Charentais cantaloupe melon is a highly aromatic fruit in which esters represent the major class of volatile compounds. Introduction of the long shelf-life character by traditional genetics or inhibition of ethylene production by biotechnology has resulted in a great loss of aroma production in melon. The two last steps of esters biosynthesis is mediated by alcohol dehydrogenases (ADH) and alcohol acyl-transferase (AAT). We report here that AAT is encoded by a gene family of at least three members that account for the great diversity of esters formed in melon, including sulphur esters that are responsible for the typical aroma of cantaloupes. Site-directed mutagenesis demonstrated the crucial role of some amino acids in AAT activity and substrate specificity. ADHs participate in the biosynthetic pathway of esters by providing substrates to the AATs. We have isolated two highly divergent *ADH* genes that are expressed specifically in fruit and up-regulated during ripening. The encoded proteins show differential substrate specificity. The expression of all *AAT* and *ADH* genes is up-regulated during fruit ripening and inhibited in antisense ACC oxidase melons or wild type fruit treated with 1-MCP. These data suggest that each member of the *AAT* and *ADH* gene family plays

A. Ramina et al. (eds.), Advances in Plant Ethylene Research: Proceedings of the 7th International Symposium on the Plant Hormone Ethylene, 189–195.

a specific role in the biosynthesis of aroma in melon fruit and that ethylene is a major regulator of their expression.

Keywords: Alcohol acyltransferase, alcohol dehydrogenase, aroma biosynthesis genes, esters

1. Introduction

The sensory quality of fruit encompasses a range of properties such as sweetness, acidity, aroma, firmness and colour that are associated with specific metabolic pathways and are generally coordinated during the ripening process. The development of good quality attributes depends on many factors including variety, culture conditions, harvest date and storage conditions. In climacteric fruit, the plant hormone ethylene is controlling most of the ripening events and therefore most of the metabolic pathways responsible for the development of sensory quality (Lelièvre *et al.*, 1997; Giovannoni, 2004). In addition, ethylene controls the rate of ripening and the duration of shelf life. Increasing storability has been one of the major goals of the breeders in the last decades and this has generally been accompanied by a loss of flavour. In the case of cantaloupe Charentais melons, for example, traditional fast ripening genotypes with a sharp climacteric phase have been replaced on the market by mid- or long-shelf-life varieties, which are considered of lower sensory quality by most consumers. More specifically aromas, which represent a major quality parameter of Charentais melon, are substantially reduced in long keeping genotypes. In this paper we present (i) the role of ethylene in the biosynthesis in aromas in relation with the storage life of melon, (ii) the molecular characteristics of members of the alcohol acyl transferase and alcohol dehydrogenase gene family and (iii) the biochemical characteristics of the encoded proteins.

2. Ethylene and Aroma Volatile Production

The aroma volatiles evolved by cantaloupe melons are mainly made of a complex mixture of esters, of saturated and unsaturated aldehydes and alcohols and of sulphur compounds (Homatidou *et al.*, 1992; Wyllie *et al.*, 1995; Beaulieu and Grimm, 2001). The aliphatic and branched esters represent the largest portion of volatiles and are essential contributors to the aroma. Long- or mid-shelf-life commercial Charentais melons genotypes (*Cucumis melo* var. *cantalupensis)* generated by conventional breeding are available. Some of them have been obtained using a non-ripening melon named 'Vauclusien'. In these hybrids, the development of abscission zone is impaired or delayed, which renders the determination of harvest time difficult. Aubert and Bourger (2004) have studied the production of aroma volatiles in 15 Charentais melon cultivars having differential ripening rates and storability. Aubert and Bourger (2004) have found a considerable reduction (49–87%) of aroma volatiles production in long shelf-life cultivars as compared to the original type and mid shelf-life. In particular, most of the esters with low odour threshold (potent odourants) were reduced by 2–30-fold. Interestingly, Bauchot *et al* (1998) came to similar conclusions using hybrids of Charentais melons obtained by crossing with an antisense ACO line. The transgenic hybrids evolved 60 to 85% less total volatiles than the non-transformed hybrids (Bauchot *et al.*, 1998). Volatiles with low odour values, such as ethyl, 2-methylpropyl and 2-methylbutyl acetates, were half to one-fifth lower than in controls whereas potent odourants, such as ethyl-2-methylpropanoate and ethyl-2-methylbutanoate, were about 3% that of controls. Together, these data suggest that ethylene is controlling preferentially the synthesis of the most potent odourants.

In order to determine which step in the ester biosynthetic pathway is under the control of ethylene. Using fruit disks incubated in the presence of various precursors, the steps at which ester formation was inhibited in ethylene-suppressed fruit was the reduction of fatty acids and aldehydes. As for the acyl transfer to alcohols to form esters, it was inhibited only partly, indicating that this step had both ethylene-dependent and -independent components (Flores *et al.*, 2002).

3. Molecular Characteristics and Expression of Genes of the Alcohol Acyl Transferase Gene Family

The formation of esters is catalysed by alcohol acyl transferase (AAT) enzymes that transfer an acyl-CoA to an alcohol (Harada *et al.*, 1985). These enzymes are capable of combining different alcohols and acyl-CoAs resulting in the synthesis of a wide range of esters accounting for the diversity of esters emitted by the fruit. We have demonstrated that, in Charentais melon, AAT are encoded by a gene family of at least four members with amino acid identity ranging from 84% (Cm-AAT1/Cm-AAT2) and 58% (Cm-AAT1/Cm-AAT3) to only 22% (Cm-AAT1/Cm-AAT4). Sequence alignment showed that the four putative Cm-AAT genes include conserved motifs that are common to the plant BAHD (benzylalcohol acetyl-, anthocyanin-*O*-hydroxycinnamoyl- anthranilate-*N*-hydroxycinnamoyl/benzoyl-, deacetyl-vindoline acetyltransferase) *O*-acetyltransferases gene family (St Pierre and De Luca, 2000). The *Cucumis melo AAT1 to 4* encode proteins of 462, 461, 459 and 479 amino acids, respectively which corresponds to a deduced molecular mass of 51.5, 51.8, 51.1 and 55.0 kDa (El-Sharkawy *et al.*, 2005).

Real time PCR studies demonstrated that the expression of all *Cm-AAT* genes is fruit specific and up-regulated during ripening (El-Sharkawy *et al.*, 2005). In addition, their expression is inhibited in antisense ACC oxidase melons and in fruit treated with the ethylene antagonist 1-methylcyclopropene (1-MCP). Our data therefore indicate that ethylene is a major regulator of the accumulation of all Cm-AAT transcripts.

4. Biochemical Properties of the Alcohol Acyl Transferase Proteins

All encoded AAT proteins, except Cm-AAT2, were enzymatically active upon expression in yeast and show differential substrate preferences (Yahyaoui *et al.*, 2002; El-Sharkawy *et al.*, 2005). Cm-AAT1 protein produces a wide range of short- and long-chain acyl esters but has strong preference for the formation of *E*-2-hexenyl acetate and hexyl hexanoate. Cm-AAT3 also accepts a wide range of substrates but with very strong preference for producing benzyl acetate. Cm-AAT4 is almost exclusively devoted to the formation of acetates, with strong preference for cinnamoyl acetate. Cm-AAT1 was much more active than Cm-AAT3 and Cm-AAT4 in the formation of sulphur esters by acylation of 3-(methylthio)-1-propanol and 2-(methylthio) ethanol (unpublished data), indicating that Cm-AAT1 plays a major role in the production of volatiles that give the characteristic scent of cantaloupe melons (Homatidou *et al.*, 1992).

The molecular mass of all recombinant AATs as well as of AATs extracted from melon fruit has been determined by gel filtration. It appears that all proteins are under tetrameric form and are active only under such conformation. We have found that the product of the reaction, CoA SH is a stimulator of the reaction at low concentrations until 2.5 μM and is inhibitor at higher concentrations. The introduction of a phosphotransacetylase capable of removing CoA SH during the reaction resulted in a lower K_m values for the activity of the recombinant proteins. For instance the K_m for acetyl CoA of Cm-AAT4 was decreased from 150 μM in the absence of phosphotransacetylase to 45 μM in its presence.

Site directed mutagenesis demonstrated that the failure of Cm-AAT2 to produce volatile esters is related to the presence of a 268-alanine residue instead of threonine as in all active AAT proteins. Mutating 268-A into 268-T of Cm-AAT2 restored enzyme activity, while mutating 268-T into

268-A abolished activity of Cm-AAT1. Activities of all three proteins measured with the preferred substrates sharply increase during fruit ripening. These results suggest that the multiplicity of *AAT* genes accounts for the great diversity of esters formed in melon.

5. Isolation, Molecular Characteristics and Expression of Genes of the Alcohol Dehydrogenases Family

Alcohol dehydrogenases (ADH) participate in the biosynthetic pathway of aroma volatiles in fruit by interconverting aldehydes to alcohols and providing substrates for the formation of esters (Speirs *et al*., 1998). We have isolated two highly divergent ADH genes (<8% identity at the amino acid level) of cantaloupe Charentais melon. Cm-ADH1 belongs to the medium chain zinc-containing class of ADH and is highly similar to all *ADH* genes expressed in fruit isolated so far (Chase, 1999). Cm-ADH2 belongs to the short chain classes of ADH that has never been described in fruit (Manríquez *et al*., 2006).

6. Biochemical Properties of Two Fruit-Specific Alcohol Dehydrogenases of Melon

The two encoded proteins are enzymatically active upon expression in yeast and purification (Manríquez *et al*., 2006). Cm-ADH1 has strong preference for NAPDH as a cofactor whereas Cm-ADH2 preferentially uses NADH. Both perform better as reductases with Kms 10 to 20 times lowers for the conversion of aldehydes to alcohols than for the dehydrogenation of alcohols to aldehydes. Both also show strong preference for aliphatic aldehydes. Cm-ADH1 is capable of reducing branched aldehydes such as 3-methylbutyraldehyde, whereas Cm-ADH2 cannot. Both *Cm-ADH* genes are expressed specifically in fruit and up-regulated during ripening. Gene expression is strongly inhibited in antisense ACC oxidase (AS) melons and in fruit treated with the ethylene antagonist 1-methylcyclopropene (1-MCP). Similarly total ADH activity of ripening fruit is also inhibited by 1-MCP and in AS fruit, indicating that ethylene plays a major role in regulating ADH activity. In addition, ADH activity in ripening fruit is mostly NADPH-dependent indicating that ADH1 is more active than ADH2. These data suggest that the two ADHs play specific roles in the regulation of aroma biosynthesis in melon.

Acknowledgements

L. Lucchetta and D. Manríquez have received doctoral fellowships from the Ministry of Education of Brazil and from the Government of Chile (CONICYT), respectively. F.B. Flores and I. El-Sharkawy had postdoctoral fellowships from the Ministry of Education of Spain and from INRA France, respectively. Part of the research was funded by the Midi Pyrénées regional council (grants 01008920 and 03001146)

References

Aubert, C. and Bourger, N. 2004. Investigation of volatiles in Charentais Cantaloupe melons (*Cucumis melo* Var *Cantalupensis*). Characterization of aroma constituents in some cultivars. J. Agric. Food Chem. 52: 4522–4528.

Bauchot, A.D., Mottram, D.S., Dodson, A.T. and John, P. 1998. Effect of aminocyclopropane-1-carboxylic acid oxidase antisense gene on the formation of volatile esters in Cantaloupe Charentais melon (cv. Védrantais). J. Agric. Food Chem. 46:4787–4792.

Beaulieu, J.C. and Grimm, C.C. 2001. Identification of volatile compounds in cantaloupe at various developmental stages using solid phase microextraction. J. Agric. Food Chem. 49:1345–1352.

Chase, T. 1999. Alcohol dehydrogenases: Identification and names for gene families. Plant Mol. Biol. Rep. 17: 333–350.

El-Sharkawy, I., Manríquez, D., Flores, F.B., Regad, F., Bouzayen, M., Latché, A. and Pech, J.C. 2005. Functional characterization of a melon alcohol acyltransferase gene family involved in the biosynthesis of ester volatiles. Identification of the crucial role of a threonine residue for enzyme activity. Plant Mol. Biol. 59: 343–360.

Flores, F., El Yahyaoui, F., De Billerbeck, G., Romojaro, F., Latché, A., Bouzayen, M., Pech, J. C. and Ambid, C. 2002. Role of ethylene in the biosynthetic pathway of aliphatic ester aroma volatiles in Charentais Cantaloupe melons. J. Exp. Bot. 53:201–206.

Giovannoni, J.J. 2004. Genetic regulation of fruit development and ripening. Plant Cell 16:170–180.

Harada, T., Sunako, T., Wakasa, Y., Soejima, J., Satoh, T. and Niizeki, M. 1985. An allele of the 1-aminocyclopropane-1-carboxylate synthase gene accounts for the low level of ethylene production in climacteric fruits of some appple cultivars. Theor. Appl. Genet. 101:742–746.

Homatidou, V.I., Karvouni, S.S., Dourtoglou, V.G. and Poulos, C.N. 1992. Determination of total volatile components of *Cucumis melo* L. variety *cantalupensis*. J. Agric. Food Chem. 40:1385–1388.

Lelièvre, J.M., Latché, A., Jones, B., Bouzayen, M. and Pech, J.C. 1997. Ethylene and fruit ripening. Physiol. Plant 101:727–739.

Manríquez, D., El-Sharkawy, I., Flores, F.B., Regad, F., Bouzayen, M., Latché, A. and Pech, J.C. 2006. Fruit-specific gene expression and biochemical characteristics of two highly divergent alcohol dehydrogenases of melon. Plant Mol. Biol. (in press)

Speirs, J., Lee, E., Holt, K., Yonk-Duk, K., Steele Scott, N., Loveys, B., Schuch, W. 1998. Genetic manipulation of alcohol dehydrogenase levels in ripening tomato fruit affects the balance of some flavor aldehydes and alcohols. Plant Physiol., 117:1047–1058.

St Pierre, B. and De Luca, V. 2000. Evolution of acyltransferase genes: origin and diversification of the BAHD superfamily of acyltransferases involved in secondary metabolism. In: John, R.I., Romeo, R., Varin, L. and De Luca V., Eds, Recent advances in Phytochemistry. Evolution of metabolic pathways, vol 34, Elsevier Science Pub. Oxford, UK, pp 285–315.

Wyllie, S.G., Leach, D.N., Wang, Y. and Shewfelt, R.L. 1995. Key aroma compounds in melons. p. 248–257. In: R.L. Roussef and M.M. Leahy (eds), Fruit flavors. ACS Symposium series 596. American Chemical Society, Washington D.C.

Yahyaoui, F.El, Wongs-Aree, C., Latché, A., Hackett, R., Grierson, D. and Pech, J.C. 2002. Molecular and biochemical characteristics of a gene encoding an alcohol acyl-transferase involved in the generation of aroma volatile esters during melon ripening. Eur. J. Biochem. 269:2359–2365.

Creating climacteric melon fruit from nonclimacteric parentals: postharvest quality implications

Obando, J.[1], Miranda, C.[1], Jowkar, M.M.[1,3], Moreno, E.[2], Sour, M.K.[4], Martínez, J.A.[1], Arús, P.[2], García-Mas, J.[2], Monforte, A.J.[2] and Fernández-Trujillo, J.P.[1]*

[1]Technical University of Cartagena, Paseo Alfonso XIII 48. ETSIA & Institute of Plant Biotechnology. 30203 Cartagena, Murcia, Spain. Tel. 34- 96832 5436. Fax 34-968325433. [2]Department Genética Vegetal. Laboratori de Genètica Molecular Vegetal CSIC-IRTA. Carretera de Cabrils s/n. 08348 Cabrils (Barcelona). Spain. Tel. 34-93 750 7511. Fax: 34-937533954 antonio.monforte@irta.es. [3]Present address: Islamic Azad Univ., Kermanshah Branch, Baghe-ney, Kermanshah, Iran. Tel. 98- 9188393066. Fax. 98- 8318356321. [4]Plant Nutrition Institute. Univ of Hohenheim. Institute 330. D-70593 Stuttgart. Germany. Tel. 49- 711459 3711. (*Corresponding author: jf68es@terra.es, juanp.fdez@upct.es)

1. Introduction

Physiologists have historically divided ripening into two general classifications, climacteric and nonclimacteric, based on the presence or absence of dramatic ripening-related increases in respiration, respectively (Biale and Young, 1981). This concept is fundamental to any discussion of ethylene in postharvest systems. Two reviews of this topic (Biale and Young, 1981; Goldschmidt, 1997) concluded that both categories represent an oversimplification. In certain species such as pepper, Asian pear, tomato, and melons climacteric cultivars or mutants of both categories can be found (Watkins, 2002; Downs *et al.*, 1991; Villavicencio *et al.*, 1999). Consequently, nonclimacteric or climacteric group patterns have been proposed. Both terms are commercially used to understand postharvest shelf-life (Watkins, 2002). The nonclimacteric group would have ripening-related responses (or not) to exogenous ethylene (Goldschmidt, 1997), while climacteric fruit would have an autocatalytic regulatory mechanism of ripening (Kubo *et al.*, 2003). Climacteric melon for example, would belong to the second group since it shows autocatalytic type regulation of ethylene having a marked increase in ethylene when treated with 1-MCP at the climacteric (Kubo *et al.*, 2003). Both ethylene-dependent and ethylene-independent

A. Ramina et al. (eds.), Advances in Plant Ethylene Research: Proceedings of the 7th International Symposium on the Plant Hormone Ethylene, 197–205.

pathways of gene regulation coexist in climacteric melon fruit (Hadfield *et al.*, 2000). Data concerning the lack of a climacteric rise in muskmelons attached to the plant (Shellie and Saltveit, 1993), the climacteric-like behavior of young citrus (Katz *et al.*, 2005), and the role of ethylene biosynthesis in the ripening of nonclimacteric fruit (Alonso *et al.*, 1995; Cazzonelli *et al.*, 1998; Chervin *et al.*, 2004; Trainotti *et al.*, 2005) challenge the definition of climacteric as the opposite of nonclimacteric.

Fruit shelf-life is a very important factor for ensuring fruit of suitable quality reaches consumers. In melon fruit, high ethylene production has been associated with reduced shelf-life (Zheng and Wolff, 2000). The quantitative genetic control of ethylene production and respiration rates has been poorly studied in fruits because the genetic control of such traits is complex and involves a large number of genes or quantitative trait loci (QTLs) (Zheng *et al.*, 2002). A very efficient way to dissect complex traits is to analyze introgression lines (IL) containing a single homozygous chromosome segment from a donor parent in the genetic background of an elite cultivar (Eshed and Zamir, 1995). These lines have, intentionally, a high percentage of the recurrent parent genome, so they have been defined as near isogenic lines (NILs). A NIL population can be defined as a genomic library where inserts are chromosome fragments of the donor parental and vector is the recurrent parent genome. NILs from tomato and melons are useful for dissecting complex traits and for detecting the QTLs involved in fruit quality (Eduardo *et al.*, 2006; Overy *et al.*, 2005; Rousseaux *et al.*, 2005), including sensitivity to chilling injury (Goldstal *et al.*, 2005) and aroma volatiles (Lewinsohn *et al.*, 2005).

In melons, accessions with good shelf-life are mostly found for vars. *Saccharinus* and *Inodorus* (Liu *et al.*, 2004), particularly when fruit are stored at temperatures of 12–18°C, without risk of chilling injury (CI) (Miccolis and Saltveit, 1995). CI symptoms in melons include skin scald and pitting (Miccolis and Saltveit, 1995), both of which have been associated with reduced sensitivity to ethylene during cold storage (Ben-Amor *et al.*, 1999). Périn *et al.* (2002) demonstrated increased climacteric behavior when crossing PI161375 and the ′Védrantais′ melon.

The aim of this work was to study fruit quality traits, including ripening. An unexpected result lead us to study how it was possible to create climacteric near isogenic lines of melon from two nonclimacteric parentals, and how the climacteric lines showed the typical postharvest behavior of these kinds of fruit. Differences between quality and metabolic profiles can identify those pathways affected by the introgression and allow genetic maps for metabolic alterations to be established.

2. Materials and Methods

2.1 Plant materials

An introgression line population consisting of near isogenic lines (NILs) derived from a cross between the *Cucumis melo* L. Spanish cultivar ′Piel de Sapo′ (PS), and an exotic Korean accession 'Shongwan Charmi' (PI 161375), each of them having a single introgression from PI 161375 into the PS background, was used to detect alleles of genes or QTLs controlling fruit morphological traits (Eduardo *et al*., 2005). The NIL SC3.5 with introgressions in the linkage group (LG) III (Eduardo *et al*., 2005) was used to obtain subNILs with smaller introgressions (encoded as NILs 5M*x*, x = 1 to 10) to improve the resolution of the mapping. The experimental design consisted of seven replicates of three plants each (n = 9 for SC3.5, 5M9, 5M10, and PI161375; n = 20 for the parental PS). Two or three fruits per replicate were evaluated at harvest. Respiration rate and ethylene production at 20°C was measured in five fruits per NIL or parentals of different replicates.

2.2 Physiological measurements

Fruit weight and density (as measured in $kg \cdot m^{-3}$ by water displacement measurements to calculate gaseous exchange on a fruit weight basis) and other quality traits (soluble solids, titratable acidity, flesh firmness) were calculated according to Fernández-Trujillo *et al*., (2005a, 2005b). The fruit respiration rate and ethylene production were determined by using the static method. Following this method, individual fruit were enclosed in a hermetic container for 1 h at 20 ± 1°C, and C_2H_4 was sampled at 30 min and CO_2 after 1 h. Both gases were measured by gas chromatography (Fernández-Trujillo *et al*., 2005b). For CI sensitivity, at least five fruit per replicate (n = 7) were stored in boxes covered by plastic liners (Plásticos del Segura SL, Murcia, Spain) at 8 ± 0.6°C and 87 ± 2% RH (both measured below the films). Fruit with skin scald symptoms (alone or associated to netting) were scored into five classes according to Fernández-Trujillo and Artés (1998) depending on the surface affected by CI: 0 = absent; 1 = very slight (0–5%), 2 = slight (5–10%); 3 = moderate (10–25%); 4 = severe (>25%). Moderate to severe-injured fruits and decay caused by necrotrophic fungi and other fungal diseases were considered as losses.

3. Results and Discussion

Parental melon lines PS and PI161375 showed nonclimacteric and nonabscising behavior, though PI161375 showed higher levels of respiration and ethylene production (Fig. 1). The NIL SC3.5 and seven of the nine derived subNILs showed climacteric behavior accompanied by a peak of ethylene production (Fig. 1 and data not shown) and a typical aroma (data not shown), with the exception of line 5M7.

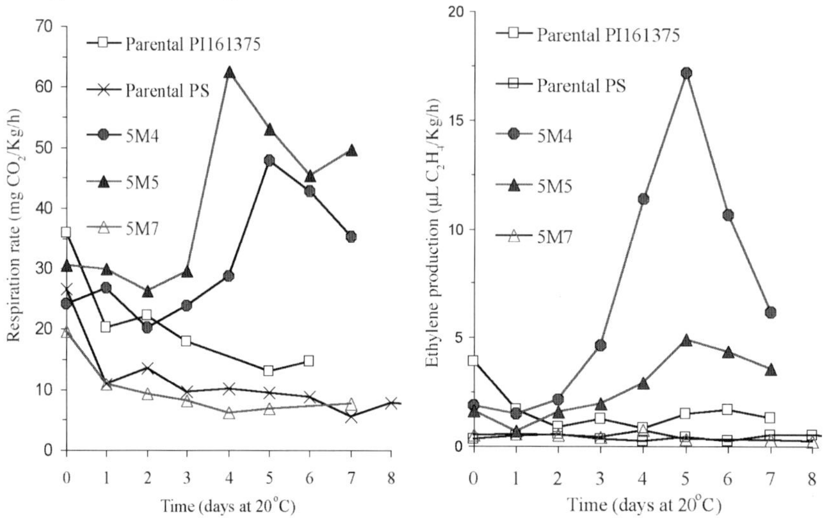

Fig. 1. Changes in respiration rate and ethylene production during postharvest ripening at 20°C in representative individual fruit from a collection of near isogenic lines of melon, and the parental lines ′Piel de Sapo′ and PI161375.

Line 5M4 and PI161375 were very sensitive to CI symptoms (netting injury, scald, and pitting) accompanied by decay. In general, NILs were more sensitive to CI than the PS parental, except 5M7 and 5M5 (Fig. 2). Decay caused by *Cladosporium* sp. affected 24% PS fruit and in the climacteric NILs CI was 1.6–3.7-fold higher than in PS. *Cladosporium* sp. developed in 1% of the PS fruit, alone or associated with *Alternaria* sp., and was also more predominant in the climacteric NILs (2–29-fold higher). Both fungi usually grow on areas suffering CI or in the fruit peduncle, because climacteric NILs were dehiscent to a different extent (particularly those showing ripening very early – earliness), such as 5M4 and 5M5). Line 5M7 suffered half the incidence of *Cladosporium* sp. decay as PS (data not shown), and 2% 5M7 fruit were also affected by *Alternaria* sp. Our results in general support the hypothesis of an increase in sensitivity to CI associated with higher ethylene production levels (Zheng and

Wolff, 2000; Ben-Amor *et al.*, 1999), with the exception of the nonnetted line 5M5. Mealy flesh, a typical problem of ripe fruit of the Korean PI161375 fruit at harvest (Périn *et al.*, 1999), affected PI161375 and several NILs (34% in SC3.5, 61% in 5M5, 43% in 5M9, and 10% in 5M10).

The climacteric lines presented noticeable differences in quality traits compared with the PS. The lines 5M4, 5M5, 5M9, and 5M10 were harvested 1 week earlier than the PS. Comparing other quality traits of the NILs versus PS, the climacteric NILS were softer than the nonclimacteric line 5M7 or the PS (both with hardness and flesh firmness above 60 $N{\cdot}mm^{-1}$ or 7 N, respectively). Only part of the softening is an ethylene-dependent process in melon (Flores *et al.*, 2001). The lines 5M4, 5M6, 5M9, and 5M10 showed higher total soluble solids above 12°Brix, and 5M4 and 5M5 also a fruit weight below 1900 g per fruit. Line 5M7 showed 2°Brix less than the PS (10.5 ± 0.5°Brix). Some NILs developed yellow to light orange flesh color (i.e., SC3.5, 5M4, 5M5, 5M9, and 5M10) compared with the white-flesh PS. Lines 5M5, 5M8, or 5M9 had titratable acidity levels above the 19 ± 2 mmol $H^{+}{\cdot}L^{-1}$ usually found in the PS.

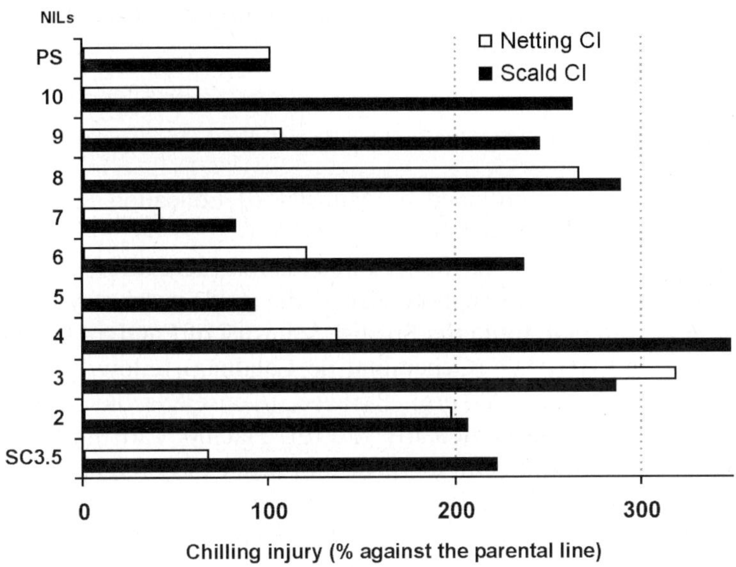

Fig. 2. Relative sensitivity to chilling injury (skin scald and scald associated to netting) compared with the parental line ´Piel de Sapo´ in the collection of near isogenic lines of melon derived from the line SC3.5 after 30 days of storage at 8°C under plastic liners.

In summary, analysis of melon NILs containing PI161375 introgressions in the LG III showed one NIL and seven subNILs with different degrees of climacteric behavior with an accompanying peak of ethylene production. QTLs associated to ethylene production, respiration rate, or sensitivity to mealiness or chilling injury can be located in the genetic map of melon. The subNILs improved our knowledge about the control of both characters and their effects on fruit quality, which seem to be controlled by at least one QTL. The candidate genes reported by Périn *et al.* (2002) are not linked to this climacteric QTL. Therefore, this QTL could induce a cascade of responses affecting fruit quality, slightly different from those reported as being ethylene-dependent (Flores *et al.*, 2001). NILs are a powerful tool for studying genetic epistatic effects and nonevident genetic variability (Esher and Zamir, 1995), and for confirming previous results from other populations (Monforte *et al.*, 2004). The NILs will allow in the future to test the hypothesis of climacteric behavior as a secondary response to fruit detachment (Shellie and Saltveit, 1993; Bower *et al.*, 2002), and are useful for determining QTLs affecting melon fruit quality, including shelf-life attributes. Melon fruit with these quantitative trait loci can be designed in order to satisfy consumer demands, reduce chilling injury, and increase shelf life in the agri-food chain.

Acknowledgments

Financial support was provided by Ministry of Education & Science of Spain and FEDER (coordinated projects AGL2003-09175-C02-01 and AGL2003-09175-C02-02, the Seneca Foundation (00620/PI/04) and funds donated by Prof. F.-T. to support part of the predoctoral stay of M.K.S. J.O. and E.M. are grateful to the Spanish Ministry of Foreign Affairs for a MAE-AECI and AGAUR (Generalitat de Cataluña) fellowships, respectively. We thank CIFEA-Torre Pacheco (Consejería de Agricultura, Región de Murcia) and particularly to Prof. Plácido Varó for crop management. J. Obando is grateful to Spanish Ministry of Foreigner Affairs for a MAE-AECI grant. Plastic liners were provided by Plásticos del Segura S.L. (Murcia) and PS seeds were donated by Semillas Fitó.

References

Alonso JM, Chamarro J, Granell A (1995) Evidence for the involvement of ethylene in the expression of specific RNAs during maturation of the orange, a nonclimacteric fruit. Plant Mol Biol 29:385–390.

Ben-Amor M, Flores B, Latche A, Bouzayen M, Pech JC, Romojaro F (1999) Inhibition of ethylene biosynthesis by antisense ACC oxidase RNA prevents chilling injury in Charentais cantaloupe melons. Plant Cell Environ 22:1579–1586.

Biale J, Young R (1981) Respiration and ripening in fruits-retrospect and prospect. In: Friend J and Rhodes M (Eds.). Recent advances in the biochemistry of fruits and vegetables. Academic Press, NY, pp. 1–39.

Bower J, Holford P, Latche A, Pech JC (2002). Culture conditions and detachment of the fruit influence the effect of ethylene on the climacteric respiration of melon. Postharvest Biol Technol 26:135–146.

Cazzonelli CI, Cavallaro AS, Botella JR (1998) Cloning and characterisation of ripening-induced ethylene biosynthetic genes from non-climacteric pineapple (*Ananas comosus*) fruits. Aust J Plant Physiol 25:513–518.

Chervin C, El Kereamy A, Roustan JP, Latche A, Lamon J, Bouzayen M (2004) Ethylene seems required for the berry development and ripening in grape, a non-climacteric fruit . Plant Sci 167:1301–1305.

Downs CG, Brady CJ, Campbell J, McGlasson WB (1991) Normal ripening cultivars of *Pyrus serotina* are either climacteric or nonclimacteric. Sci Hortic 48:213–221.

Eduardo I, Arús P, Monforte AJ (2005) Development of a genomic library of near isogenic lines (NILs) in melon (*Cucumis melo* L.) from the exotic accession PI161375. Theor Appl Genet 112:139–148.

Eduardo I, Obando J, Weltty N, Martínez JA, Alarcón A, Arús P, Álvarez JM, van der Knaap E, Fernández-Trujillo JP, Monforte AJ (2007) Estimating the genetic architecture of fruit quality traits in melon (*Cucumis melo* L.) using a genomic library of Near-Isogenic Lines. J Am Soc Hortic Sci 132:80–89.

Eshed Y, Zamir D (1995) An introgression-line population of *Lycopersicon pennelli* in the cultivated tomato enables the identification and fine mapping of yield-associated QTLs. Genetics 141:1147–1162.

Fernández-Trujillo JP, Artés F (1998) Chilling injury in peaches during conventional and intermittent warming storage. Int J Refrig 21:265–272.

Fernández-Trujillo JP, Obando J, Martínez JA, Alarcón A, Eduardo I, Monforte AJ, Arús, P (2005a) Statistical multivariate analysis of melon shape: A case study using near isogenic lines. Acta Hortic 674:537–544.

Fernández-Trujillo JP, Obando J, Martínez JA, Alarcón A, Eduardo I, Arús P, Monforte AJ (2005b) Gestión de la calidad de experimentos con frutos de una colección de líneas casi isogénicas de melón (in Spanish). In: Atienza J and Rabasseda J (coord.). III Virtual Iberoamer Cong Qual Lab Manag., Madrid, pp. 149–158.

Flores F, Ben-Amor, M., Jones, B., Pech, J.C., Bouzayen, M., Latche, A., Romojaro, F. (2001) The use of ethylene-suppressed lines to assess differential sensitivity to ethylene of the various ripening pathways in Cantaloupe melons. Physiol Plant 113:128–133.

Goldschmidt EE (1997) Ripening of citrus and other non-climacteric fruits: A role for ethylene. Acta Hortic (ISHS) 463:335–339.

Hadfield KA, Dang T, Guis M, Pech JC, Bouzayen M, Bennett AB (2000) Characterization of ripening-regulated cDNAs and their expression in ethylene-suppressed Charentais melon fruit. Plant Physiol 122:977–983.

Katz E, Riov J, Weiss D, Goldschmidt EE (2005) The climacteric-like behaviour of young, mature and wounded citrus leaves. J Exp Bot 56:1359–1367.

Kubo Y, Owino WO, Liu X, Hiwasa K, Nakano R., Inaba A (2003) Differential regulation of ripening ethylene biosynthesis in several climacteric fruits. In: Vendrell M, Klee H, Pech JC, Romojaro F (Eds.). Biology & Biotechnology of the Plant Hormone Ethylene III, pp. 239–240.

Lewinsohn E, Sitrit Y, Bar E, Azulay Y, Ibdah M, Meir A, Yosef E, Zamir D, Tadmor Y (2005) Not just colors – carotenoid degradation as a link between pigmentation and aroma in tomato and watermelon fruit. Trends Food Sci Technol 16: 407–415.

Liu L, Kakihara F, Kato M (2004) Characterization of six varieties of *Cucumis melo* L. based on morphological and physiological characters, including shelf-life of fruit. Euphytica 135:305–313.

Miccolis V, Saltveit ME (1995). Influence of storage period and temperature on the postharvest characteristics of six melon (*Cucumis melo* L., Inodorus Group) cultivars. Postharvest Biol Technnol 5:211–219.

Monforte AJ, Oliver M, Gonzalo MJ, Álvarez JM, Dolcet-Sanjuan R, Arús P (2004). Identification of quantitative trait loci involved in fruit quality traits in melon (*Cucumis melo* L.). Theor Appl Genet 108:750–758.

Overy SA, Walker HJ, Malone S, Howard TP, Baxter C J, Sweetlove LJ, Hill SA, Quick WP (2005) Application of metabolite profiling to the identification of traits in a population of tomato introgression lines. J Exp Bot 56:287–296.

Périn C, Dogimont C, Giovinazzo N, Besombes D, Guitton L, Hagen L, Pitrat M (1999). Genetic control and linkages of some fruit characters in melons. Cucurbit Genetics Coop Rept 22:16–18.

Périn C, Gómez-Jiménez M, Hagen L, Dogimont C, Pech JC, Latche A, Pitrat M, Lelievre JM (2002) Molecular and genetic characterization of a non-climacteric phenotype in melon reveals two loci conferring altered ethylene response in fruit. Plant Physiol 129:300–309.

Rousseaux MC, Jones, CM, Adams D, Chetelat R, Bennett A, Powell A (2005) QTL analysis of fruit antioxidants in tomato using *Lycopersicon pennellii* introgression lines. Theor Appl Genet 111:1396–1408.

Shellie KC, Saltveit ME (1993). The lack of respiratory rise in muskmelon fruit ripening on the plant challenges the definition of climacteric behavior. J Exp Bot 44:1403–1406.

Trainotti L, Pavanello A, Casadoro G (2005) Different ethylene receptors show an increased expression during the ripening of strawberries: does such an increment imply a role for ethylene in the ripening of these non-climacteric fruits? J Exp Bot 56:2037–2046.

Villavicencio L, Blankenship SM, Sanders DC, Swallow WH (1999) Ethylene and carbon dioxide production in detached fruit of selected pepper cultivars. J Am Soc Hort Sci 124:402–406.

Watkins CB (2002) Ethylene synthesis, mode of action, consequences and control. In: Knee, M. (Ed.). Fruit quality and its biological basis. Sheffield Academic Press Ltd, UK.

Zheng XY, Wolff, DW (2000) Ethylene production, shelf-life and evidence of RFLP polymorphisms linked to ethylene genes in melon (*Cucumis melo* L.). Theor Appl Genet 101:613–624.

Zheng, XY, Wolff DW, Crosby KM (2002) Genetics of ethylene biosynthesis and restriction fragment length polymorphisms (RFLPs) of ACC oxidase and synthase genes in melon (*Cucumis melo* L.). Theor Appl Genet 105:397–403.

Effect of selenium addition on phenylalanine ammonia lyase (PAL) activity and ethylene production in leafy vegetables

Diaz, K.E.[1, a], Mensuali-Sodi, A.[1], Serra, G.[1], Malorgio, F.[2]*, Tognoni, F.[2], Ferrante, A.[3] and Pezzarossa, B.[4]

[1]Scuola Superiore S. Anna, Pisa, Italy, [2]Dip. Biologia delle Piante Agrarie, Università di Pisa, Italy, [3]Dip. Produzione Vegetale, Università di Milano, Italy, [4]CNR, Istituto per lo Studio degli Ecosistemi, Sede di Pisa, Italy; [a]Present address: Dip. Producción Vegetal. Facultad Agronomía. University of Centro de la Provincia de Buenos Aires, Argentina.
(*Corresponding author: fmalorgio@agr.unipi.it)

1. Introduction

Minimally processed vegetables (MPV) are prepared and handled to maintain their fresh nature while providing convenience to the consumer, as ready-to-eat products. Cut operations during the MPV preparation induce activation of several biochemical processes leading, for example, to an increase of ethylene production and tissue browning. Enzymatic browning is the result of the oxidation of phenols that are produced starting by phenylalanine ammonia lyase (PAL). In plants, selenium (Se) is well known for its high potential of protecting biomembranes, eradicating free particles, and delaying senescence (Pezzarossa *et al.*, 1999).

The aim of the work was to study the effect of Se on vegetables production and the accumulation ability in leaves of minimally processed leafy vegetables. The effect of Se treatments on the shelf life of leafy vegetables during storage was also evaluated.

2. Methods

Chicory (*Chicorium intybus* L.) and green salad bowl lettuce (*Lactuca sativa* L. var. Acephala) plants were grown in a floating system under greenhouse conditions. Se was added to the nutrient solution as sodium selenate at rate of 0 (control), 0.5, and 1 ppm Se. Experiments were carried out in December 2004 and April 2005. Leafy vegetables were harvested and

A. Ramina et al. (eds.), Advances in Plant Ethylene Research: Proceedings of the 7th International Symposium on the Plant Hormone Ethylene, 207–209.

stored at 4–5°C in dark conditions at UR 90–95%. PAL activity, ethylene evolution, and total Se content were determined at harvest time and during the storage.

3. Results and Discussion

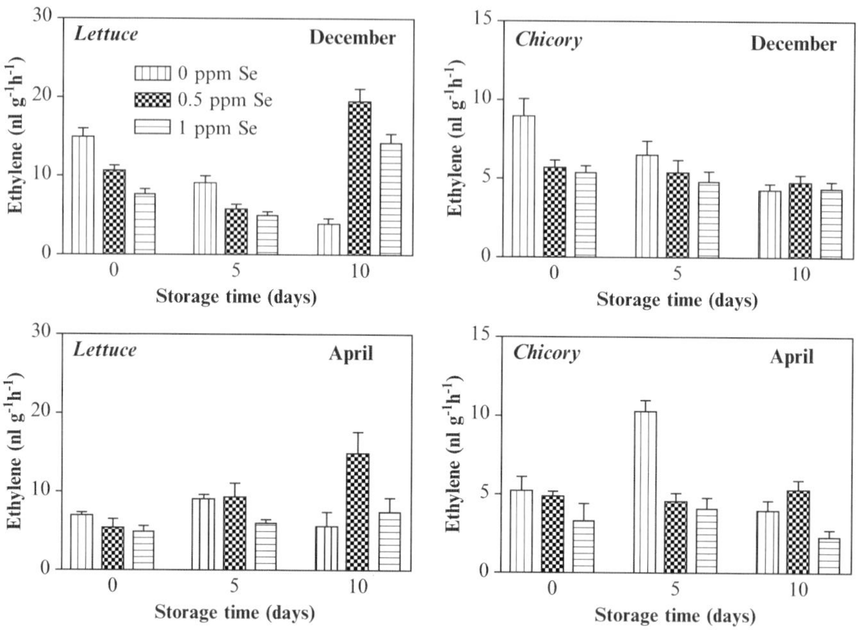

Fig. 1. Ethylene production in leafy vegetables grown with different Se concentration in nutrient solution (0, 0.5, and 1 ppm) at harvest and during storage. Data are means with standard error ($n = 6$).

PAL activity in wounded tissue produces phenolic compounds responsi ble for tissues browning (Soliva-Fortuny and Martín-Belloso, 2003). PAL activity was higher in *Lactuca sativa* than in *Chicorium intybus* plants and it was significantly reduced by the Se treatments in both species only at harvest time. The addition of 0.5 ppm and 1 ppm Se reduced the PAL activity of 18 and 38%, respectively in lettuce, and of 28 and 40%, respectively in chicory. Lettuce leaves showed the lower ethylene production compared to chicory in both experiments. The higher Se treatment (1 ppm) reduced the ethylene both at the harvest time and after 5 days of storage (Fig. 1). Leaf Se concentration in both species was directly related to the Se added. At 0, 0.5, and 1 ppm Se in the nutrient solution, the concentration at the harvest time increased from 0.44 to 11.9 and 26.11 ppm respectively in lettuce, and from 0.34 to 14.60 and 29.14 ppm respectively in chicory. Ethylene accumulation in boxes during storage showed a significant negative correlation with the Se concentration in leaves ($r = -0.80$ in lettuce and –0.86 in chicory). Se in plants is converted to Se-methionine and reduces free methionine, which is the initial substrate of the ethylene biosynthesis (Tapiero *et al.*, 2003).

Acknowledgements

The work was supported by MIUR-PRIN project “Prodotti orticoli di IV gamma: strategie di produzione e miglioramento della qualita`” (2004).

References

Pezzarossa, B., Malorgio, F., Tonutti, P., 1999. Effects of selenium uptake by tomato plants on senescense, fruit ripening and ethylene evolution (Santorini, Grecia, 5–8 settembre 1998). Kluwer Academic Publishers, Dordrecht, Netherlands, pp. 275–276.

Soliva-Fortuny, R.C., Martín-Belloso, O., 2003. New advances in extending the shelf-life of fresh cut fruits: a review. Trends in Food Science and Technology. 14(9), 341–353.

Tapiero H., Townsend D.M., Tew, K.D., 2003. The antioxidant role of selenium and seleno compounds. Biomedicine & Pharmacotherapy. 57, 134–144.

Development of a new palladium-based ethylene scavenger

Ilkenhans, T.[1], Poulston, S.[1], Rowsell, L.[1], Smith, A.W.J.[1]* and Terry, L.A.[2]

[1]Johnson Matthey Technology Centre, Blount's Court, Sonning Common, Reading, Berkshire, RG4 9NH, UK. [2]Plant Science Laboratory, Cranfield University, Bedfordshire, MK45 4DT, UK. (*Corresponding author: smithawj@matthey.com)

Abstract

A new palladium-promoted material was discovered with significant ethylene adsorption capacity. The material effectively scavenged exogenously administered and/or endogenously produced ethylene by climacteric fruit. Corresponding inhibition of ethylene-induced ripening was observed.

1. Introduction

Uncontrolled levels of ethylene can result in premature ripening of fruit and vegetables, wilting of cut flowers and loss of green colour and bitterness in vegetables. However, there has been a paucity of research in recent years on developing novel and more effective ethylene scavenging materials. This work has identified a new palladium-promoted material with significant ethylene adsorption capacity at room temperature. The material is compared with another ethylene scavenger, potassium permanganate (Shorter *et al.*, 1992; Howard *et al.*, 1994).

2. Results and Discussion

Initial screening was carried out in a plug flow reactor with 200 $\mu l\ l^{-1}$ ethylene, 10% (v/v) O_2 balanced with He at *ca.* 100% relative humidity (RH). A palladium (Pd)-promoted material that has significant adsorption capacity at room temperature has been discovered (4500 $\mu l\ g^{-1}$ adsorber).

A. Ramina et al. (eds.), Advances in Plant Ethylene Research: Proceedings of the 7th International Symposium on the Plant Hormone Ethylene, 211–213.

Materials were tested for ethylene scavenging ability in the presence of fruit. A banana was monitored by GC analysis for 24 h to measure changes in the gas composition in a 1-l vessel. After this time the ethylene level rose to ~5 μl l^{-1}. When the experiment was repeated with the Pd-based adsorber, no ethylene was detected over 24 h (the minimum detection limit was ~0.5 μl l^{-1} ethylene).

The effect of moisture was investigated by pretreating scavenging materials with water vapour. Materials were placed in a container with ~100% RH (21°C) prior to analysis (Table 1).

Table 1. Effect of moisture on ethylene adsorption capacity.

Material	Pretreatment	C_2H_4 ads. cap. μl g^{-1}
Pd material	None	4162
Pd material	Water vapour, 100 h	3753
5 wt % $KMnO_4/Al_2O_3$	Dried 110°C	750
5 wt % $KMnO_4/Al_2O_3$	Dried 110°C, water vapour 72 h	0

The Pd-based scavenger was also evaluated in 3-l sealed jars containing three pre-climacteric banana cv Cavendish fruit initially treated with or without 100 μl l^{-1} ethylene to attest efficacy and effects of ethylene removal. Ethylene and CO_2 levels were monitored daily (Table 2).

Table 2. Change in C_2H_4[a] (μl l^{-1}) and CO_2[b] (%) concentrations over 3 days with and without the addition of 100 μl l^{-1} ethylene.

Days	No scavenger				Pd-based Scavenger			
	+ Ethylene		– Ethylene		+ Ethylene		– Ethylene	
	C_2H_4	CO_2	C_2H_4	CO_2	C_2H_4	CO_2	C_2H_4	CO_2
0	98.19	---	0.00	---	94.10	---	0.00	---
1	84.15	2.12	0.04	2.02	0.00	2.31	0.00	1.94
2	85.23	6.49	0.05	3.61	0.00	4.63	0.01	3.33
3	76.97	13.92	0.05	5.23	0.00	6.50	0.01	4.72

[a]LSD ($P = 0.05$) = 10.266;.[b]LSD ($P = 0.05$) = 0.494

The Pd-based material effectively reduced exogenous ethylene concentrations to sub-physiologically active levels in banana. A concomitant reduction in CO_2 production and control of colour change from green to yellow was also observed. The Pd-based scavenger had no adverse affect on fruit quality or subsequent ripening when removed. Moreover, the material has been shown to have significant advantages over supported $KMnO_4$.

References

Shorter A. J., Scott K. J., Ward G. and Best D. J. (1992) Effect of ethylene adsorption on the storage of Granny Smith apples held in polyethylene bags. Postharvest Biology and Technology, 1(3):189–194.

Howard L. R., Yoo K. S., Pike L. M. and Miller Jr. G. H. (1994) Quality changes in diced onions stored in film packages. Journal of Food Science 59(1): 110–117.

Ethylene behavior in arazá fruit (*Eugenia stipitata* Mc Vaugh) during ripening and storage at different temperatures

Carrillo, M.P.[1], Hernández, M.S.[1]*, Barrera, J.A.[1], Martínez, O.[2] and Fernández-Trujillo, J.P.[3]

[1]Instituto Amazónico de Investigaciones Científicas SINCHI. Calle 20 # 5-44, Bogotá; [2]Universidad de los Andes. Dep. Biología. Bogotá, Colombia; [3]Universidad Politécnica de Cartagena. Dep. Agric. & Food Engin. Paseo Alfonso XIII, 48. ETSIA. 30203 Cartagena (Murcia). Spain. (*Corresponding author: shernandez@sinchi.org.co, cocona@colomsat.net.co)

1. Introduction

The postharvest shelf-life of arazá fruit is approximately 72 h at 20°C due to exacerbated dehydration, softening, and eventual anthracnose infection (Hernández and Trujillo, 2004). Softening and skin color changes can be closely associated with the pattern of ethylene production (Flores *et al*., 2001; Rogez *et al*., 2004). Here we determine ethylene production in arazá fruits when stored at three different temperatures, and in relation to the onset of respiration rate and several quality traits.

2. Materials and Methods

Green mature arazá fruit were harvested in Florencia (Caquetá, Colombia) and stored at 7 ± 1°C, 12 ± 1°C, and at 20 ± 1°C and 95% RH. The quality traits were evaluated after 1 week, with or without a shelf-life period (3 days at 20°C and 70% RH). CO_2 and C_2H_4 were measured by gas chromatography using an Agilent HP 4890D gas chromatograph (with a thermal conductivity detector for CO_2 and a flame ionization detector for C_2H_4, with a Shincarbon ST column). Flesh firmness was determined using a Magness Taylor force gauge on two opposite equatorial points on the fruit. The skin color was measured by *L** *C** *H** coordinates with a Hunter Lab colorimeter (D65 illuminant; 2° observer). Total soluble solids (TSS) were measured with a Fisher hand refractometer at 20°C. Data was subjected

A. Ramina et al. (eds.), Advances in Plant Ethylene Research: Proceedings of the 7th International Symposium on the Plant Hormone Ethylene, 215–217.

to analysis of variance with SAS 6.3 statistic program using temperature and time as factors.

3. Results and Discussion

Arazá fruits showed a climacteric peak at 20°C (170 mg $CO_2 \cdot kg^{-1} \cdot h^{-1}$) concomitant with maximum ethylene production (15 μl $C_2H_4 \cdot kg^{-1} \cdot h^{-1}$). Ethylene production peak levels were 8.6 and 5.9 μL $C_2H_4 \cdot kg^{-1} \cdot h^{-1}$ at 12°C or 7°C, respectively. A temperature-dependent behavior was identified, similar to that reported in guava (Bron *et al.*, 2005). During the shelf-life period of the fruit previously stored at 12°C or 7°C, ethylene production and respiration rates doubled the levels reached by fruit that had been stored at 20°C (data not shown).

Skin color coordinates (L^* = 54, C^* = 35, and H^* = 107°, at harvest) were significantly influenced by temperature and storage time ($P<0.05$) with faster color changes taking place at 20°C (L^* = 67.7, C^* = 55.3, H^* = 83.9) than at 12°C (L^* = 65.4, C^* = 49.1, H^* = 83.8) or 7°C (L^* = 58.3, C^* = 38.9, H^* = 95.3). In general, loss of green skin color is an ethylene-dependent process (Flores *et al.*, 2001). Arazá softening progressed faster at the highest storage temperature (Txt significant at $P<0.05$). The slower softening at 7°C can be partly associated to development of chilling injury (Bron *et al.*, 2005; Hernández *et al.*, 2003). TSS was unrelated to the behavior of ethylene production (data not shown).

In summary, the storage of arazá fruits at 20°C exhibited a moderate ethylene production concomitant with a climacteric respiration pattern, and a fast softening and color changes from green to yellow. The storage at 12 or 7°C of arazá fruit avoided climacteric patterns and delayed ripening changes but the upsurge in ethylene production and the respiration rate during the subsequent shelf-life can be associated with the development of chilling injury, particularly at 7°C.

References

Bron DU, Ribeiro RV, Cavalini FC, Jacomino AP, Trevisan MJ (2005) Temperature-related changes in respiration and Q(10) coefficient of guava. Sci. Agric. 62:458–463.

Flores F, Ben Amor M, Jones B, Pech JC, Bouzayen M, Latche A, Romojaro F (2001) The use of ethylene-suppressed lines to assess differential sensitivity to ethylene of the various ripening pathways in Cantaloupe melons. Physiol. Plant 113:128–133.

Hernández MS, Barrera J, Fernández-Trujillo JP, Arjona HE, Martínez O (2003) Efecto de la temperatura de almacenamiento en la fisiología y calidad del fruto de araza (*Eugenia stipitata* Mc Vaugh). Actas de Horticultura 37:1074–1081.

Hernández MS, Fernández-Trujillo JP (2004) Arazá fruit. In: Gross KC, Saltveit ME and Wang CY (eds) USDA Agricultural Handbook No. 66. http://www.ba.ars.usda.gov/hb66/027arazá.pdf. USDA. USA.

Rogez H, Buxan R, Mignolet E, Souza JNS, Silva EM (2004) Chemical composition of the pulp of three typical Amazonian fruits: araca-boi (*Eugenia stipitata*), bacuri (*Platonia insignis*) and cupuacu (*Theobroma grandiflorum*). Eur. Food Res. Technol. 218:380–384.

Ethylene production in nectarine fruit of different maturity as measured by time-resolved reflectance spectroscopy

Vanoli, M.[1]*, Eccher Zerbini, P.[1], Grassi, M.[1], Jacob, S.[1], Rizzolo A.[1], Torricelli, A.[2], Spinelli, L.[2] and Cubeddu, R.[2]

[1]CRA-IVTPA Istituto Sperimentale per la Valorizzazione Tecnologica dei Prodotti Agricoli, via Venezian, 26 - Milan, Italy. [2]CNR-INFM and CNR-IFN, Politecnico di Milano-Dipartimento di Fisica, piazza Leonardo da Vinci, 32 – Milan, Italy. (*Corresponding author: m.vanoli@ivtpa.it)

1. Introduction

Maturity at harvest influences all aspects of postharvest life and quality of fruit. In climacteric fruits, such as nectarines, ripening is characterized by increasing ethylene production (EP) and loss of firmness (F), but great variability was observed between fruit (Tonutti *et al.*, 1996). Recently, it has been shown that the absorption coefficient at 670 nm (μ_a), measured by time-resolved reflectance spectroscopy (TRS) on 'Spring Bright' nectarines, is a valuable tool to assess the maturity, i.e. the biological age, of individual fruit without disrupting its structure (Eccher Zerbini *et al.*, 2006). The aim of this research was to study EP in relation to both fruit maturity at harvest, as measured by TRS, and to softening at two postharvest ripening temperatures.

2. Material and Methods

'Spring Bright' nectarines were picked on July 15, 2005 and selected by size (A and B). Fruits (330 per size) were individually weighed, measured on two sides by TRS (Eccher zerbini *et al.*, 2006), ranked by decreasing μ_a averaged on the two fruit sides (i.e. from the less to the more mature fruit) and divided into 30 groups corresponding to 30 levels of μ_a. Eleven sample sets (one for analysis at harvest, five for analysis during ripening at 10°C and five for analysis during ripening at 20°C), each one representing the whole range of μ_a, were prepared by randomly choosing one fruit from each of the 30 groups. F was measured at harvest (Eccher Zerbini *et al.*, 2006);

A. Ramina et al. (eds.), Advances in Plant Ethylene Research: Proceedings of the 7th International Symposium on the Plant Hormone Ethylene, 219–221.

F and EP (Rizzolo *et al.*, 2006) were assessed during postharvest ripening after 36, 56, 84, 152 and 158 h at 20°C and after 23, 71, 142, 188 and 238 h at 10°C. Sampling at 158 h at 20°C is missing, due to decay of fruit.

3. Results

EP and F were strongly affected by temperature and time of ripening (Fig. 1). On average, EP was always higher at 20°C than at 10°C and increased exponentially with time, while F quickly decreased, especially at 20°C. F loss began when EP was still low and EP increased remarkably when fruits were already very soft (Tonutti *et al.*, 1996; Trainotti *et al.*, 2003). Fruits reached the melting stage (F = 15N) after 142 h at 10°C and earlier (after 56 h) at 20°C, when EP was about 50 pmol/kg/s at both temperatures.

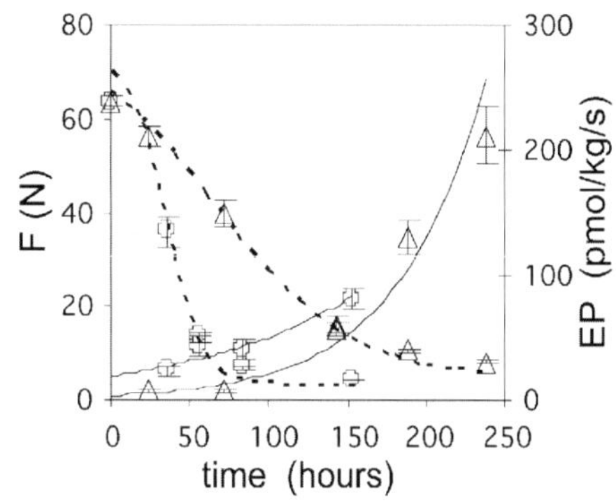

Fig. 1. EP (-) and F (--) at 10°C (Δ) and at 20°C (O).

Considering μ_a and neglecting size, all fruits can be grouped into five μ_a classes: 0.35–0.31, 0.30–0.26, 0.25–0.21, 0.20–0.16 and 0.15–0.10 cm^{-1}. Every class showed a characteristic and different trend in EP and F. At both temperatures, the lower the μ_a, the earlier the onset of EP. At 10°C EP rate increased with decreasing μ_a. At 20°C (Fig. 2) four different trends were found: 0.35–0.26 cm^{-1} nectarines showed an exponential increase in EP with low starting values; 0.25–0.21 cm^{-1} fruits had higher EP than the previous classes but with a lower EP rate; 0.20–0.16 cm^{-1} nectarines displayed a quick increase in EP until 84 h and then a decrease of EP rate; in the more mature fruits (μ_a<0.15 cm^{-1}), EP peaked at 84 h. In the five μ_a classes, softening followed the logistic pattern at both temperatures with a time shift according to μ_a: fruit with high μ_a softened later than those having low μ_a (Eccher Zerbini *et al.*, 2006).

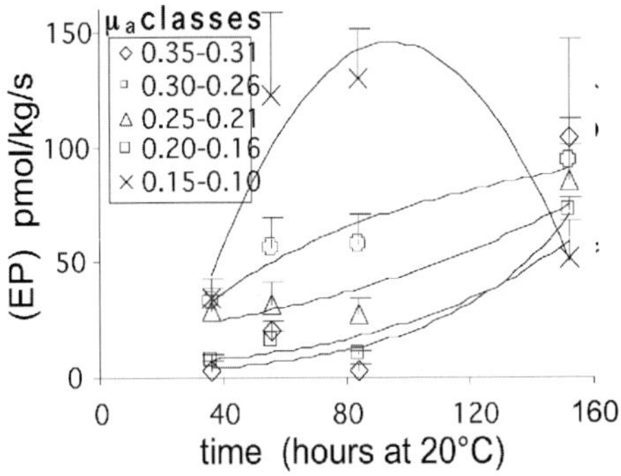

Fig. 2. EP at 20°C according to μ_a classes.

4. Discussion and Conclusions

Maturity at harvest (μ_a) affects EP and softening in a different way: with increasing maturity (i.e. decreasing μ_a), the EP curve is accelerated, reaching earlier the climacteric, while softening is shifted towards earlier time but with the same rate. The different EP curves of the different μ_a classes can all be considered as normal climacteric curves with different time scales, from the longest (less mature, high μ_a fruit; 10°C) to the shortest (more mature, low μ_a fruit; 20°C). The low μ_a fruit at 10°C behave like the high μ_a fruit at 20°C, but at 20°C the useful time window before fruit decay is shorter than at 10°C (150 vs 250 h).

In conclusion, our results confirm that μ_a can be regarded as an effective index of the biological age of fruit: it can be used to predict the softening rate during postharvest ripening, as well as to explain the different trends in ethylene production.

References

Eccher Zerbini P., Vanoli M., Grassi M., Rizzolo A., Fibiani M., Cubeddu R., Pifferi A., Spinelli L., Torricelli A. (2006). Postharvest Biol. Technol. 39: 223–232.

Rizzolo A., Eccher Zerbini P., Grassi M., Cambiaghi P., Bianchi G. (2006). J. Food Qual 29: 184–202.

Tonutti P., Bonghi C., Ramina A. (1996). J. Hort. Sci. 71(1): 141–147.

Trainotti L., Zanin D., Casadoro G. (2003). J. Exp. Bot. 54: 1821–1832.

Ethylene production by '*Prunus domestica*' plums during storage at different temperatures

Rato, A.E.*, Campos, D., Barroso, J.M. and Agulheiro, A.C.

Universidade de Évora – Instituto de Ciências Agrárias Mediterrânicas. Depto. De Fitotecnia - Apt 94 – 7002-554 Évora. (*Corresponding author: aerato@uevora.pt)

Abstract

'Raínha Claudia verde' plums fruits are usually stored at 0–2°C. This cultivar has a small commercial period because of its short postharvest life. In cold chambers fruits became soft very quickly and were not adequate for sale.

The aim of this study was to compare the effect of two different storage temperatures (1°C and 7 ±1°C) during 2, 5, 8 and 14 days in ethylene production upon rewarming. We also studied the evolution of fruits softening rate, % acidity and SSC (soluble solids content) in response to cold storage conditions.

1. Introduction

Ethylene is necessary for a normal ripening process in all climacteric fruits. The magnitude of the ethylene peak production in normal ripening conditions can vary enormously between fruit species. Low temperatures interact with ethylene biosynthesis and ripening in many species. Exposure of pears to low temperatures promotes ethylene synthesis (Gerasopoulos and Richardson, 1997) some apple cultivars exhibited a similar behaviour (Johnston *et al.*, 2002).

In '*Prunus domestica*' cv 'Raínha Claudia verde' plums, it was observed after a cold storage period a decrease in ethylene production upon rewarming. The longer the storage period the stronger is the depression in ethylene biosynthesis (Rato *et al.*, 2005). In nectarines it was observed that a decrease in ethylene production occurs in fruits stored at 0°C for 35 days after transfer to 20°C. This lack of ethylene production promotes severe chilling symptoms in this species (Zhou *et al.*, 2001).

A. Ramina et al. (eds.), Advances in Plant Ethylene Research: Proceedings of the 7th International Symposium on the Plant Hormone Ethylene, 223–226.

2. Material and Methods

After collecting the fruits a group of eight was separated to evaluate ethylene production without cool storage. Other fruits were split in two groups and stored at 1°C and 7 ±1°C. Two subsamples of fruits, from each temperature (1°C and 7°C), were taken periodically (2, 5, 8 and 14 days). Fruits were analysed for firmness, % acidity and SSC and ethylene production.

One ml gas samples were withdrawn from the headspace and injected into a gas chromatograph fitted with a FID detector on a daily basis.

3. Results and Discussion

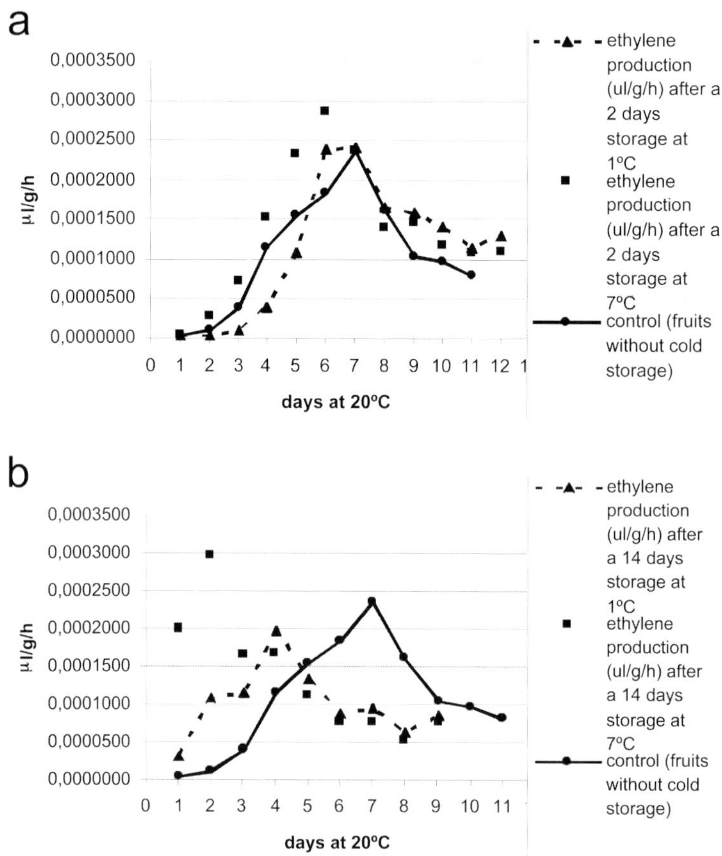

Fig. 1. **(a)** Ethylene fruit production after 2 days of cold storage. **(b)** Ethylene fruit production after 14 days of cold storage.

SSC and % acidity showed an expected evolution rate. Fruits held at 7°C showed a higher SSC and a lower % acidity than fruits held at 1°C (data not shown). Because this specie doesn't have starch as a primary carbohydrate reserve, the higher SSC observed is due to fruit dehydration. Upon rewarming, fruits stored during 14 days at 1°C produce less ethylene than control (Fig. 1). Storage temperature of 7°C didn't affect the ethylene production rate (0.3 nl/g/h) in all storage periods.

All fruits stored for 14 days exhibited an ethylene maximum production earlier than those stored for shorter periods. A short period of cold storage (2 days at 7°C) induced an enhancement of ethylene production relative to fruits held at 20°C.

Fruit firmness decreased during the storage period no matter the temperature (Fig. 2). Fruits held at 7°C for 14 days, remained significantly firmer ($p<0.0011$) than fruits held at 1°C. Similar results were obtained with pepino (*Solanum muricatum*) when fruits held at lower temperatures became softer than fruits held at higher temperatures due to chilling injury (Martinez-Romero *et al*., 2003). In nectarine a delayed storage prevented woolliness; this treatment allows normal ethylene production. Fruits subjected to immediate storage developed severe woolliness (Zhou *et al*., 2001).

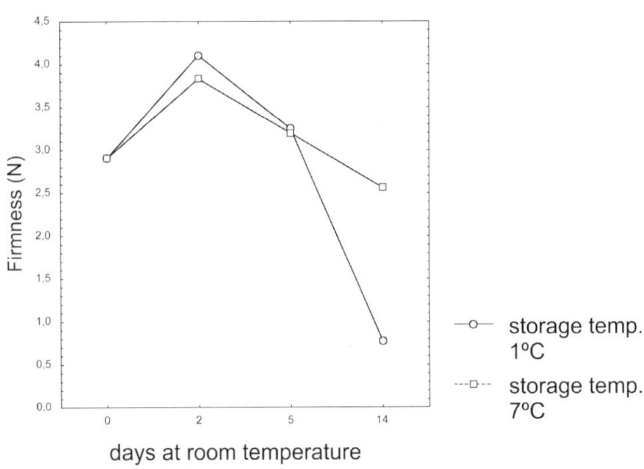

Fig. 2. Fruit firmness after 14 days of cold storage for two different temperatures.

Despite data of SSC and % acidity evolution suggesting that fruit storage conditions are better at lower temperatures, fruits held at 7°C showed a lower softening rate and a higher ethylene production rate than fruits held at 1°C which may indicate some chilling injury in this variety.

Ethylene-related cold induction and fruit softening should be clarified in this variety by future research.

References

Gerasopoulos, D.; Richardson, D.G. 1997. Ethylene production by 'd'Anjou' pears during storage at chilling and nonchilling temperature. Postharvest Biology and Technology, 32(6):1092–1094.

Johnston, J.W.; Hewett, E.W.; Hertog, M.; Harker, F.R. 2002. Temperature and ethylene affect induction of rapid softening in 'Granny Smith' and 'Pacific Rose TM' apple cultivars. Postharvest Biology and Technology, 25:257–264.

Martinez-Romero, D.; Serrano, M.; Valero, D. 2003. Physiological changes in pepino (*Solanum muricatum* Ait.) fruit stored at chilling and non-chilling temperatures. Postharvest Biology and Technology, 30:177–186.

Rato, A.E.; Marreiros, H.I.; Santos, A.C.; Barroso, J.M. 2005. Effects of different fruit calcium levels on the postharvest physiology of plums (*Prunus domestica* L.). proceedings of 5th International Postharvest Symposium (682) pp 171–175. Verona, Italy.

Zhou, H.; Dong, L.; Ben-Arie, R.; Lurie, S. 2001. The Role of Ethylene in the Prevention of chilling injury in nectarines. Journal of Plant Physiology, 158:55–61.

The peach 1-aminocyclopropane-1-carboxylic acid synthase isogene, *Pp-ACS1*, is required for fruit softening

Tatsuki, M.*, Haji, T. and Yamaguchi, M.

National Institute of Fruit Tree Science, NARO, Tsukuba 305-8605, Japan.
(*Corresponding author: tatsuki@affrc.go.jp)

1. Introduction

The fruit of melting-flesh peach (*Prunus persica* (L). Batsch) produces increasing levels of ethylene, and the flesh firmness softens rapidly during the ripening stage. On the other hand, stony hard peaches barely soften on the tree or after harvest, although the fruit changes colour normally, contains highly soluble solids and has good flavour (Haji *et al.*, 2001). It has been assumed that a low level of ethylene production by stony hard peach is responsible for the inhibition of fruit softening, because exogenous ethylene softens them effectively (Hayama *et al.*, 2003; Haji *et al.*, 2003). Since ethylene production occurs and the fruit softens by the application of ACC, ACC oxidase activity and ethylene sensing are normal in stony hard peach (Haji *et al.*, 2003). For this reason, it has been considered that ACC synthesis is the key in the stony hard peach fruit. In this study, ethylene production and the expression patterns of ACC synthase genes and ACC oxidase gene were examined in the stony hard peach cultivars 'Yumyeong', 'Odoroki', and 'Manami' and the melting-flesh cultivar 'Akatsuki'.

2. Results and Discussion

In the fruit of 'Akatsuki', a melting-flesh peach, an increase in ethylene production and a decrease in flesh firmness began within 1 day after harvest. By contrast, the flesh of the stony hard peaches 'Yumyeong', 'Odoroki', and 'Manami' remained hard, and ethylene evolution did not increase during storage. The application of 5000 $\mu L\ L^{-1}$ propylene promoted softening of 'Yumyeong', 'Odoroki', and 'Manami' fruits.

In 'Akatsuki', *Pp-ACS1* mRNA was detected 1 day after harvest, then transcript became abundant (Fig. 1). No expression of ACC synthase

A. Ramina et al. (eds.), Advances in Plant Ethylene Research: Proceedings of the 7th International Symposium on the Plant Hormone Ethylene, 227–228.

isogenes could be detected in ‘Yumyeong’, ‘Odoroki’, and ‘Manami’ during fruit storage in air (Fig. 1). On the other hand, *Pp-ACO1* mRNA was expressed at a constant level in stony hard peaches and was increased by propylene treatment. These results indicate that ethylene evolution did not occur after harvest in stony hard peaches because of a lack of expression of *Pp-ACS1* in the fruit.

Since *Pp-ACS1* mRNA was induced normally in senescing flowers, wounded leaves and wounded immature fruit of ‘Yumyeong’, *Pp-ACS1* is suppressed only at ripening stage, and not a defect in *Pp-ACS1*. DNA gel blot analysis of *Pp-ACS1* showed no marked differences in polymorphism in ‘Akatsuki’ and ‘Yumyeong’; it was assumed that there might be no insertion or deletion in the regulatory domain of *Pp-ACS1* in ‘Yumyeong’.

In conclusion, the suppression of fruit softening in stony hard peach cultivar is caused by a low level of ethylene production, which depends on suppressed expression of *Pp-ACS1*.

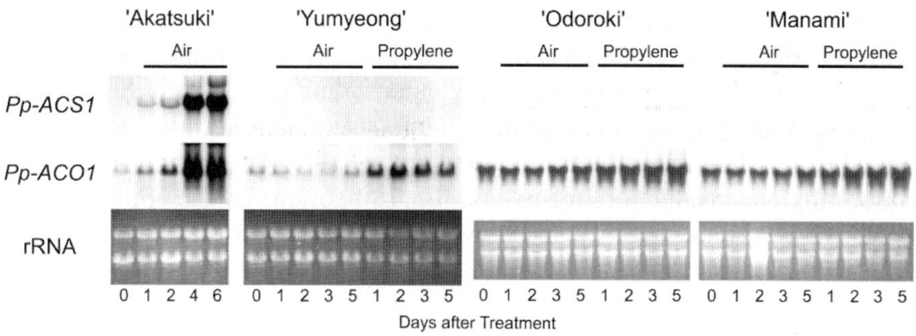

Fig. 1. Expression levels of *Pp-ACS1* and *Pp-ACO1* in peach fruit.

References

Haji T, Yaegaki H, Yamaguchi M (2001) Changes in ethylene production and flesh firmness of melting, nonmelting and stony hard peaches after harvest. J Jap Soc Hortic Sci 70:458–459

Haji T, Yaegaki H, Yamaguchi M (2003) Softening of stony hard peach by ethy lene and the induction of endogenous ethylene by 1-aminocyclopropane-1-carboxylic acid (ACC). J Jap Soc Hortic Sci 72:212–217

Hayama H, Ito A, Moriguchi T, Kashimura Y (2003) Identification of a new expansin gene closely associated with peach fruit softening. Postharvest Biol Technol 29:1–10

Monitoring genes whose expression is related to ethylene biosynthesis genes in tomato fruit by cDNA macroarray

Imanishi, S.*, Noguchi, A., Matsuo, E. and Nagata, M.

National Institute of Vegetable and Tea Science, Department of Physiology and Quality Science, Postharvest Research Team, NARO, 360 Kusawa, Ano, Tsu, Mie, 514-2392, Japan. (*Corresponding author: phppublish@ml.affrc.go.jp)

1. Introduction

Fruit ripening is a complex, genetically programmed process. It has been shown that many ripening processes of climacteric fruit are regulated by ethylene (Alexander and Grierson, 2002). Despite the understanding of ethylene biosynthesis and perception, the mechanism regulating fruit ripening, including factors for the ethylene climacteric, remain obscure. To obtain information on the genetic mechanism which underlies the pleiotropic effects of ethylene, we screened 10,911 tomato cDNA clones for responsiveness to hormone treatment and difference in genetic background using cDNA macroarray technology.

2. Methods

The macroarray used in this work was generated by the "*research project for utilizing advanced technologies in agriculture, forestry and fisheries no. 1424*". The cDNA clones used for generation of the macroarray had been prepared from four developmental stages fruits (mature green, breaker, yellow and red) and mature leaves of tomato (cv Micro-Tom) (Yamamoto *et al.* 2005). Self-organized maps (SOM) were employed for analysis of the expression patterns of 10,911 genes. In the 20 × 20 SOM, we picked up the categories in which ripening-related ethylene biosynthesis genes, ACC synthase 2, ACC synthase 4 and ACC oxidase 1 were classified, and the putative functions of the genes in each category were assigned through annotation (Imanishi *et al.* 2005).

A. Ramina et al. (eds.), Advances in Plant Ethylene Research: Proceedings of the 7th International Symposium on the Plant Hormone Ethylene, 229–231.

3. Results and Discussion

To better understand the role of the clones in each category, expression patterns for these genes during fruit development were analyzed (Fig. 1). 18 genes in the category (20, 20), in that ACC oxidase 1 gene arranged and that induced by jasmonate both in wild type and *nor* mutant tomato, were up-regulated during tomato fruit ripening. Considering with the ratio of "metabolism" genes which up-regulated during tomato fruit ripening is higher than that of "defense/stress", this result suggests that jasmonates play some role in fruit ripening through regulating "metabolism" genes expression. 13 genes were found in the LeACS4 category whose expression level was not changed significantly by any treatment or genetic background, but was up-regulated in the fruit ripening. 8 of these 13 genes could not assign their functions ("unclassified" or annotated with "unknown"). Further studies of their expression patterns might be able to help us understand the mechanism of fruit ripening.

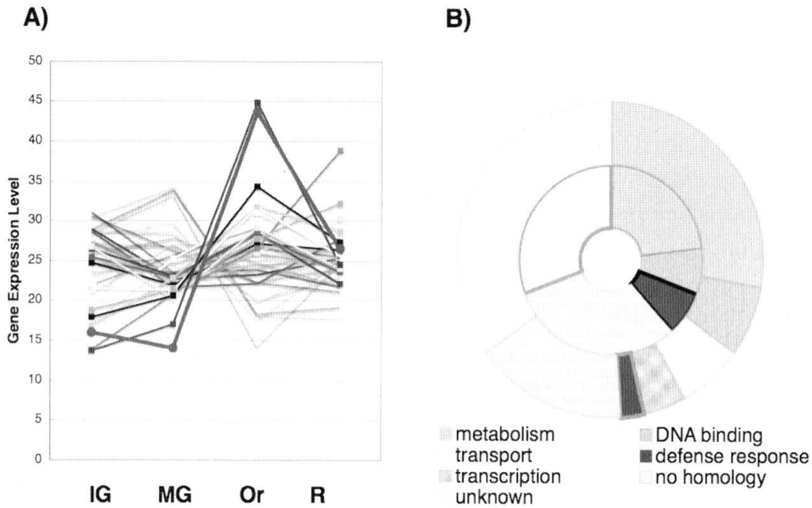

Fig. 1. Expression pattern in tomato fruit ripening **(A)** and deduced functions of the genes in the LeACS4 category. Genes that up-regulated in fruit ripening are marked by squares (A) and whose functions are indicated as inner circle **(B)**.

References

Alexander L, Grierson D (2002) Ethylene biosynthesis and action in tomato: a model for climacteric fruit ripening. J Exp Bot 53:2039–2055

Imanishi S, Noguchi A, Hatakeyama R, NagataM (2005) Monitoring the effect of jasmonates on the expression of ripening related genes in tomato fruit disks by cDNA macroarray. ISHS Acta Hortic 682:149–154

Yamamoto N, Tsugane T, Watanabe M, Yano K, Maeda F, Kuwata C, Torki M, Ban Y, Nishimura S, Shibata D (2005) Expressed sequence tags from the laboratory-grown miniature tomato (Lycopersicon esculentum) cultivar Micro-Tom and mining for single nucleotide polymorphisms and insertions/deletions in tomato cultivars. Gene 356:127–134

Ethylene is required for the progression of fruit softening and a continuous mRNA accumulation of softening-related genes in peach

Hayama, H.*, Ito, A., Sakamoto, D., Nakamura, Y. and Kashimura, Y.

National Institute of Fruit Tree Science, Tsukuba, Ibaraki 305-8605, Japan
(*Corresponding author: hhiroko@affrc.go.jp)

1. Introduction

Peach (*Prunus persica* (L.) Batsch) fruit are climacteric and an increase in ethylene production is observed during fruit ripening. However, the relationship between ethylene and fruit softening is unclear, because peach fruit softening begins significantly earlier than the observed increase in ethylene production (Tonutti *et al*., 1996). The stony hard peach is an ideal mutant for studying the relationship between fruit softening and ethylene. In this mutant, the ripe fruit produces little ethylene and firm flesh is maintained during ripening, but exogenous ethylene induces a rapid loss of firmness (Haji *et al*., 2003), which accompanied by increases in both endo-PG and exo-PG activities (Hayama *et al*., 2006).

2. Results and Discussion

In order to investigate the role of ethylene in peach fruit softening during ripening, stony hard peach fruit 'Manami' were treated with various concentrations of ethylene. There was no noticeable decrease in flesh firmness without ethylene treatment, while the degree of softening depended on the amount of applied ethylene in a range of 0.1–100 $\mu L\ L^{-1}$. When ethylene treatment was interrupted, the degree of softening was greatly reduced. These results indicated that continuous ethylene treatment was required for the initiation and progression of fruit softening and that ethylene concentration is also an important factor in regulating the rate of softening in peach.

Eight genes, which putatively encode cell wall metabolism-related proteins, were investigated for mRNA accumulation patterns in the two different softening phenotypes of melting and stony hard peaches. All of the

A. Ramina et al. (eds.), Advances in Plant Ethylene Research: Proceedings of the 7th International Symposium on the Plant Hormone Ethylene, 233–234.

mRNAs investigated accumulated in fruit of the melting-flesh 'Akatsuki' during ripening. By contrast, in the stony hard-flesh 'Manami', the mRNAs for a putative endopolygalacturonase (*PpPG*), an α-L-arabinofuranosidase/ β-xylosidase (*PpARF/XYL*), and an expansin (*PpExp3*) showed either much lower levels or did not accumulate, and were identified as softening-related genes. The other five mRNAs, including *PpExp1* and *PpExp2* as shown in Hayama *et al.* (2000, 2003), were detected at similar extents as in normal softening 'Akatsuki'. Interruption of ethylene treatment indicated that softening-related genes were regulated at the transcriptional level and quickly responded to the presence or absence of ethylene before the softening response occurred, suggesting that ethylene directly regulates the transcription of these softening-related genes.

References

Haji T, Yaegaki H, Yamaguchi M (2003) Softening of stony hard peach by ethy lene and the induction of endogeneous ethylene by 1-aminocyclopropane-1-carboxylic acid (ACC). J Jpn Soc Hortic Sci 72: 212–217.

Hayama H, Shimada T, Haji T, Ito A, Kashimura Y, Yoshioka H (2000) Molecular cloning of a ripening-related expansin cDNA in peach: evidence for no relationship between expansin accumulation and change in fruit firmness during storage. J. Plant Physiol 157: 567–573.

Hayama H, Ito A, Moriguchi T, Kashimura Y (2003) Identification of a new expansin gene closely associated with peach fruit softening. Postharvest Biol Technol 29: 1–10.

Hayama H, Tatsuki M, Ito A, Kashimura Y (2006) Ethylene and fruit softening in the stony hard mutation in peach. Postharvest Biol Technol 41: 16–21.

Tonutti P, Bonghi C, Ramina A (1996) Fruit firmness and ethylene biosynthesis in three cultivars of peach (*Prunus persica* L. Batsch). J Hortic Sci 71: 141–147.

Changes in the autocatalytic ethylene production in transgenic tomato expressing the *iaaM* gene from *Agrobacterium tumefaciens*

Vioque, B.[1]*, Castellano, J.M.[1] and Chamarro, J.[2]

[1]Inst. de la Grasa-CSIC, Dept. Fisiología y Tecnología de Productos Vegetales. Avda. Padre García Tejero 4 E-41012, Sevilla, Spain; [2]Inst. de Biología Molecular y Celular de Plantas-UPV-CSIC, Dept. Biología del Desarrollo, Avda. de los Naranjos s/n E-46022, Valencia, Spain.
(*Corresponding author: bvioque@cica.es)

1. Introduction

Fruit ripening and the role that ethylene plays in this process are complex. Many studies support that the interaction of ethylene with other plant hormones determines the onset and progress of the ripening process (Brady, 1987). Recent results in tomato suggest that ethylene and auxin cross-talk throughout fruit ontogeny (Khoja *et al.*, 2003). Our approach to investigate on this interaction is based in the use of transgenic tomato (*Lycopersicum esculentum* Mill. cv Ailsa craig) transformed with the *iaaM* gene from *Agrobacterium tumefaciens*. This gene codes for the enzyme catalysing the first step in the auxin biosynthesis and increase the endogenous auxin levels. In a preliminary study, we advanced that *iaaM* tomato transformed with the strong constitutive CaMV *35S* promoter further than to overproduce auxin also overproduces ethylene during the climacteric crisis (Castellano *et al.*, 1999). In this work, we analyse in a more detailed study, the changes in the autocatalytic ethylene production in three lines of transgenic tomato expressing the *iaaM* gene under the control of different gene promoters (*CaMV35S*, *HMG1* and *PG*). The *HMG1* promoter induces the *iaaM* expression in the earlier phases of fruit setting and growth (Jelesko *et al.*, 1999); whereas the expression of the polygalacturonase (PG) promoter coincides with the fruit ripening (DellaPenna *et al.*, 1986).

A. Ramina et al. (eds.), Advances in Plant Ethylene Research: Proceedings of the 7th International Symposium on the Plant Hormone Ethylene, 235–237.

2. Results and Discussion

The Northern analysis of the three lines of transgenic tomato confirmed the correct insertion of the *iaaM* gene as well as the expected expression patterns. Ailsa Craig is a climacteric fruit characterised by a moderate–low rate of ethylene production. At the climacteric maximum, the fruit produced as mean 0.77 ± 0.41 nmol ethylene g^{-1} h^{-1}. A 30% of the fruits released less than 0.50 nmol g^{-1} h^{-1} and only 5% of them produced more than 1.50 nmol g^{-1} h^{-1}. However, transgenic tomato with the construction *35S::iaaM* released as mean 144% more ethylene at the climacteric than the wild-type fruit. Only a scanty fraction of fruits (2%) produced less than 0.50 nmol ethylene g^{-1} h^{-1} and a significant percentage (16%) produced more than 1.50 nmol C_2H_4 g^{-1} h^{-1}. Fruit expressing the *HMG1::iaaM* transgene presented a range of ethylene productions intermediate between wild-type and *35S::iaaM*. Transgenic *PG::iaaM* fruit exhibited however similar behaviour as the wild type. A detailed study of the profile of ethylene production indicated that the expression of the *iaaM* gene induced substantial changes that affected not only the magnitude of the climacteric maximum but also the moment at which this maximum takes place. Wild-type fruits reached maximal ethylene production at 68 ± 23 h after the onset of the colour change. Tomato expressing the *35S::iaaM* transgene achieved it almost one day later (88 ± 25 h); whereas *HMG1::iaaM* tomato did not show alterations in this sense. In the *PG::iaaM* fruit, the maximum was advanced 10 h (58 ± 22 h) with respect to the WT. The results support the involvement of auxin in the regulation of tomato ripening and suggest an active role of the hormone as a component of the developmental cue that triggers the autocatalytic ethylene production.

Acknowledgements

This work was supported by the research project AGL2001-2288 (Plan Nacional I+D+I 2000-2003, Spain).

References

Brady CJ (1987) Fruit ripening. Annu. Rev. Plant Physiol. 38:155–178

Castellano JM, Chamarro J, Vioque B (1999) Ethylene biosynthesis in transgenic auxin-overproducing tomato plants. Biology and Biotechnology of the Plant Hormone Ethylene II. Kluwer Academic Publishers. Dordrecht

DellaPenna D, Alexander DC, Bennet AB (1986) Molecular cloning of tomato fruit polygalacturonase. PNAS USA 83:6420–6234

Jelesko JG, Jenkins SM, Rodríguez-Concepción M, Gruissem W (1999) Regulation of tomato *HMG1* during cell proliferation and growth. Planta 208:310–318

Khoja HA, Kalil Ahmad Q, Jones B, Leclercq J, Sanchez MT, Latche A, Pech JC, Bouzayen M. (2003). Ripening-associated regulation of gene expression in the tomato. Biology and Biotechnology of the Plant Hormone Ethylene III. IOS Press

Expression profile of ripening-related genes during ethylene evolution and fruit softening in apple (*Malus × domestica* Borkh.)

Stella, S., Costa, F.* and Sansavini, S.

Dipartimento di Colture Arboree, Universita' degli Studi di Bologna, 40127 Bologna, Italy. (*Corresponding author: fcosta@agrsci.unibo.it)

1. Introduction

The physiological and biochemical processes that fruit undergo during development and maturation change the fruit's general composition. Besides the important evolution of chemical compounds involved in determining color, flavor/aroma formation, and sugar/acid content, modification of flesh firmness is one of the major changes occurring in fruit at the onset of ripening. While softening starts at the beginning of maturation, it peaks most notably during the last stage of fruit ripening in concomitance with ethylene burst.

Fruit softening is a process triggered by the synergic and coordinated action of several cell-wall enzymes acting to loosen polysaccharides, a mechanism that in climacteric fruits such as apple is generally ethylene-dependent (Giovannoni, 2004; Powell *et al.*, 2003; Brummel and Harpster, 2001). Ethylene is a hormone that in apple is known to have a direct relationship to fruit softening (Yang and Hoffman 1984), and both ethylene production and softening are factors limiting fruit storability and shelf life since the greater the firmness, the better fruit resistance is to typical postharvest diseases and, hence, the more appealing and healthy in appearance. We thus analyzed the expression profiles of five genes, Md-ACS, Md-ACO1, and Md-ERS1 involved in ethylene production and perception and Md-PG1 and Md-Exp3 in fruit softening, during the ripening of the apple cv Mondial Gala and herewith report the results.

2. Materials and Methods

Mondial Gala ripening was monitored over 5 weeks from post full-bloom to overripe, i.e., postharvest, stage in an orchard at Bologna University's experiment station. The five stages recorded throughout the ripening

A. Ramina et al. (eds.), Advances in Plant Ethylene Research: Proceedings of the 7th International Symposium on the Plant Hormone Ethylene, 239–242.

process represent fruit development from preclimacetrium to full ripening (ethylene production–climacterium phase). Ethylene production and fruit softening were measured using at each stage using a gas chromatographer and an Effegi penetrometer. One fruit subsample was treated at harvest with 1 ppm of 1-MCP for 12 h at 24°C, and ethylene production was monitored three times a week over 30 days in the treated sample and the control, both stored at room temperature. This overall approach made it possible to characterize the functional evolution of five candidate genes via real-time PCR.

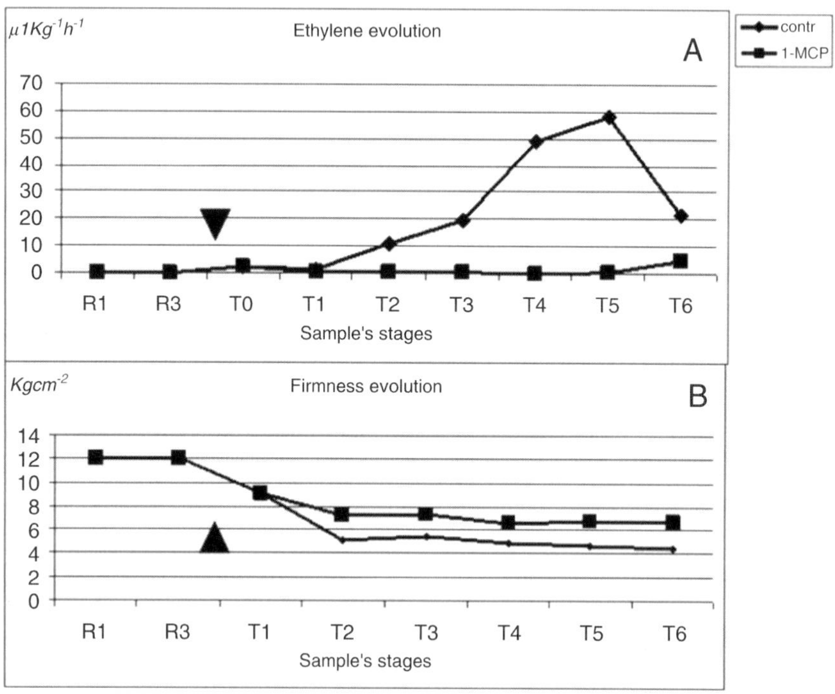

Fig. 1. Comparison between ethylene production (Fig. 1A) and fruit softening evolution (Fig. 1B) in Mondial Gala apple, control vs. 1-MCP-treated. In both figures the black arrow indicates treatment time coincident with harvest.

3. Results and Discussion

Mondial Gala is a cultivar that initiates ethylene production in postharvest, reaching a peak output of 58 μL Kg^{-1} h^{-1} 20 days after picking (Fig. 1A). In the climacterium phase coincident with the ethylene burst (maximum ethylene), fruit ripening is also marked by a pronounced softening, with firmness dropping to a value of 4.7 Kg cm^{-2} by the hormone peak (Fig. 1B). The Md-ACO1 and Md-ERS1 profiles show maximum transcript accumulation at the onset of ethylene burst whereas Md-ACS, though exhibiting the same profile, registered its maximum expression at the highest ethylene production rate measured over the 30 days of observation. The expression of these three genes was down-regulated by 1-MCP, underscoring that they are ethylene-related and specifically active during the climacterium.

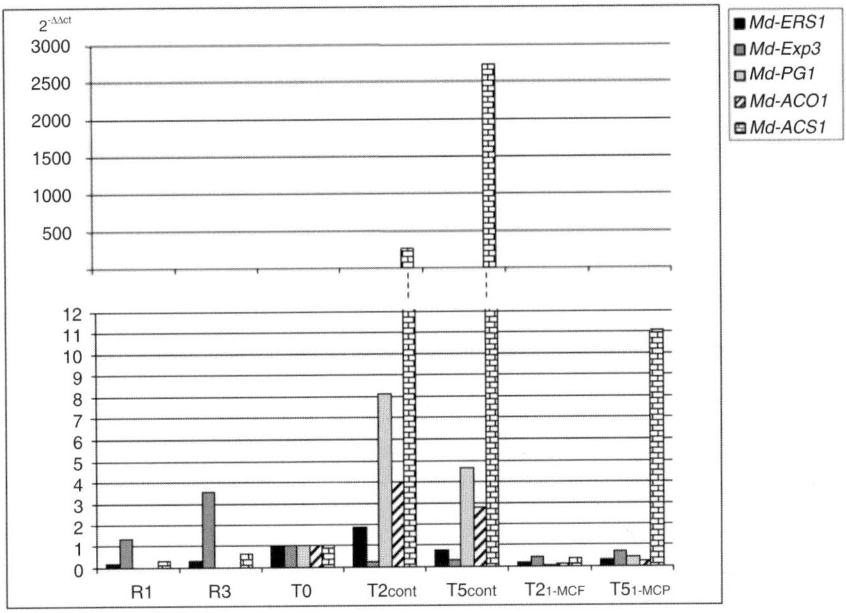

Fig. 2. Expression profile of the five genes performed via real-time PCR. To corresponds to harvest date as well as 1-MCP application.

The expression profile of Md-PG1, a gene typical of the softening process, also proved to be ethylene-dependent, its peak being up-regulated at ethylene burst, i.e., similar to those of Md-ERS1 and Md-ACO1, and down-regulated by 1-MCP (Fig. 2). Md-Exp3, an expansin involved in fruit ripening, evinced its maximum transcript accumulation before harvest, i.e., earlier than Md-PG1's expression. These latter two expression profiles support in apple the theory of early and late phases in softening process respectively controlled by expansin and polygalacturonase that was initially posited in tomato (Rose *et al.*, 1997).

References

Brummell D.A. and Harpster M.H. 2001. Cell wall metabolism in fruit softening and quality and its metabolism in transgenic plants. Plant Mol. Biol. 47, 311–340.

Giovannoni J.J. 2004. Genetic regulation of fruit development and ripening. Plant Cell 16, S170–S180.

Powell A.L.T., Kalamaki M.S., Kurien P.A., Guerrieri S. and Bennett A.B. 2003. Simulteneous transgenic suppression of LePG and LeExp1 influences fruit texture and juce viscosity in a fresh market tomato variety. J. Agric. Food Chem. 51, 7450–7455.

Rose J.K.C., Lee H.H. and Bennett A.B. 1997. Expression of a divergent expansin gene is fruit – specific and ripening – regulated. Proc. Natl. Acad. Sci. 94, 5955–5960.

Yang S.F. and Hoffman N.E. 1984. Ethylene biosynthesis and its regulation in higher plants. Annu. Rev. Plant Physiol. 35, 155–189.

Preliminary investigation into the uneven ripening of banana peel after 1-MCP treatment

De Martino, G.[1]*, Mencarelli, F.[1] and Golding, J.B.[2]

[1]Department of Food Science and Technology, University of Tuscia, Via De Lellis, 01100, Viterbo, Italy. [2]New South Wales Department of Primary Industries, Gosford Horticultural Institute, Locked Bag 26, Gosford, NSW, Australia. (*Corresponding author: demartino@unitus.it)

1. Introduction

Peel appearance of banana fruit is an important quality criteria that influences consumer acceptability. Many studies have shown promising results with the use of the ethylene antagonist 1-methylcyclopropene (1-MCP) on extending green life and shelf life of bananas (Macnish *et al.* 1997, 2000; Golding *et al.* 1998; Jiang *et al.* 1999; Harris *et al.* 2000). However, uneven peel ripening following some different timings of 1-MCP treatment has also been observed (Golding *et al.* 1998; Harris *et al.* 2000). This uneven peel degreening appears when, after ethylene treatment, 1-MCP is applied to bananas which have not completely lost the green colour and results in a dull-grey appearance with some visual symptoms similar with chilling injury. In this preliminary experiment we investigated the hypothesis that the uneven peel ripening could be related to an imbalance in respiration pattern, with accumulation of volatile compounds such as ethanol and acetaldehyde. Different post-ripening treatments (1-MCP, nitrogen, ethanol, chilling temperatures) were used to assess if dysfunctional ripening behaviour could lead to the appearing of these symptoms.

2. Materials and Methods

Five bunches of mature green bananas (*Musa* sp., AAA group, Cavendish subgroup, cultivar 'Williams') from Coffs Harbour, NSW, were used for this experiment. After ethylene treatment (200 μL L^{-1} for 24 h at 20°C) bananas were allocated different treatment units: 1-MCP treatment (200 ppb, 20°C, 24 h), nitrogen treatment (O_2 < 0.1%, 20°C, 24 h), ethanol treatment (0.3 mL L^{-1}, 20°C, 24 h), chilling treatment (0–2°C, 24 h) and

A. Ramina et al. (eds.), Advances in Plant Ethylene Research: Proceedings of the 7th International Symposium on the Plant Hormone Ethylene, 243–246.

untreated control (20°C). In each treatment were allocated bananas from five different bunches. Ethylene production, ethanol and acetaldehyde production, skin colour, chlorophyll fluorescence, chlorophyll content and firmness were assessed at regular intervals.

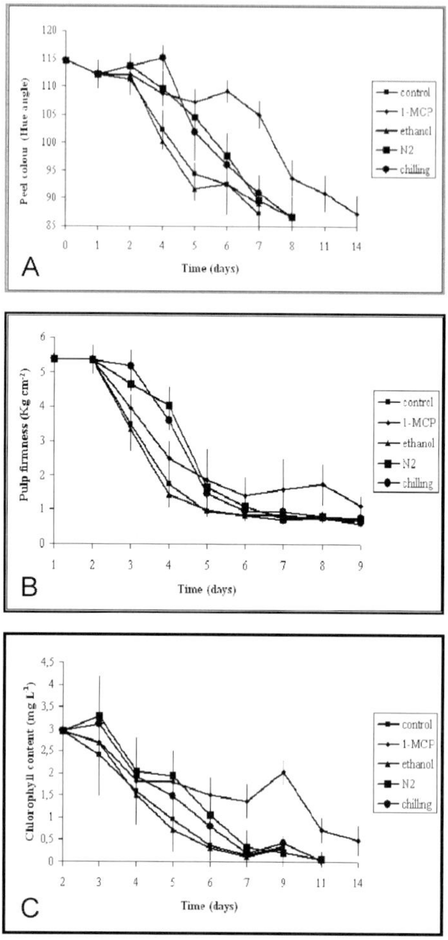

Fig. 1. Peel colour (**A**), pulp firmness (**B**) and chlorophyll content (**C**) of banana fruit treated with 200 μL L^{-1} of ethylene then treated with 200 nL L^{-1} of 1-MCP for 24 h at 20°C (1-MCP) or with 0.3 mL L^{-1} of ethanol (ethanol) or kept for 24 h in nitrogen atmosphere (O2 < 0.1%) at 20°C (N_2) or stored for 24 h at chilling temperature (0–2°C) (chilling) or maintained at 20°C and not treated (control). Day 2 = 48 h after ripening initiation with ethylene/24 h after various treatments. Bars indicate standard deviation.

3. Results

Different accumulation of volatile compounds was observed between 1-MCP-treated and untreated fruits but not significant difference in ethanol and acetaldehyde was noted between 1-MCP-treated greyish bananas and fruits from other treatments.

Ethylene production 24 h after 1-MCP, ethanol, nitrogen or chilling treatment and 48 h after ethylene treatment generally increased during ripening. Ethylene production was not affected by the 1-MCP treatment which in contrast delayed peel degreening (Fig. 1). The peel colour of 1-MCP fruit at the end of the experiment (Day 14) was similar to the peel colour of the other treatments at Day 7–8 (Fig. 1); anyway, 1-MCP-treated fruit maintained a general uneven light-green colour moving from the middle towards the tips. The presence of general greying within the peel was evident in 1-MCP treated fruit. All other post-ripening treatments did not lead to the appearance of the 'grey' colour on the peel. It appears the 1-MCP treated peel may still undertake some normal senescence events that occur during banana ripening. Considering the effect of the other experimental conditions tested, we observed that the 24 h post-climacteric nitrogen and chilling treatments led to a temporary inhibitory effect on ripening after Day 2, but this delay disappeared with the continuation of shelf life (Fig. 1). These results are likely to be related to the effect of initial low O_2 and low temperature storage conditions (24 h) on slowing down fruit metabolism or altering normal pulp softening.

4. Conclusions

These results suggest 1-MCP treatment 24 h after the initiation of ripening can still delay peel degreening by decreasing in chlorophyll content and maintaining chlorophyll fluorescence. However the 1-MCP treatment may not prevent other peel aging events, such as peel browning. Peel aging events and uneven peel degreening and ripening in 1-MCP treated fruit maybe affected by other factors such as water content. This hypothesis is currently being investigated.

References

Golding JB, Shearer D, Wyllie SG, McGlasson WB (1998) Application of 1-MCP and propylene to identify ethylene-dependent ripening processes in mature banana fruit. Postharvest Biology and Technology 14: 87–98.

Harris DR, Seberry JA, Wills RBH, Spohr LJ (2000) Effect of fruit maturity on efficiency of 1-MCP to delay the ripening of bananas. Postharvest Biology and Technology 20: 303–308.

Jiang Y, Joyce DC, Macnish AJ (1999) Responses of banana fruit to treatment with 1-methylcyclopropene. Plant Growth Regulation 28: 77–82.

Macnish AJ, Joyce DC, Hofman PJ (1997) 1-Methylcyclopropene delays ripening of 'Cavendish' banana fruit. In: Proceedings of the Australasian Postharvest Conference, University of Western Sydney, Australia, pp. 282–284.

Macnish AJ, Joyce DC, Hofman PJ, Simons DH (2000) 1-MCP gas can be used to control banana ripening. Good Fruit and Vegetables 10: 56.

Effect of combined 1-MCP and cold storage on the shelf life and postharvest quality of tomato

Martínez-Madrid, M.C.[1], Egea, M.I.[2], Sánchez-Bel, P.[2], Valdenegro, M.[2], Flores, F.[2] and Romojaro, F.[2]*

[1]Dpto. Agroquímica y Medio Ambiente, Universidad Miguel Hernández, Ctra. Beniel km 3.2, 03312-Orihuela, Alicante, Spain; [2]Centro de Edafología y Biología Aplicada del Segura (CEBAS-CSIC), Apdo. Correos 164, 30100-Espinardo, Murcia, Spain. (*Corresponding author: felix@cebas.csic.es)

1. Introduction

Tomato is a very important fruit due to the high consumption of it in the world. From the nutritional point of view, it also presents interesting characteristics such as the high content of substances with antioxidant activity. It has been assumed that regular intake of dietary antioxidants from fruits and vegetables would lead to a decrease in oxidative damage to key structures in the body, like lipids, proteins and DNA.

Tomato is a climacteric fruit and, therefore, its ripening is controlled by the plant hormone ethylene. The shelf life of tomato depends on post-harvest conditions and also on the time of harvest.

1-Methylcyclopropene (1-MCP) inhibits ethylene action by blocking the ethylene receptors and furthermore it blocks the ethylene physiological action for a period of time. As a result, MCP delays fruit ripening and extends shelf life.

The aim of this work has been the study of the effect of treatment with 1-MCP and storage at a non-chilling temperature on sensory and anti-oxidant properties of tomato varieties Rambo and Raf.

2. Experimental Design

Tomato fruits (*Lycopersicon esculentum* Mill) were provided by the Farming Cooperative Agroaguilas (Mazarrón, Murcia, Spain). After harvesting, only fruits having uniform size and at ripening stage "breaker" were selected. They were divided into three groups for each cultivar: (1) control

A. Ramina et al. (eds.), Advances in Plant Ethylene Research: Proceedings of the 7th International Symposium on the Plant Hormone Ethylene, 247–248.

fruits at the moment of harvest and that were analysed at day zero; (2) treated fruits with 1-MCP 500 ppb at 12°C during 24 h and then stored at 12°C; (3) untreated control fruits that were maintained at 12°C. After storage at 12°C, analytical determinations were made and also after 4 days and 7 days, respectively, at 20°C.

The analytical determinations were ethylene, by gas chromatography; texture, as pulp firmness, by penetrometry; colour, using a numerical scale of ripening stages, being 1 = green, 2 = breaker, 3 = turning, 4 = pink, 5 = light red and 6 = red; lycopene, by spectrophotometry; hydrophobic and hydrophilic antioxidant activity, as trolox equivalent antioxidant capacity.

3. Conclusions

Although a direct action on ethylene production has not been observed, the results of parameters reflecting ethylene-dependent processes point out that it has been stated a delay in ripening of approximately 10 days for variety Raf and 15 days for Rambo.

No relationship was observed between lipid soluble TEAC and lycopene content, suggesting that other hydrophobic substances could significantly contribute to the global value of this activity.

Due to the increase of oxidative reactions during the progress of ripening and the evolution to senescence, the increase of the levels of antioxidant substances is induced as a defence mechanism of the plant tissues in order to neutralize the generated free radicals. At the final stage, this strategy is not effective due to the high level of cell disorganization.

The effect of chilling and 1-MCP on quality attributes and physicochemical aspects of cell wall components of Passe-Crassane pears

Spinardi, A.M.*, Giovenzana, V. and Mignani, I.

Dipartimento di Produzione Vegetale, University of Milan, via Celoria 2, 20133 Milan, Italy, Phone: +39-0250316567; Fax: +39-0250316553 (*Corresponding author: anna.spinardi@unimi.it)

Most winter pear cultivars, including Passe-Crassane, do not ripen at warm temperatures. Postharvest exposure to chilling temperatures is required to synchronize the onset of the climacteric increase in ethylene production and the ripening of individual fruit (Lelièvre *et al.*, 1997). During ripening, pears softened, developing a melting texture after a period of shelf life. Softening and textural changes result principally from primary cell wall modifications. This study was initiated to evaluate the changes in cell wall fractions throughout the postharvest life of Passe-Crassane, comparing them with loss of firmness and with the production of ethylene during cold storage, ripening and failure to ripen.

Pears (*Pyrus communis* L. 'Passa Crassana') were picked at commercial ripening stage; one group was kept at 20°C, another was stored at 0°C and a third group was treated with 400 ppb of 1-MCP for 12 h before cold storage. Fruits were sampled after 30 days and cold-stored fruits after 135 days. Flesh firmness was evaluated using a hand penetrometer. Internal ethylene concentration was determined with a gas chromatograph equipped with an alumina column. Isolation of cell wall material (alcohol insoluble solids, AIS) was obtained by extraction of fruit tissue in boiling ethanol (Rose *et al.*, 1998). Samples of water-soluble and water-insoluble fractions of AIS were assayed spectrophotometrically for total uronic acid (UA) (Ahmed and Labavitch, 1977) and by GLC for noncellulosic neutral sugar composition (Albersheim *et al.*, 1967).

Pears maintained for 30 days at 20°C did not produce ethylene (Fig. 1). Fruit softened but did not develop a buttery and juicy texture. Cell wall total pectin content showed a decrease without any change in water-soluble pectin fraction (Fig. 2). After 30 days, cold storage ethylene synthesis already began (Fig. 1). Fruit softening did not occur, in spite of a consistent decrease in cell wall total pectin and a higher UA content of the

A. Ramina et al. (eds.), Advances in Plant Ethylene Research: Proceedings of the 7th International Symposium on the Plant Hormone Ethylene, 249–251.

water-soluble fraction (Fig. 2). Fruits sampled after 135 days cold storage showed a marked increase in internal ethylene associated with a loss of firmness and pectin degradation, indicating that ripening proceeds. In 1-MCP-treated pears, ethylene synthesis was as expected, strongly inhibited (Fig. 1). Fruit softening did not show any variation, pectin solubilisation was reduced and a higher total pectin content was maintained throughout the storage period (Fig. 2). Neutral sugar composition was affected by cold storage, increasing galactose and decreasing arabinose levels in the water-soluble fraction, while no effect was caused by the 1-MCP treatment (data not shown).

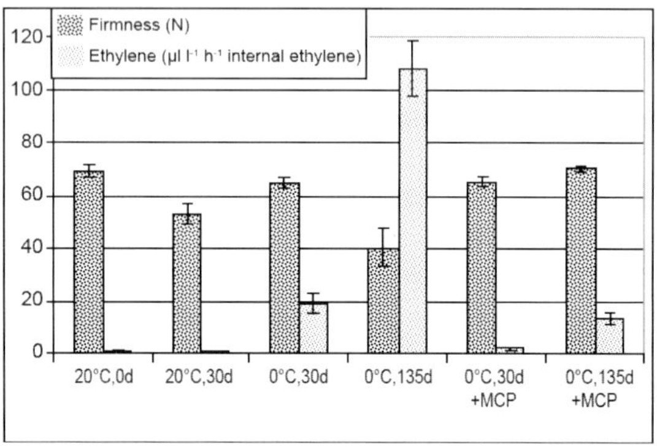

Fig. 1. Effect of cold storage and MCP treatment on firmness and ethylene production of Passe-Crassane pears. Vertical bars indicate SE.

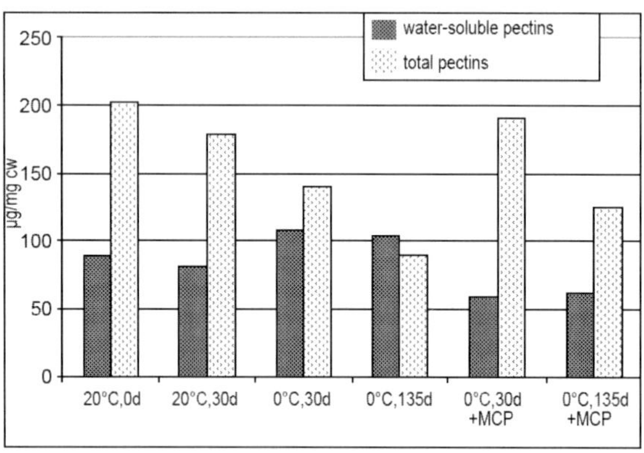

Fig. 2. Effect of cold storage and MCP treatment on cell wall total and water-soluble polyuronides of Passe-Crassane pears.

References

Ahmed AER, Labavitch JM (1977) A simplified method for accurate determination of cell wall uronide content. J Food Biochem 1:361–365

Albersheim P, Nevins DJ, English PD, Karr A (1967) A method for the analysis of sugars in plant cell wall polysaccharides by gas-liquid chromatography. Carbohydr Res 5:340–345

Lelièvre JM, Tichit L, Dao P, Fillion L, Nam YW, Pech JC, Latché A (1997) Effects of chilling on the expression of ethylene biosynthetic genes in Passe-Crassane pear (*Pyrus communis* L.) fruits. Plant Mol Biol 33:847–855

Rose JKC, Hadfield KA, Labavitch JM, Bennet AB (1998) Temporal sequence of cell wall disassembly in rapidly ripening melon fruit. Plant Physiol 117:345–361

1-MCP treatment affects tomato ethylene and polyamine metabolism

Tassoni, A.[1,2], Watkins, C.B.[1], Bagni, N.[2] and Davies, P.J.[1]*

[1]Cornell University, Ithaca NY, USA; [2]University of Bologna, Italy.
(*Corresponding author: pjd2@cornell.edu)

1. Introduction

In fruit such as the tomato (*Solanum lycopersicum*), ethylene initiates ripening and senescence, whereas polyamines have been considered senescence inhibitors. Putrescine is elevated in long-keeping Alcobaca tomato (Dibble *et al.*, 1988), and polyamine applications delay tomato ripening (Law *et al.*, 1991). Ethylene and polyamine biosynthetic pathways share S-adenosyl-methionine (SAM) as a common intermediate. The effect of 1-MCP (1-Methylcyclopropene) treatment (20 μL L^{-1}, 4 h, 23°C) on ethylene and polyamine metabolism and associated gene expression was investigated during tomato ripening to determine whether its effect in delaying ripening (Watkins, 2002) could be via polyamines in addition to 1-MCP binding to ethylene receptors (Klee, 2002).

2. Results and Discussion

1-MCP temporarily delayed ripening for 8 days (Fig. 1A), delaying ethylene production, and the expression of ACC synthase and ethylene receptor genes (ETR4,5,6; less ETR1,2, no effect on NR), though not ACC oxidase. Resumption of ripening was associated with a renewed expression of the genes encoding the ethylene receptors ETR4,5,6 (Fig. 1B).

In 1-MCP-treated fruit, free putrescine did not increase during the period of ripening inhibition, but increased as fruit started to ripen; spermidine and spermine were unaffected. The activity of the putrescine biosynthetic enzyme arginine decarboxylase (ADC) was higher in treated fruit. SAM-decarboxylase (SAMDC), which metabolizes putrescine to spermidine, peaked at the same time as putrescine levels in control and treated fruit. Gene expression for ADC peaked early in non-treated fruit and coincident with the delayed peak in putrescine in treated fruit (Fig. 1C) (Tassoni

A. Ramina et al. (eds.), Advances in Plant Ethylene Research: Proceedings of the 7th International Symposium on the Plant Hormone Ethylene, 253–255.

et al., 2006). A coincident peak in the gene expression for arginase, SAMDC, and spermidine and spermine synthases was also seen in treated fruit. No effect was observed on ornithine decarboxylase.

In summary, the content of polyamines does not rise with the prevention of ripening by 1-MCP, indicating that polyamines do not play role during the period of delayed ripening. However, the polyamine level increased as ripening started. We suggest that polyamines, rather than regulating ripening per se, may modulate the rate of ripening and over-ripening by counteracting the ripening promotion by ethylene.

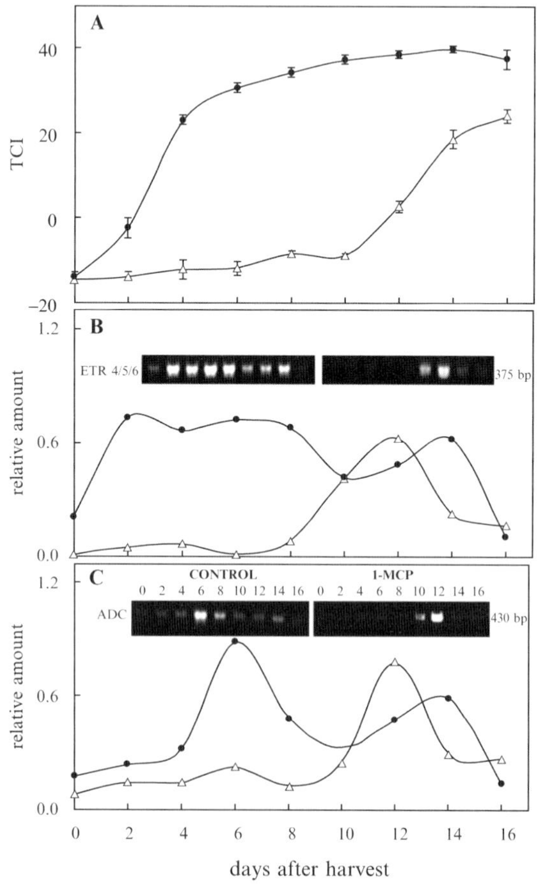

Fig. 1. (**A**) Colour (tomato colour index, TCI) of tomato fruit untreated or treated with 1-MCP. (**B**) RT-PCR analysis of ethylene receptor expression levels, *LeETR4,5,6*. (**C**) RT-PCR analysis of arginine decarboxylase. (•), Untreated; (Δ), 1-MCP-treated fruit.

References

Dibble ARG, Davies P, Mutschler MA. (1988) Polyamine content of long-keeping Alcobaca tomato fruit. Plant Physiol 86, 338–340.

Klee HJ. (2002). Control of ethylene-mediated processes in tomato at the level of receptors. J Exp Bot 53, 2057–2063.

Law DM, Davies PJ, Mutschler MA. (1991) Polyamine-induced prolongation of storage in tomato fruits. Plant Growth Reg 10, 283–290.

Tassoni A, Watkins CB, Davies PJ. (2006) Inhibition of the ethylene response by 1-MCP in tomato suggests that polyamines are not involved in delaying ripening, but may moderate the rate of ripening or over-ripening. J Exp Bot 57: 3313–3325.

Watkins CB. (2002) Ethylene synthesis, mode of action, consequences and control. In: Knee M, ed. Fruit quality and its biological basis, Sheffield Academic Press, pp. 180–224.

Ethylene stimulates emission of terpenoids and aliphatic esters in citrus fruits

Herrera, A.[1], Rodrigo, M.J.[2], Gil, J.[2] and Zacarías, L.[2]*

[1]Universidad Nacional de Colombia - Facultad de Agronomía -Bogotá; [2]Instituto de Agroquímica y Tecnología de Alimentos, IATA-CSIC, Valencia, Spain. (*Corresponding author: lzacarias@iata.csic.es)

1. Introduction

The aroma is an important parameter in fruit quality and is due to hundreds of volatile organic compounds including alcohols, aldehydes, terpenes, esters, ketones, lactones, sulfur-containing compounds among others.

In citrus fruit, the main volatile compounds are terpenoids which accumulate in the gland oils of the fruit, being the monoterpene limonene the major volatile compound in the essential oil of citrus peel (Ruberto, 2002).

Exogenous treatment with ethylene is commercially used worldwide to promote peel degreening in many citrus species (Salveit, 1999), but the effect of this treatment on other parameters of citrus fruit quality such as volatile emission has not been yet investigated. Recent studies on the effect of ethylene on volatile emission in climateric and non-climateric fruits showed that ester biosynthesis can be both dependent and independent of ethylene (Sarni-Manchado *et al.*, 1997; Aharoni *et al.*, 2000; Flores *et al.*, 2002; Echeverría *et al.*, 2004).

The aim of this study was to characterize the effect of postharvest ethylene treatment on the emission of volatile organic compounds in Fortune mandarins.

2. Results and Discussion

Headspace gas chromatography–mass spectrometry analysis of aroma emission of mature-green fruits of the hybrid Fortune (*Citrus reticulata* × *Citrus clementina*) stored for 72 h in air at 25°C showed the presence of more than 20 terpenoids emitted from whole mature-green fruit. Two alcohols, linalol and -citronellol were also identified. In ethylene-treated fruits,

A. Ramina et al. (eds.), Advances in Plant Ethylene Research: Proceedings of the 7th International Symposium on the Plant Hormone Ethylene, 257–259.

most of these terpenoids were also identified, but interestingly ten aliphatic esters were also detected that were not emitted from the fruit incubated in air.

The emission of limonene, the most abundant monoterpene in the peel of citrus fruit, sharply declined after fruit detachment, irrespective of the presence of ethylene. The monoterpene ocimene and the sesquiterpene trans-caryophyllene transiently increased by ethylene treatment. Ethylene markedly increased the emission of β-elemene, valencene and α-panasinsene that reached a maximum after 7 days and then remained steady. In fruits previously treated with 1-MCP, the ethylene-induced emission of terpenoids was initially reduced. As the effect of 1-MCP was progressively lost, the emission of these three terpenoids was progressively recovered, being after 2 weeks similar to that of ethylene-treated fruits.

Hexyl acetate, hexyl hexanoate, ethyl octanoate and butyl octanoate were the main aliphatic esters emitted from the whole mature-green Fortune mandarin treated with ethylene. This effect was totally ethylene-dependent. Esters emission reached a maximum 3–6 days after ethylene exposure and remained high during the whole period. As with terpenoids, the effect of 1-MCP reducing esters emission was progressively lost after a week of treatment. No ester emission was detected in mandarin fruit incubated in air. The ethylene-induced terpenoids and esters emission required the continuous presence of ethylene. Removal of ethylene after 7 days of treatment resulted in a rapid decline of volatile emission. Collectively, these results illustrate the role of the ethylene on the emission of volatiles in citrus fruits, and that its effect on the emission of aliphatic esters may be a common phenomena occurring in both climacteric and non-climacteric fruits.

References

Aharoni A, Keizer L, Bouwmeester H (2000) Identification of the SAAT gene involved in strawberry flavor biogenesis by use of DNA microarrays. Plant Cell 12: 647–661

Echeverría G, Fuentes J, Graell J, Lara I, López M (2004) Aroma volátile compounds of 'Fuji' apples in relation to harvest date and cold storage technology a comparison of two seasons. Postharvest Biol Technol 32: 29–44

Flores F, Yahyaoui F, De Billerbecck G, Romojaro F, Latche A, Bouzayen M, Pech J, Ambid (2002) Role of ethylene in the biosynthesis pathway of aliphatic ester aroma volatiles in Charentais cantaloupe melons. J Exp Bot 53: 201–206

Ruberto G, (2002) Analysis of volatile components of Citrus Fruit Essential Oils. In: JF Jackson and JF Linskens (Eds). Analysis of Taste and Aroma. Springer Verlag. Molecular Methods of Plant Analysis Series Vol 21 Berlin pp 123–157

Salveit M (1999) Effect of ethylene on quality of fresh fruits and vegetables. Postharvest Biol Technol 15: 279–292

Sarni-Manchado P, Verries C, Tesnìere C (1997) Molecular characterization and structural analysis of one alcohol dehydrogenase gene (*GV-Adh1*) expressed during ripening of grapevine berry. Plant Sci 125: 177–187

Ethylene evolution and its relationship with fruit maturation and ripening in late season peaches cultivars O'Henry and Late Nos

Luchsinger, L.E.*, Silva, E. and Lizana, L.A.

Postharvest Study Center (CEPOC), Faculty of Agricultural Sciences, University of Chile, Casilla 1004, Santiago, Chile. (*Corresponding author: lluchsin@uchile.cl)

1. Introduction

Maturity is only one aspect of the quality of perishable products, but it has great influence on their postharvest behavior during marketing, as well as on their ultimate organoleptic quality. Consequently, it has been the subject of considerable research. Characteristics that change with advancing maturity are valuable as harvest indicators. The objectives of the present study were to characterize the quality and maturity changes of late season peach (*Prunus persica* L.) fruit cvs. O'Henry and Late Nos during maturation and ripening, and to identify harvest maturity indices relating the different variables.

2. Materials and Methods

After fruit set, 400 similar fruits (grown in Chile) were tagged in order to follow maturation and ripening on the tree. During harvest, 48 fruits were randomly harvested every 3–6 days. The first harvest was January 25th for O'Henry fruit and March 14th for Late Nos. The ethylene evolution rate (EER) at 20°C, the fresh weight, the peel ground and cover color (L*, a*, b*, C*, and H_{ab}) were measured in all 48 fruits. Flesh color, firmness at several points on the fruit, soluble solids (SS), pH, titratable acidity (TA), and SS/TA ratio were measured in only 24 fruits, and the remaining fruits were held at 20°C for up to 7 days, after which the same parameters were measured again (ripening period). Ethylene concentration in a jar was measured with a gas chromatograph (Perkin Elmer Autosystem) using nitrogen carrier gas and equipped with an FID, 1 mL sample loop, and a Poropak column 80/100 GR. 3.2 at 90°C. Quantification was done by

A. Ramina et al. (eds.), Advances in Plant Ethylene Research: Proceedings of the 7th International Symposium on the Plant Hormone Ethylene, 261–263.

measuring peak area with a PE Nelson model 120 integrator. EER was log transformed to meet the assumption of homogeneity of variance. Individual fruit comprised the experimental unit. Pearson correlation coefficients were determined between variables to explore possible relations with logEER. The SAS stepwise procedure was used for model building. In this analysis, logEER was used as the dependent variable (maturity index). To test the validity of criteria used in the model and the significance of coefficients of independent variables the SAS Regression Procedure was performed.

3. Results and Discussion

For fruit of both cultivars, good Pearson correlation coefficients (*R*) were found between ethylene evolution rate logarithm and the pulp resistance to pressure at the cheek, suture and shoulder of the fruit, and SS/TA ratio (Table 1). Regression test using ethylene production rate logarithm as an indicator for ripeness (dependable variable) and other quality and ripen parameters as independent variables gave no satisfactory results (R^2 of 0.06) for either of the two cultivars.

4. Conclusions

No models (with $R^2 \geq 0.6$), to predict EER in terms of a nondestructive parameter were found for both cultivars. Therefore, no reliable harvest maturity indices could be established for these two late season cultivars.

Table 1. Pearson correlation coefficients ($r^2 \geq 0.6$) at harvest between ethylene evolution rate logarithm and different fruit characteristics at harvest for late season peaches cvs. O'Henry and Late Nos grown in Chile.

Log Ethylene Evolution Rate vs			
Firmness			SS/TA
Cheeks	Suture	Shoulder	
cv O'Henry			
• 0.60	• 0.61	• 0.69	0.64
cv Late Nos			
• 0.62	• 0.61	• 0.67	0.63

References

Luchsinger, L. and Walsh, C. 1998. Development of an objective and non-destructive harvest maturity index for peaches and nectarines. Acta Horticulturae 465(2):679–687.

Luchsinger, L.E. 1996. Quantification of peach fruit maturity, chilling injury and changes in cell wall composition during storage. PhD Diss., University of Maryland, College Park, USA. 135 p.

Luchsinger, L. and Walsh, C. 1993. Change in ethylene rate and ground color in peaches (cv. Redhaven and Marqueen) and nectarines (cv. Fantasia) during maturation and ripening. Acta Horticulturae 343:70–72.

Tonutti, P., Casson, P. and Ramina, A. 1991. Ethylene biosynthesis during peach fruit development. Journal of American Society for Horticultural Science 116(2):274–279.

The use of μPEACH 1.0 to investigate the role of ethylene in the initiation of peach fruit ripening

Begheldo, M., Ziliotto, F., Rasori, A. and Bonghi, C.*

Department of Environmental Agronomy and Crop Science, University of Padova, Viale dell'Università 16, I-35020 Legnaro (Padova), Italy. (*Corresponding author: claudio.bonghi@unipd.it)

1. Introduction

The use of natural or transgenic mutants is one of the ways to investigate specific physiological pathways in model plants such as Arabidopsis and tomato. In peach, a natural mutant (named slow ripening, *slr*), originated from a free pollination of the cv Fantasia, a yellow-fleshed nectarine, is characterized by a block of fruit development and an altered ripening pattern (no softening, pale yellow color). The *slr* fruit growth curve stops at S3 stage (after pit hardening) and the fruit remains smaller and ethylene evolution at basal levels. Propylene treatments, performed at preclimacteric stage induce a similar slight up-regulation of ethylene biosynthesis in Fantasia and *slr* fruits, but no further evolution of the hormone was observed in the latter. Northern analysis and semiquantitative RT-PCR have shown an up-regulation of the ethylene biosynthetic (*Pp-ACO1* and *Pp-ACS1*) and *PG* genes in Fantasia, while no significant changes have been observed in *slr* (Ziliotto, unpublished data). The lack of softening in *slr* might be related to the absence of PG transcripts supporting the role of this enzyme in the process. To better characterize the *slr* phenotype, changes in transcriptome profiles in relation to propylene treatment have been evaluated using the μPEACH 1.0 microarray.

2. Material and Methods

Fruits of Fantasia and *slr* at S3, preclimacteric and climacteric stages were collected from trees grown at the experimental farm of the agricultural faculty at the Padova University. Due to the absence of parameters indicating the onset of the process, S3slr and S4slr fruits were harvested at the same time of Fantasia. A portion of the fruits harvested at the preclimacteric stage

A. Ramina et al. (eds.), Advances in Plant Ethylene Research: Proceedings of the 7th International Symposium on the Plant Hormone Ethylene, 265–267.

were flushed with propylene (500 ppm) in humidified air (6 l/h) for 48 h. The following stages were compared in terms of transcript profiles: (a) S4 vs S3, (b) S4*slr* vs S3*slr*, (c) S3 vs S3*slr*, (d) S4 vs S4*slr*, (e) S4P vs S4T0, and (f) S4*slr*P vs S4*slr*T0 by using μPEACH 1.0 microarray. Microarray construction, hybridization and analysis were carried out as described by Trainotti *et al.*, (2006) using a replicated dye swap experimental design.

3. Result and Discussion

Three-hundred seventy-six genes are differentially expressed during the transition from immature to mature Fantasia fruit, while only 54 genes show changes in expression during the S3slr/S4*slr* transition. This difference is confirmed by comparing Fantasia and *slr* fruits at immature (S3 vs S3*slr*) and ripening (S4 vs S4*slr*) stages: in fact in the first comparison only 29 were differentially expressed, while in the second, 322 have been showing significant differences in terms of transcript level. Propylene treatment at preclimacteric stage in both fruit types, up-regulates 65 genes in Fantasia and 49 in *slr*, but only five are shared by the two gene sets; so in fantasia and *slr* fruits ethylene may act within a different genetic background.

Considering ethylene physiology, besides the upregulation of genes to be known as ripening-related and ethylene-dependent (Trainotti *et al.*, 2006), propylene induces an up-regulation of genes involved in the ethylene signal perception (*Pp-ERS*, *Pp-ETR2*) and transduction (*Pp-EIN2*, *Pp-ERF5*) in Fantasia (S4 vs S3), but not in *slr*. The treatment causes a significant downregulation of a tomato *E8*-like gene. The same gene appears to be more expressed in S4*slr* fruit in comparison to S4 Fantasia. The higher expressing *E8*-like gene might be related to the lower ethylene level detected in *slr* fruit considering that tomato fruit over-expressing E8 were characterized by ethylene biosynthesis at basal levels (Kneissl and Deikman, 1996). The presence in *slr* fruit of an interference on ethylene biosynthesis and action is confirmed by the analysis of transcription factors actively transcribed during the climacteric phase as those belonging to AUX-IAA family (Trainotti *et al.*, 2006). In Fantasia propylene-treated fruits an increase of transcripts corresponding to spotted probes showing high similarity to Arabidopsisis *IAA4*, *6*, *9*, *13*, *16*, *17* occurs, while in *slr* no changes were detected. In tomato, a downregulation of one of these proteins (IAA9) causes hypersensitivity to IAA and irregular fruit development (Wang *et al.*, 2005). A different behavior is shown by a member of the MADS-box family showing high identity to *Arabidopsis* SEP3: this gene is highly induced during ripening only in Fantasia fruit (Trainotti *et al.*, 2006), but its expression is unaffected by ethylene in Fantasia and *slr* demonstrating that it is developmentally-regulated.

The alteration of expression of genes encoding elements controlling ethylene pathway and transcription factors suggest that *slr* fruit could be a suitable model to study the regulation of dependent- and independent-ethylene processes occurring at ripening.

References

Kneissl M.L., Deikman J. 1996. The tomato E8 gene influences ethylene biosynthesis in fruit but not in flowers. Plant Physiol 112:537–547.

Trainotti L., Bonghi C., Ziliotto F., Zanin D., Rasori A., Casadoro G., Ramina A., Tonutti P. 2006. The use of microarray μPEACH 1.0 to investigate transcriptome changes during transition from pre-climacteric to climacteric phase in peach fruit. Plant Sci 170:606–613.

Wang H., Jones B., Li Z., Frasse P., Delalande C., Regad F., Chaabouni S., Latché A., Pech J-C., Bouzayen M. 2005. The tomato Aux/IAA transcription factor IAA9 is Involved in fruit development and leaf morphogenesis. Plant Cell 17:2676–2692

5. ABSCISSION AND SENESCENCE

Ethylene responses in abscission and other processes of cell separation in Arabidopsis

Patterson, S.[1]*, Butenko, M.[2] and Kim, J.[1]

[1]University of Wisconsin, Madison; Department of Horticulture, 1575 Linden Dr., Madison, WI, USA; [2]University of Oslo, Department of Molecular Biosciences, University of Oslo, N-0316 Oslo, Norway. (*Corresponding author: spatters@wisc.edu)

Abstract

Early abscission or detachment of an organ from the main body of a plant can result in unwanted crop loss and decrease in the quality of the fruit harvested. For many decades it has been observed that gassing plants with ethylene will often hasten abscission, and greenhouse flower growers are careful to avoid any unwanted ethylene. As new genes regulating abscission are identified, scientists have also begun to try to understand how these genes interact with genes regulating ethylene and other hormone responses. Gene expression and genetic analysis indicates that ethylene can affect the abscission process at two distinct phases during the abscission process.

1. Introduction

Abscission, or loss of organs from the main body of the plant, has been studied by plant physiologists for many decades. Even more interesting is to consider the fact that this process was probably one of the earliest traits that early humans selected for when they began to cultivate plants. Crop plants such as wheat (*Triticum monococcum*) and rice (*Oryza sativa*) from plants with premature seed shatter would have been selected against as the early gatherers most likely collected food for harvest and retaining a small amount for planting the following year. Not only can premature abscission result in significant decreases in yield, but it can also lead to reduced quality. For example, although canola should be harvested when the moisture levels are reduced from 8 to 10%, in dry and hot conditions shatter occurs easily and so the siliques are harvested before full ripening thus reducing the quality of the oil. Recent focus on understanding abscission in Arabidopsis

A. Ramina et al. (eds.), Advances in Plant Ethylene Research: Proceedings of the 7th International Symposium on the Plant Hormone Ethylene, 271–278.

may assist in increased understanding of the genes regulating this process and lead to improved crop practices as well as potential genetic modification.

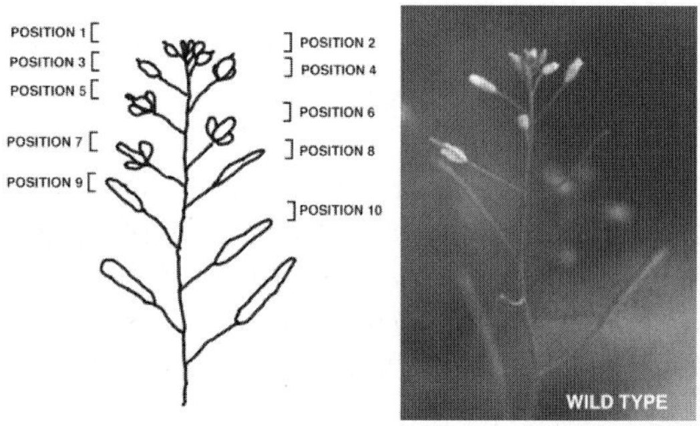

Fig. 1. Arabidopsis inflorescence showing position of flowers.

For more than a century, it has been recognized that the process of abscission is developmentally regulated and hastened by stress or disease responses. These stresses are most often associated with increased production of ethylene, and in the early 1900s it was determined by the Russian scientist Neljubov that early leaf abscission on trees grown near leaky gas lines was the result of exposure to higher than normal levels of ethylene (described in Abeles *et al.*, 1992). Subsequent studies showed that the abscission process could be delayed by application of auxin at certain stages of development (Rubenstein and Leopold, 1963; Abeles and Rubenstein, 1964). Most of these early studies were conducted on bean *Phaseolus vulgaris* and demonstrated that in young leaves auxin delayed abscission while in older leaves auxin had little effect in delaying abscission and in some cases may even contribute to increased sensitivity to ethylene. Other studies suggested a role for abscisic acid in regulating abscission and even gibberellins and cytokinins (Abeles, 1992). It is evident that the interactions of hormones are quite complex and perhaps these early studies were correct in implicating all the hormones in regulation of abscission. Using floral organ abscission in Arabidopsis as a model system, we have been able to developmentally follow the process of abscission by floral position (Fig. 1) and also ask specific questions about genes regulating ethylene responses and abscission. We propose that there are four major steps governing the process of abscission (Fig. 2) and that ethylene can affect this process at two major steps as designated by ETR1 and EIN2.

In our model of abscission, we propose that during the first step or Pre-abscission there is formation or establishment of a zone of cells (the abscission zone) that do not differentiate into distinct organs or tissue types and that the abscission zone cells are characterized as small isodiametric densely cytoplasmic cells with increased vesicular trafficking. Early studies on tomato *Lycopersicon esculentum*, elderberry *Sambucus nigra*, and bean *Phaseolus vulgaris* have shown that these abscission zone cells are easily distinguishable from neighboring cells and consist of four to fifty layers of cells (Jensen and Valdovinos, 1967; Addicott, 1982; Wright and Osborne, 1974). In Arabidopsis, the abscission zone of floral organs is composed of four to six cell layers and similar in features to those previously described in other plant species (Patterson and Bleecker, 2004). Phase II is represented as the dissolution of the middle lamella, and at this stage of development the cells within the abscission zone are responsive to both ethylene and auxin. During Phase II, we predict further expansion of the cells within the abscission zone and cell wall loosening of the proximal cells. We also predict that these cells are still responsive to ethylene; however, no longer responsive to auxin. Last, in stage IV or postabscission, we predict cell separation and suberization and repair of the proximal layer of cells. Genes responsible for regulating ethylene responses ETR1 and EIN2 as well as several additional genes associated with delayed abscission are placed in the pathway.

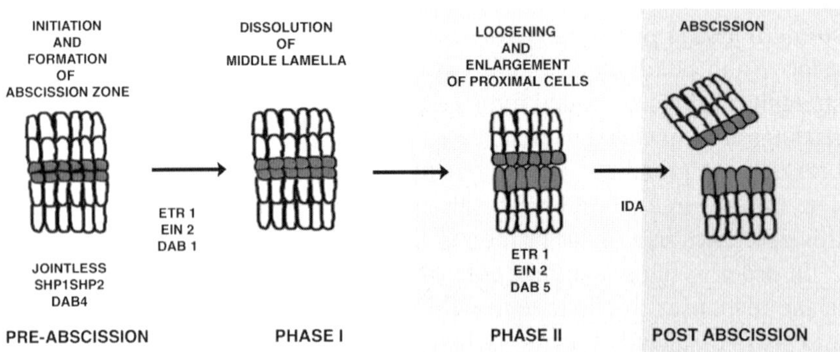

Fig. 2. Model of abscission.

2. Results

There are distinct morphological changes that occur during the different stages of abscission. If the floral organs in Arabidopsis are forcibly removed different developmental stages, the condition of the abscission zone

cells correlates with the stage of development. In young flowers that have their petals removed shortly after anthesis the revealed petal abscission zone is ragged and broken cells are exposed, an indication that no dissolution of the middle lamella or loosening has occurred (Fig. 3, left panel). At slightly later stages such as position 3 or 4, the cells revealed are flattened rather than broken (Fig. 3, middle panel) and at later stages we observe varying degrees of cell rounding (Fig. 3, right panel). As petals and other floral organs abscise, the abscission zone cells reach full expansion as indicated in the model in Fig. 2.

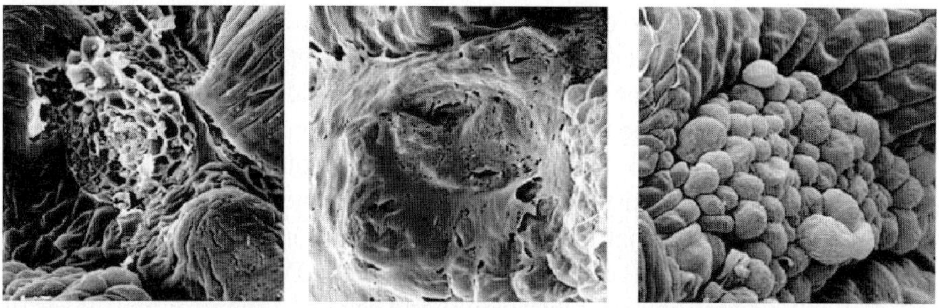

Fig. 3. Scanning electron micrographs of petal abscission zones at three developmental positions: left – early, position 1/2; middle – position 3/4; and right – position 7/8.

The morphological changes observed during the phases of abscission are also collaborated by breakstrength measurements (Fig. 4). The breakstrength assays provide a quantitative measurement of the force required to remove the petals at the different developmental stages and high breakstrength readings are found for stages revealing broken cells, while decreased readings are associated with increased cell rounding. In wild type-Arabidopsis, breakstrength readings are similar for all ecotypes tested (COL, WS, C24). The initial force to remove petals is approximately 1.7 g equivalents and this decreases in a linear fashion until petals are abscised.

In order to understand the role of ethylene more clearly, we also looked at the responses to ethylene in wild type, ethylene response mutants *etr1-1* and *ein 2-1*, and several delayed abscission mutants (Patterson and Bleecker, 2004; Butenko *et al.*, 2003). We observed that ethylene clearly hastened abscission as previously reported in several other plant species, but that the ability to perceive ethylene was not essential. Ethylene response mutants *etr1-1* and *ein2-1* both demonstrated a delay in floral organ abscission, but also both mutants ultimately displayed floral organ abscission despite the inability to perceive ethylene. SEM observations on the revealed abscission zones as well as breakstrength measurements also showed that booth *etr1-1* and *ein2-1* followed the same basic processes

during abscission as wild type aside from a developmental delay. Alternatively, delayed floral organ abscission mutants such as *dab1-1, dab3-1, dab4* and *dab5* both showed normal responses to ethylene but delays in abscission. While *dab1-1* and *dab5-1* showed a similar pattern of breakstrength to wild type, *dab5-1* and *dab3-1* showed an altered progression of breakstrength. Two additional mutants, *nev* (nevershed) and *ida* (indehiscent in abscission) demonstrated no abscission despite normal ethylene responses. It is interesting to note that the breakstrength pattern for these mutants is similar to wild type until position eight when it begins to increase again. We believe that this reflects a change in regulation of repair processes during abscission that we predict is normally repressed.

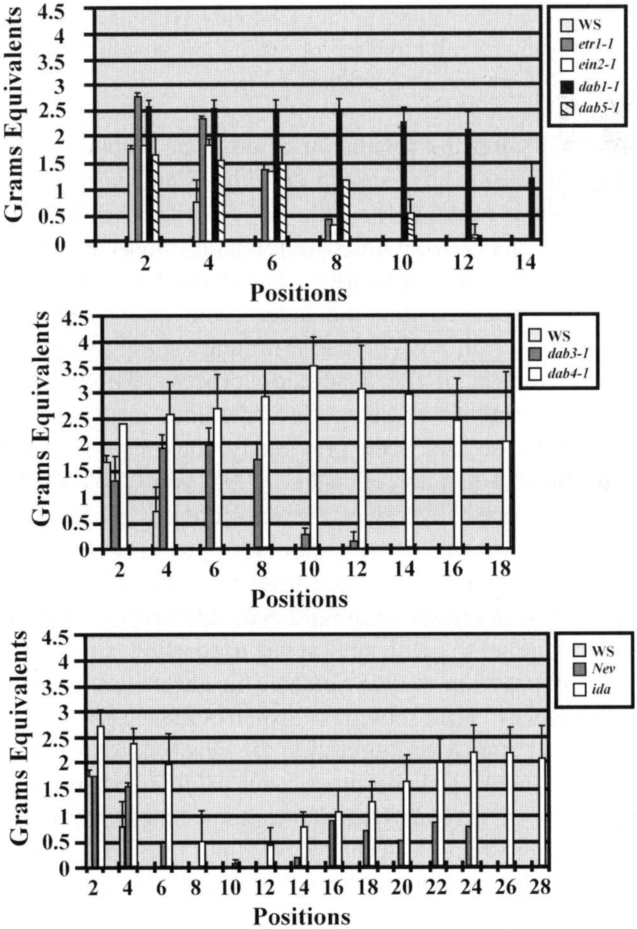

Fig. 4. Breakstrength measurements on wild type and selected ethylene response mutants and delayed abscission mutants.

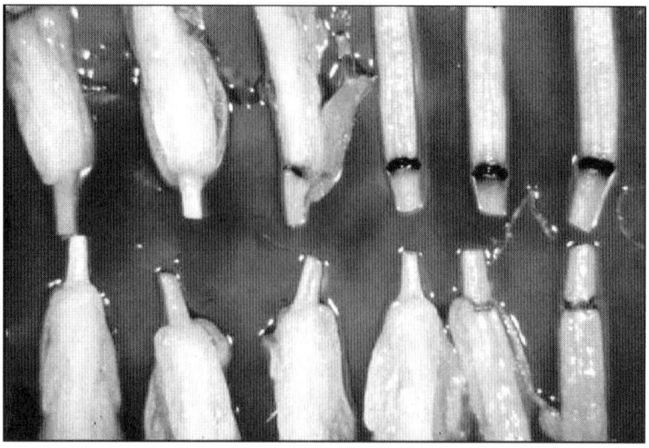

Fig. 5. (Color figure in the Annex, p.457) Chitinase GUS expression in wild type and *etr1-1*. *etr1-1* shows similar pattern of expression to wild type, but is delayed developmentally. WT :above; *etr1-1* below.

To further understand the regulation of ethylene responses during abscission, delayed abscission genes such as *dab4-1* and *ida,* were crossed to the ethylene response mutant *etr1-1*. In *dab4-1*, *etr1-1* double mutants, the delayed abscission phenotype and strong apical dominance characteristics of *dab4-1* were displayed indicating that *dab4-1* is epistatic to *etr1-1*. Alternatively, we observed that the indehiscent abscission phenotype of *ida* was displayed in *ida*, *etr1-1* double mutants suggesting multiple pathways for the regulation of the abscission process. This observation was also confirmed by analysis of promoter GUS lines in the different mutant backgrounds. Chitinase has been previously shown to be upregulated in response to ethylene (Chen and Bleecker, 1995) and CHIT:GUS transgenic lines of *etr1-1*, *ida*, and several of the *dab* mutants all showed GUS expression upregulated in the floral organ abscission zone as the abscission process occurs (Fig. 5). IDA:GUS transgenic lines were also evaluated and altered expression was observed in wild type and *etr1-1*. Specifically, IDA is expressed in filament, petal, and sepal abscission zones in wild type, while in *etr1-1* expression is also detected in nectarines and onset is later and terminates sooner. This provides further evidence for the existence of multiple converging pathways regulating the abscission process.

In an effort to understand the interaction of other hormones during the abscission process, we screened the *dab* mutants and *ida* at a seedling level for altered hormone responses. Although seedling responses (triple response) to ethylene in *dab 4-1* were normal at all levels of ethylene tested (0.1 ppm to 100 ppm), in the presence of 10 μM jasmonic acid ethylene responses were magnified (Fig. 6. *dab4-1* is insensitive to jasmonic acid, and thus the reduction of seedling growth in response to ethylene on 10 μM jasmonic acid, provides further evidence that jasmonic acid has a role in seedling elongation and responses to ethylene.

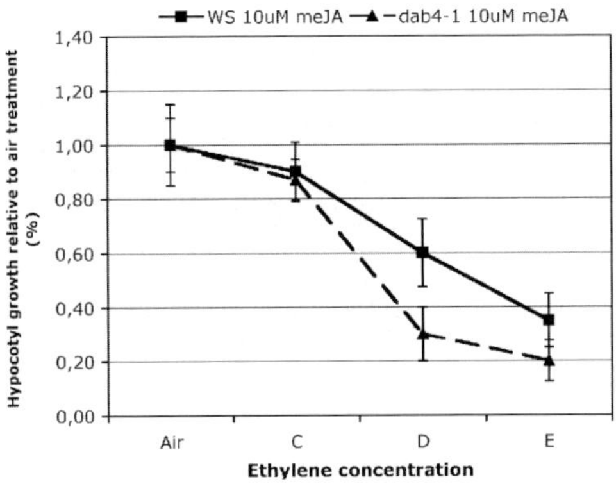

Fig. 6. Seedling responses of wild type and *dab4-1* to ethylene in the presence of 10 μM jasmonic acid (C = 0.04 ppm; D = 0.6 ppm; E = 20.5 ppm).

Altered seedling responses to other hormones have not been observed on any of the *dab* mutants or *ida*, but changes in gene expression in different tissue types using RT PCR were observed for *dab4-1* for several of the hormone response genes. Genes regulating signaling responses or hormone regulation included ARF3, IAA7, SPY, EIN2, and BRI1 (Fig. 7). In addition, levels of expression of several of these genes were also altered during the developmental process of abscission (data not shown). Preliminary results from the Cohen lab at the University of Minnesota indicated altered levels of IAA in *dab4-1* versus wild type, and additional experiments at different developmental stages of development will further address the regulation of auxin in the *dab4-1* genetic background. These changes in levels of expression lend support to the earlier proposed hypothesis that regulation of abscission involves multiple hormone pathways (Osborne, 1989).

It is an exciting time to be studying abscission, and recently, there have been many new genes discovered that have a role in regulating this process (Lewis *et al.*, 2006). We predict that in the next decade, additional attention will be directed to understanding abscission and the role of ethylene and other growth regulators in governing this process.

Acknowledgments

This work was supported by USDA grant #00-35301-9085, NSF DBI-0077719, and the Norwegian Research Council. We would like to thank

Ayala Most for excellent technical assistance and members of the Patterson lab for critical discussions.

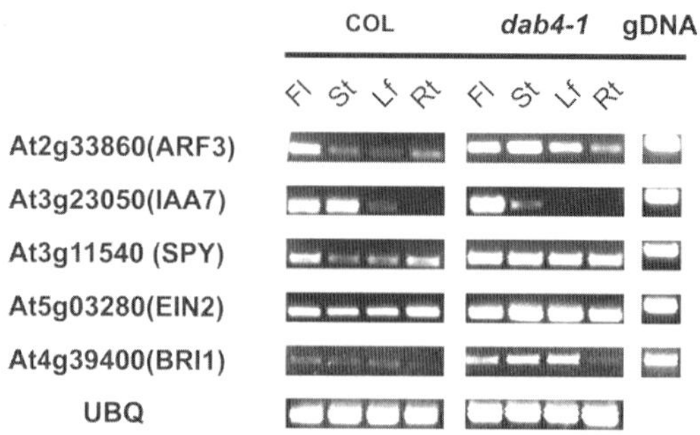

Fig. 7. Expression of selected hormone associated genes in different tissue types of *dab4-1* and wild-type Arabidopsis.

References

Abeles FB, Morgan PW, Salveit ME. (1992) *Ethylene in Plant Biology* Edition 2, Academic Press, San Diego, CA.

Abeles FB and Rubenstein B. (1964) Regulation of ethylene evolution and leaf abscission by auxin. *Plant Physiology* 39, 963–969.

Addicott FT. (1982) *Abscission.* University of California Press, Berkeley, CA.

Butenko MA, Patterson SE, Grini PE, Mandal A, Aalen RB. (2003) Inflorescence deficient in abscission controls floral organ abscission in Arabidopsis and identifies a novel family of putative ligands in plants. *Plant Cell* 15, 2296–2307.

Chen QG and Bleecker AB. (1995) Analysis of ethylene-signal transduction kinetics associated with seedling-growth response and chitinase induction in wild-type and mutant *Arabidopsis. Plant Physiology* 108, 597–607.

Jensen T and Valdovinos J. (1967) Fine structure of abscission zones I. Abscission zones of the pedicels of tobacco and tomato flowers at anthesis. *Planta* 77, 298–318.

Lewis MW, Leslie ME, Liljegren SJ. (2006) Plant separation: 50 ways to leave your mother. *Current Opinion in Plant Biology.* 9, 59–65.

Osborne DJ. (1989) Abscission. *Critical Reviews in Plant Sciences* 8, 103–129.

Patterson SE and Bleecker AB. (2004) Ethylene dependent and independent processes associated with floral organ abscission in Arabidopsis. *Plant Physiology* 134, 194–203.

Rubenstein B and Leopold A. (1963) Analysis of the auxin control of bean leaf abscission. *Plant Physiology* 38, 262–267.

Wright M and Osborne DJ. (1974) Abscission in *Phaseolus vulgaris.* The positional differentiation and ethylene-induced expansion growth of specialized cells. *Planta* 120, 163–170.

Proteomic analysis of pollination-induced senescence in Petunia flowers

Jones, M.L.*[1], Bai, S.[1], Willard, B.[2], Stead, A.[1,3] and Kinter, M.[2]

[1]The Ohio State University, Department of Horticulture and Crop Science, 1680 Madison Ave, Wooster, Ohio, 44606, USA. [2]Cleveland Clinic Foundation, Proteomics Laboratory, Department of Cell Biology, 9500 Euclid Ave, Cleveland Ohio, 44195, USA.[3] Royal Holloway University of London, School of Biological Sciences, Egham, Surrey, UK. (*Corresponding author: jones.1968@osu.edu)

Abstract

Senescence represents the last stage of flower development, ultimately culminating in the death of the petals. The senescence program is regulated by coordinated changes in gene and protein expression, and the later stages of senescence share many characteristics of programmed cell death. In some flowers, like petunias, petal senescence is associated with increased ethylene production and wilting is accelerated by treatment with ethylene. We have recently taken a proteomic approach to profile protein changes during the pollination-induced senescence of petunia petals. Two-dimensional electrophoresis comparing unpollinated corollas at 24 h, 48 h, and 72 h after flower opening to corollas at 24 h, 48 h, and 72 h after pollination has identified a number of proteins that are up- or down-regulated during corolla senescence. Electrospray ionization-tandem mass spectrometry (ESI-MS/MS) has been used to determine the identity of these differentially expressed proteins. A total of 143 proteins have been sequenced and greater than 85% have been identified. The function of the up-regulated proteins is being confirmed using virus-induced gene silencing (VIGS).

1. Introduction

The senescence program is regulated by highly coordinated changes in gene expression, and the later stages of senescence share many biochemical and morphological characteristics of programmed cell death (PCD)

A. Ramina et al. (eds.), Advances in Plant Ethylene Research: Proceedings of the 7th International Symposium on the Plant Hormone Ethylene, 279–284.

(Jones, 2004). Plant PCD also involves components that are regulated posttranscriptionally (Swidzinski *et al.* 2004), and the senescence of flowers is associated with increased enzymatic activity and increased abundance of individual proteins (Borochov and Woodson 1989; Jones 2004; Langston *et al.* 2005). Treating flowers with the protein synthesis inhibitor, cycloheximide, effectively delays petal senescence in ethylene-sensitive and insensitive flowers (Wulster *et al.* 1982; Lay Yess *et al.* 1992; Jones, unpublished). These experiments indicate that the timely progression of the senescence program in petals is dependent on *de novo* protein synthesis. Therefore, it is hypothesized that identifying those proteins that increase or decrease in abundance during petal senescence will provide insight into the biochemical pathways that control senescence.

2. Materials and Methods

Petunia × *hybrida* "Mitchell Diploid" plants were grown in the greenhouse as previously described (Langston *et al.* 2005). Corollas were collected at 24, 48, and 72 h after flower opening and 24, 48, and 72 h after self pollination for proteomics experiments. Ethylene production from pollinated and unpollinated corollas was determined as previously described (Langston *et al.* 2005). Total soluble proteins were extracted from corollas and quantified using the methods described in Coaker *et al.* 2004. Two hundred micrograms of total protein was used to rehydrate 11-cm immobilized pH gradient (IPG) strips (pH 5–8; BioRad). Isoelectric focusing using the protean IEF cell (BioRad) was according to the manufacturers recommendations. Equilibrated strips were immediately subjected to SDS-PAGE (12.5%) for second dimension separation. Following electrophoresis, gels were fixed (50% ethanol, 10% acetic acid), rinsed in water (3X), and stained with Gel Code Blue (Pierce). At least three biological and three technical replicates were performed for each sample.

Digital images of the stained gels were generated using the Syngene GeneGnome imaging system, and qualitative and quantitative protein differences between senescing and nonsenescing petals were identified using PDQuest 7.4.0 software (BioRad). Differentially expressed proteins were excised from the gel, trypsin digested, and subjected to capillary column HPLC-electrospray ionization–tandem mass spectrometry (ThermoFinnigan LTQ) (Kinter and Sherman 2000). Data was analyzed by using all generated collision-induced dissociation (CID) spectra to search the NCBI nonredundant database using the SEQUEST and MASCOT algorithms with a "green plants" taxonomy filter. For those proteins not identifiable by querying the NCBI nonredundant database, the CID spectra were manually interpreted to determine the amino acid sequence of the proteins and this sequence data was used to search the TIGR Petunia Gene Index and a translated Solanaceae EST database developed at OSU.

3. Results and Discussion

Senescence in petunia flowers is accompanied by a burst of ethylene production from the corolla (Fig. 1A). Pollination accelerates ethylene production and symptoms of petal wilting. To identify proteins involved in petal senescence, protein profiles were compared between corollas at 24, 48, and 72 h after pollination (hap) and corollas at 24, 48, and 72 h after flower opening (Fig. 1B).

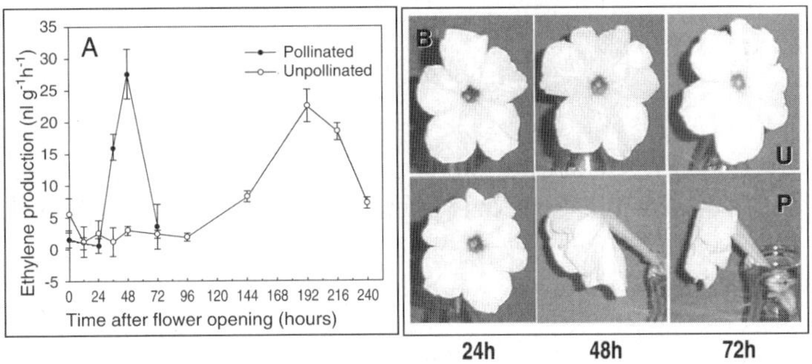

Fig. 1. *Petunia x hybrida* "Mitchell Diploid" flowers. **A**. Ethylene production from corollas of pollinated and unpollinated flowers. **B**. Unpollinated (top) and pollinated (bottom) flowers were collected at 24, 48, and 72 h after flower opening for proteomic analyses.

Two-dimensional polyacrylamide gel electrophoresis (2-DE) was used to separate the individual protein components of the corolla extracts (Fig. 2). Analyses using PDQuest 7.4.0 (BioRad) identified 90 proteins that were up-regulated and 53 proteins that were down-regulated when comparing unpollinated and pollinated corollas (greater than twofold change, significant at $p < 0.05$). Up-regulated proteins included those that first appeared in pollinated corollas or those that increased in abundance following pollination. Most of the differentially expressed proteins were detected when comparing the mid and advanced stages of senescence (i.e., 48 h unpollinated to 48 h pollinated and 72 h unpollinated to 72 h pollinated).

Protein spots were excised from the 2-D gels and proteins were sequenced by electrospray ionization–tandem mass spectrometry (Cleveland Clinic Proteomics Laboratory). Over 85% of the proteins have been identified based on sequence homology to proteins in the NCBI nonredundant protein database or a Solanaceae translated EST database created at OSU (data not shown). Classification of the annotated, up-regulated proteins revealed that unknown, energy metabolism, carbohydrate metabolism, and protein degradation made up the largest functional groups of proteins, comprising ~59% of the total annotated proteins (data not shown).

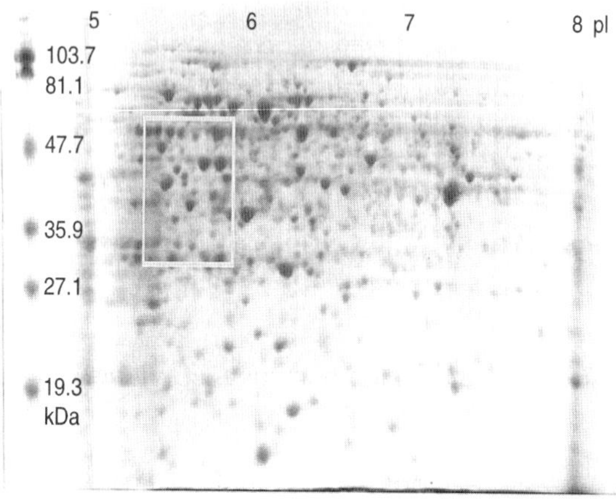

Fig. 2. Total proteins extracted from corollas of pollinated and unpollinated flowers were separated using 2-D gel electrophoresis. Gels were stained with Gel Code Blue to visualize the individual protein spots. This is an example of one of the replicate gels containing proteins from 72 h pollinated corollas. The box indicates the region enlarged in Fig. 3.

Once a flower has been pollinated, the programmed senescence of the corolla allows the plant to breakdown proteins, nucleic acids, and membranes and to remobilize their constituents (primarily N and P) to developing tissues, including the ovary. Many of the senescence up-regulated genes that have been identified from flowers encode enzymes involved in these degradative processes (Jones 2004). Similarly, the majority of the up-regulated proteins we have identified have a putative function in the dismantling and remobilization of petal constituents.

Fig. 3 shows an enlarged area of six 2-D gels comparing the protein profiles in unpollinated petals at 24, 48, and 72 h after flower opening (top) to the protein profiles of pollinated petals at 24, 48 and 72 hap (bottom). This region is indicated by a box on Fig. 2. Protein number 1 (circle) has homology to an endonuclease from potato (Accession #50657596). This protein has a MW of 43 kDa and a pI of 6.7. The endonuclease was first detectable at 48 hap and increased at 72 hap. This corresponds with petal wilting and ethylene production, and correlates with the detection of DNA fragmentation and the activity of a 43 kDa endonuclease in senescence petunia corollas (Langston *et al.*, 2005). Protein 2 (square) has homology to the 20S proteasome alpha 6 subunit from *Nicotiana benthamiana* (Accession #22947842) and has an estimated MW of 37 kDa and a pI of 5.7. This protein was up-regulated during the advanced stage of senescence, with increases first detected at 72 hap. The proteasome degrades proteins that have been targeted for destruction by ubiquitination. In support of the central role of protein degradation via ubiquitination, some of the up-regulated proteins that we have identified have peptides with MW differences that can be explained by ubiquitination (data not shown).

Two dimensional electrophoresis followed by electrospray ionization-tandem mass spectrometry has identified 143 proteins that are differentially expressed during petal senescence. The identity of these proteins provides further insight into the biochemical pathways that are executing the petal senescence program in ethylene-sensitive flowers. Future experiments will determine the expression patterns of the corresponding genes and their regulation by the plant hormone ethylene. Virus-induced gene silencing will be used to confirm their role in petal senescence.

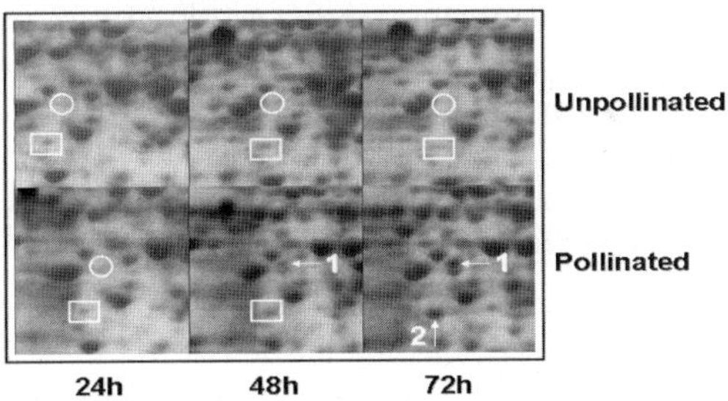

Fig. 3. Protein differences at 24, 48, and 72 h after flower opening in unpollinated and pollinated corollas. A region of the gel (boxed region from Fig. 2) is shown for one replicate gel for the six treatments. The circles and arrow #1 indicate a putative endonuclease. The rectangles and arrow #2 indicate a putative 20S proteasome alpha 6 subunit.

Acknowledgments

This research was supported by The Ohio State University D.C. Kiplinger Endowment, the OARDC competitive grants program and the Fred C. Gloeckner Foundation.

References

Borochov A and Woodson WR (1989) Physiology and biochemistry of flower petal senescence. Horticultural Reviews 11:15–43

Coaker GL, Willard B, Kinter M, Stockinger EJ, and Francis DM (2004) Proteomic analysis of resistance mediated by Rcm 2.0 and Rcm 5.1, two loci controlling resistance to bacterial canker of tomato. Molecular Plant Microbe Interactions 17:1019–1028

Jones ML (2004) Changes in Gene Expression. *In* Plant Cell Death Processes. L Nooden ed. Elsevier Academic Press, San Diego, CA. Pgs 51–72

Kinter M and Sherman NE (2000) Protein Sequencing and Identification Using Tandem Mass Spectrometry. John Wiley and Sons, Inc., New York

Langston BL, Bai S, and Jones ML (2005) Increases in DNA fragmentation and induction of a senescence-specific nuclease are delayed during the senescence of ethylene-insensitive (*etr1-1*) transgenic petunias. Journal of Experimental Botany 56:15–23

Lay Yess M, Stead AD, and Reid MS (1992) Flower senescence in daylily (*Hemerocallis*). Physiologia Plantarum 86:308–314

Swidzinski JA, Leaver CJ, and Sweetlove LJ (2004) A proteomic analysis of plant programmed cell death. Phytochemistry 65:1829–1838

Wulster G, Sacalis J, and Janes H (1982) The effect of inhibitors of protein synthesis on ethylene-induced senescence in isolated carnation petals. Journal of the American Society for Horticultural Science 107:112–115

Circadian rhythm of ethylene related genes under postharvest conditions in transgenic Arabidopsis plants

Kenigsbuch, D.[1]*, Ovadia, A.[1,2], Tabibian-Keissar, H.[3]
and Chalupowicz, D.[1]

[1]Volcani center, Postharvest Dept., P.O.B. 6, Bet-Dagan 50250, Israel.
[2]Beit-Berl Collage, Agriculture and Environment Dept., Beit-Berl 44905, Israel. [3]Bar-Ilan University, Life Science Faculty, Ramat-Gan, 52900, Israel. (*Corresponding author: davke@agri.gov.il)

1. Introduction

The circadian clock is an endogenous cellular mechanism which allows organisms to anticipate daily environmental changes resulting from the earth's rotation on its axis (Alabadi *et al.*, 2001). A wide variety of physiological and behavior responses in plants, animals, and some prokaryotes are regulated by the clock (Harmer *et al.*, 2001; Young and Kay, 2001; McClung, 2006). Endogenous circadian oscillations of gene expression resulting from auto-regulatory feedback loops underlie the physiological rhythms and form the basis of a clock central oscillator (Más, 2005). The setting of the clock is synchronized to the environment by entrainment signals, including light and temperature (Salome and McClung, 2005). Entrained rhythmic outputs under circadian clock regulation are maintained under constant conditions (Gardner *et al.*, 2006). In Arabidopsis at least 6% of the genes are expressed under the circadian clock (Harmer *et al.*, 2001; Schaffer *et al.*, 1998).

The plant hormone ethylene plays a role in the life cycle of plants from germination to ripening and senescence (reviewed by Mattoo and Suttle, 1991, and Abeles *et al.*, 1992). Ethylene's biosynthesis, metabolism, and signal transduction pathways are well known (Guo and Ecker, 2004). It has been suggested that there are strong interactions between light signaling, plant hormone activity (including ethylene), and the circadian clock (Nozue and Maloof, 2006). For example, ethylene inhibits hypocotyl elongation in the dark (Bleeker *et al.*, 1988), however under long day conditions ethylene has been shown to induce hypocotyl elongation (Smalle *et al.*, 1997).

A. Ramina et al. (eds.), Advances in Plant Ethylene Research: Proceedings of the 7th International Symposium on the Plant Hormone Ethylene, 285–291.

Ethylene plays a major role in regulating ripening and senescence (Bleecker and Kende, 2000). Thain *et al.*, (2004) have shown that ethylene emission in Arabidopsis plants is regulated by the circadian clock. Ethylene production levels are tightly correlated with ACC SYNTHASE 8 steady-state transcript levels. The expression of this gene is regulated by the circadian clock, as well as by light and by negative feedback regulation through ethylene signaling. Ethylene production was shown to be controlled by TIMING OF CAB EXPRESSION 1 (TOC1) and CIRCADIAN CLOCK ASSOCIATED 1 (CCA1) genes, which are critical for all circadian rhythms yet tested in Arabidopsis (Alabadi *et al.*, 2001; Alabadi *et al.*, 2002). While circadian expression was found for the ACC SYNTHASE gene family: ACS2, ACS5, ACS8, ACS9; mutations of ethylene signaling pathways did not affect the plants' rhythmic growth.

In this work, we studied the role of CCA1 and ethylene biosynthesis genes on postharvest senescence of Arabidopsis plants.

2. Materials and Methods

2.1 Plants and growth conditions

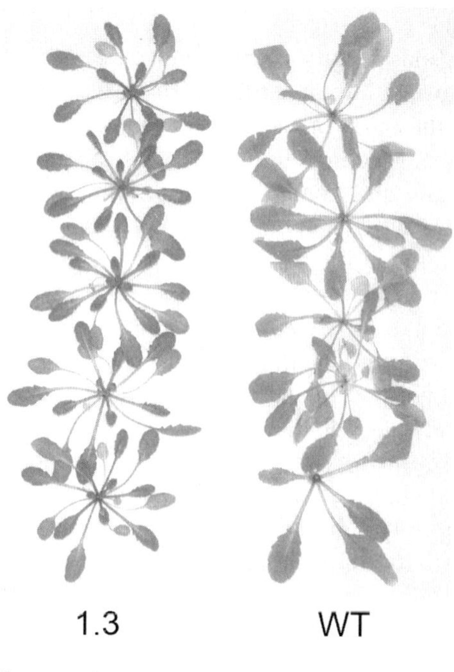

Fig. 1. Transgenic plants (1.3) and WT plants 7 days after harvest at dark (20°C).

Arabidopsis thaliana wild type (WT) colombia ecotype and transgenic plants over expressing CCA1 gene (1.3) were used in all experiments (Tabibian-Keissar, 2005). Seeds were sown in a soil/perlite mix and grown under 12:12 L:D cycles at 23°C.

2.2 Postharvest conditions

The plants were grown under 12:12 L:D cycles at 23°C. Rosettes were harvested when seedlings were 5 weeks old, packed in polyethylene bags and placed at 20°C in complete darkness for 7 days.

2.3 Semiquantitative RT-PCR

For circadian clock and ethylene gene expression, samples were collected every 4 h and frozen in liquid nitrogen. The RNA purification was done using GenElute™ Mammalian Total RNA Miniprep Kit (Sigma, USA). cDNA was created from 2 mg of the total RNA with Reverse Transcription System kit (Promega, USA). PCR was done using Reddy Mix (ABgene, UK); the primers, cycle number, and annealing temperatures varied according to the gene amplified as follows:

Circadian clock genes: CCA1 (U79156): 5′-gcggcgtgcattggactc, 5′-tcct-ccacggcctaagcg, 57°C, 27 cycles; LHY (AJ006404): 5′-ccaacgaaacagg-taagtggcg, 5′-ggtctgccacaagaatcctggc, 53°C, 27 cycles; ethylene biosynthesis genes: ACS8 (AF334712): 5′-gtccagtttcggtctaatctc, 5′-ataggtgtctcatgtcaaccc, 55°C, 28 cycles; ACS9 (AF332391): 5′-tcggtttaccaggttttcgc, 5′-acacgagttt-cttctgacgaa, 55°C, 28 cycles; ACO2 (At1g62380) 5′-ccagctacttcgcttgtcgag, 5′-gtctctacggctgctgtagga, 56°C, 28 cycles; and 18SrRNA (X16077): 5′-gtgttggcttcgggatcgg, 5′-cgctccaccaactaagaacgg, 60°C, 19 cycles. PCR products were separated on 1.4% agarose gel, stained with ethidium bromide. Semiquantitative analysis was carried out using TINA2 software.

2.4 Ethylene and 1-MCP treatments

WT leaves from 5 week old seedlings were harvested (about 4 gr) and placed in 150 ml jars. The leaves were treated with 5 ppm ethylene or 1 ppm of 1-methylcyclopropene (1-MCP, an ethylene action inhibitor) for 16 h. The leaves were maintained at 20°C in complete darkness for 7 days.

3. Results and Discussion

3.1 Postharvest transgenic plants

Fig. 2. (Color figure in the Annex, p.458) Postharvest senescence of detached leaves taken from 5 week old seedlings (7 days at 20°C, in dark).

The transgenic plants overexpressing the CCA1 gene (1.3) were found to behave differently under postharvest conditions. The rosette senescence after harvest of this line was retarded as compared to the WT (Fig. 1). The WT leaves had higher degrees of yellowing and wilting than the 1.3 leaves. The same phenomenon was found in detached leaves (data not shown).

Treating detached WT leaves with ethylene gave accelerated senescence as compared to the control. The ethylene-treated leaves showed increased yellowing and wilting. The 1-MCP treatment retarded the onset of senescence (Fig. 2). The leaves remained green and fresh in comparison to the control. These results agree with previous studies showed the involvement of ethylene in leaf senescence (Bleecker and Kende, 2000).

3.2 Gene expression under postharvest conditions

Wild type and 1.3 transgenic plants were grown under 12:12 L:D cycles, harvested and placed in complete darkness for storage. In the WT plants, the CCA1 rhythmic expression continued under D:D postharvest conditions as in L:D conditions (Fig. 3). In the 1.3 transgenic plants, the CCA1 overexpression continued under D:D postharvest conditions as in L:D conditions (Fig. 3).

Thain *et al.* (2004) showed that some of the ethylene biosynthesis genes, such as ACS8 and ACS9, are regulated by the circadian clock. We show here that the amplitude of circadian gene expression of the ACS8, ACS9, and ACO2, is diminished under D:D postharvest conditions (Fig. 4). In all of the ethylene biosynthesis genes tested, there was an expression peak on the 3rd day after harvest in the WT plants which did not occur in the 1.3 transgenic plants. It is reasonable to conclude that this reduction in expression of ethylene biosynthesis genes contributes to the retarded postharvest senescence found in the 1.3 transgenic plants.

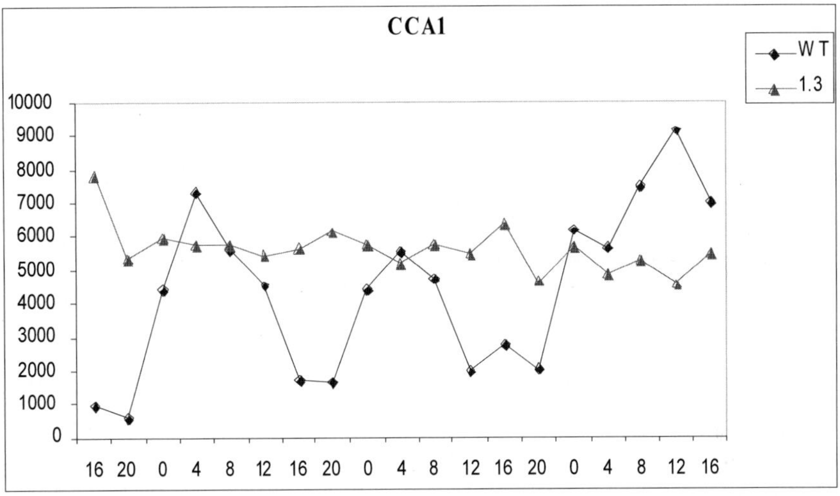

Fig. 3. Semiquantitative RT-PCR of CCA1 expression after harvest at dark in WT, and 1.3 transgenic plants.

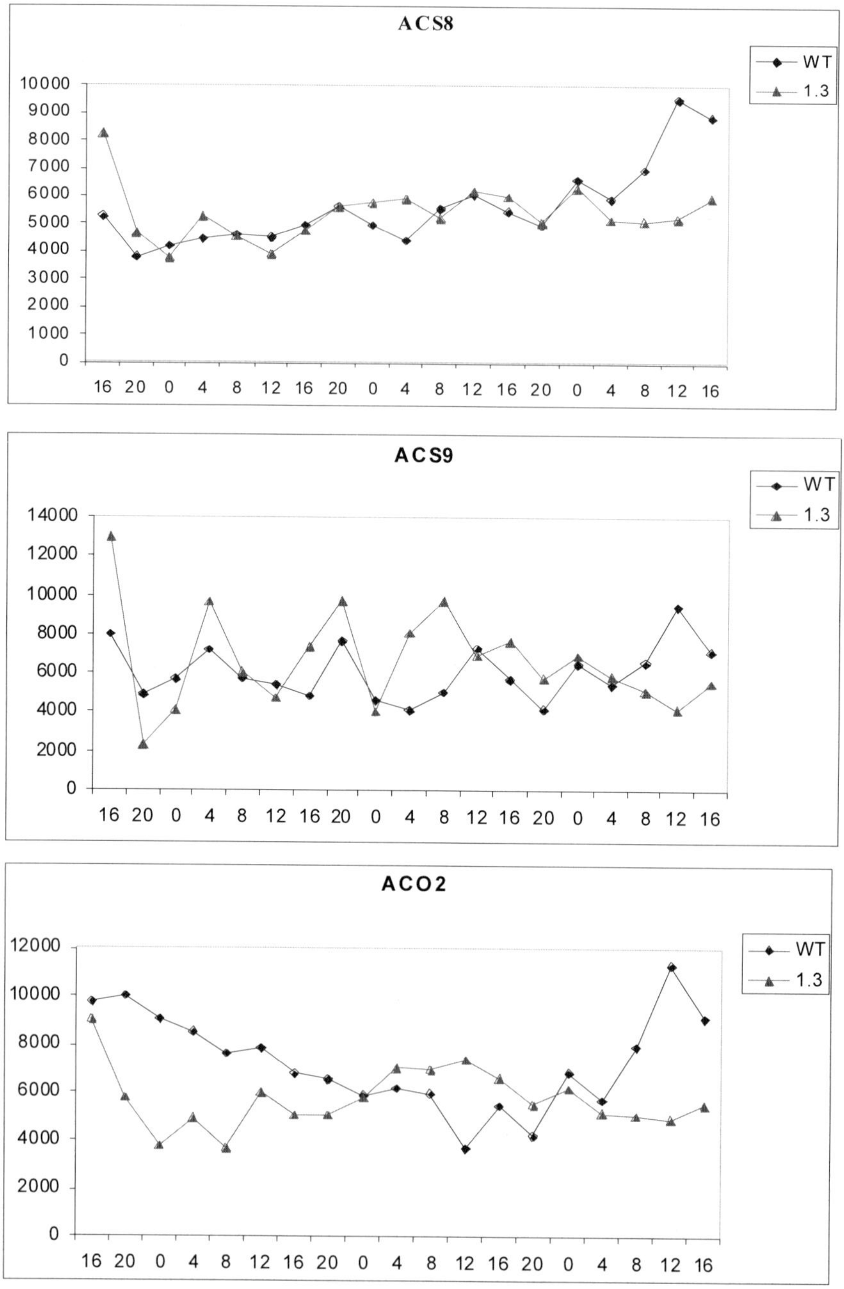

Fig. 4. Semiquantitative RT-PCR of ACS8, ACS9, and ACO2 after harvest at dark in WT and 1.3 transgenic plants.

References

Abeles FB, Morgan PW, Saltveit ME Jr. (1992) Ethylene in plant biology, 2nd Edition. San Diego: Academic Press.

Alabadi D, Oyama T, Yanovsky MJ, Harmon FG, Más P, Kay SA (2001) Reciprocal regulation between *TOC1* and *LHY/CCA1* within the Arabidopsis circadian clock. Science 293: 880–883.

Alabadi D, Yanovsky MJ, Más P, Harmer SL, Kay SA (2002) Critical role for CCA1 and LHY in maintaining circadian rhythmicity in Arabidopsis. Curr. Biol. 12: 756–761.

Bleecker AB, Kende H (2000) Ethylene: A gaseous signal molecule in plants. Ann. Rev. Cell. Dev. Biol. 16: 1–18.

Bleeker AB, Estelle MA, Somerville C, Kende H (1988) Insensitivity to ethylene conferred by a dominant mutation in *Arabidopsis thaliana.* Science 241: 1086–1089.

Gardner MJ, Hubbard KE, Hotta CT, Dodd AN, Webb AAR (2006) How plants tell the time. Biochem J. 397: 15–24.

Guo H, Ecker JR (2004). The ethylene signaling pathway: new insights. Curr. Opin. Plant Biol. 7: 40–49.

Harmer SL, Panda S, Kay SA (2001) Molecular bases of circadian rhythms. Ann. Rev. Cell Dev. Biol. 17: 215–253.

Más P (2005) Circadian clock signaling in *Arabidopsis thaliana:* from gene expression to physiology and development. Int. J. Dev. Biol. 49: 491–500.

Mattoo AK, Suttle JC (1991) The plant hormone ethylene. Boca Raton: CRC Press, Inc.

McClung CR (2006) Plant circadian rhythms. Plant Cell 18: 792–803.

Nozue K, Maloof JN (2006) Diurnal regulation of plant growth. Plant Cell Env. 29: 396–408.

Salome PA, McClung CR (2005) The *Arabidopsis thaliana* clock. J. Biol. Rhythms. 2004 (5): 425–435.

Schaffer R, Ramsay N, Smach A, Corden S, Putterill J, Carre IA, Coupland G (1998) The *late elongated hypocotyls* mutation of Arabidopsis disrupts circadian rhythms and the photoperiodic control of flowering. Cell 93: 1219–1229.

Smalle J, Haegman M, Kurepa J, Van Montagu M, Van Der Straeten D (1997) Ethylene can stimulate arabidopsis hypocotyl elongation in the Light. Proc. Natl. Acad. Sci. USA 94: 2756–2761.

Tabibian-Keissar H (2005) The role of CCA1 gene in development processes in Arabidopsis plants. Ph.D. thesis Bar-Ilan University, Ramat-Gan, Israel. (Abs. in english).

Thain SC, Vandenbussche F, Laarhoven LJ, Dowson-Day MJ, Wang ZY, Tobin EM, Harren FJ, Millar AJ, Van Der Straeten D. (2004) Circadian rhythms of ethylene emission in Arabidopsis. Plant Physiol. 136(3): 3751–61.

Young MW, Kay SA (2001) Time zones: a comparative genetics of circadian clocks. Nat. Rev. Genet. 2: 702–715.

Role of cell wall hydrolases on ethylene-induced abscission of *Dendrobium* inflorescences

Ketsa, S.* and Bunya-atichart, K.

Department of Horticulture, Kasetsart University, Bangkok 10900, Thailand. (*Corresponding author: agrsck@ku.ac.th)

Abscission is preceded by the production of an abscission zone or layer at the base of organ. Abscission zone cells are not greatly different anatomically from the surrounding cells, but typically they are smaller and less vacuolated. Anatomical studies reveal almost complete disappearance of the middle lamella and extensive swelling and disorganization of the microfibrils in the primary wall in abscission zone cells. This changes result from the combined activities of several degrading enzymes. Several of these enzymes have been shown to be regulated by ethylene (Sexton and Roberts, 1982). In our previous study, we have shown that open flowers of *Dendrobium* cv Miss Teen inflorescences did not abscise whereas flower buds considerably abscised in response to ethylene treatment (Bunya-atichart *et al.*, 2006). In the present study, we report activities of cell wall degrading enzymes in the abscission zone of of *Dendrobium* cv Miss Teen inflorescences in response to ethylene.

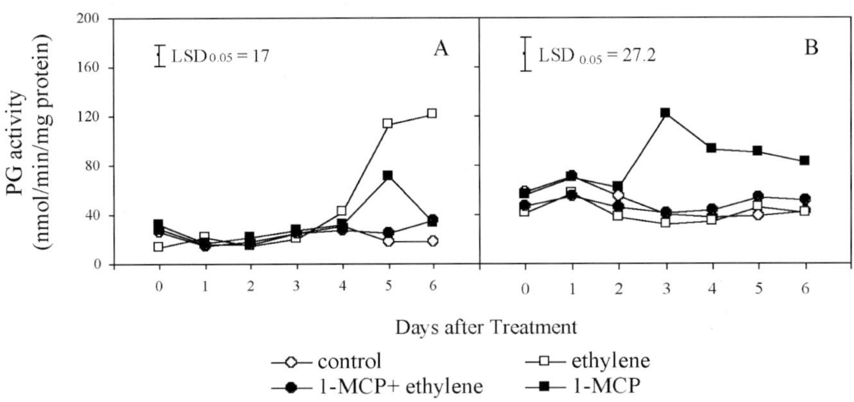

Fig. 1. PG activity of flower buds (**A**) and open flowers (**B**) in abscission zone of *Dendrobium* cv Miss Teen inflorescences following 24 h ethylene treatment. Rapid flower bud abscission occurred at day 5.

A. Ramina et al. (eds.), Advances in Plant Ethylene Research: Proceedings of the 7th International Symposium on the Plant Hormone Ethylene, 293–295.

The applied ethylene (0.4 μl/l for 24 h at 25°C) greatly hastened abscission of flower buds but had no effect on open flowers. Flower fall was completely inhibited by 1-methylcyclopropene (1-MCP, 0.5 μl/l for 3 h at 25°C) (Ketsa and Rungruchkanont, 2006). This indicated that ethylene sensitivity in orchid flowers depends on developmental stages (Abeles *et al.*, 1992). Ethylene treatment also increased activities of polygalacturonase (PG) (Fig. 1A) and cellulase (β-1,4-glucanase) (Fig. 2A) in the abscission zone of flower buds concomitant with abscission but not in open flowers (Figs. 1B, 2B). 1-MCP alone increased PG activity in the abscission zone of both flower buds and open flowers (Fig. 1) but 1-MCP inhibited an increase in cellulase activity in the abscission zone of flower buds induced by ethylene treatment (Fig. 2A). This suggested that ethylene may induce separation through cellulase (Brown, 1997). Ethylene treatment did not increase pectin methylesterase activity in the abscission zone of both flower buds and open flowers (data not shown). It is concluded that cellulase plays an important role in the separation mechanism of the abscission zone in ethylene-induced abscission of *Dendrobium* cv Miss Teen inflorescences.

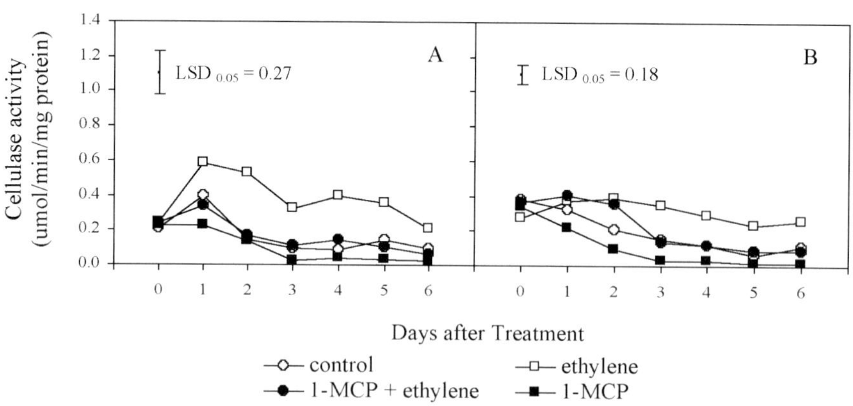

Fig. 2. Cellulase activity of flower buds (**A**) and open flowers (**B**) in abscission zone of *Dendrobium* cv Miss Teen inflorescences following 24 h ethylene treatment. Rapid flower bud abscission occurred at day 5.

Acknowledgements

The research was financially supported by the Thailand Research Fund.

References

Abeles FB, Morgan PM, Saltveit ME (1992) Ethylene in Plant Biology. Academic Press, San Diego

Brown KM (1997) Ethylene and abscission. Physiol Plant 100:567–576

Bunya-atichart K, Ketsa S, Van Doorn WG (2006) High floral bud abscission and lack of open flower abscission in *Dendrobium* cv Miss Teen: rapid reduction of ethylene sensitivity in the abscission zone. Funct Plant Biol 39:539–546

Ketsa S, Rungruchkanont K (2006) Effect of inhibitors of auxin and ethylene on abscission development of *Dendrobium* inflorescences. Published in this proceedings

Sexton R, Roberts JA (1982) Cell biology of abscission. Annu Rev Plant Physiol 33:133–162

Effect of inhibitors of auxin and ethylene on abscission development of *Dendrobium* inflorescences

Ketsa, S.* and Rungruchkanont, K.

Department of Horticulture, Kasetsart University, Bangkok 10900, Thailand. (*Corresponding author: agrsck@ku.ac.th)

Ethylene and auxin (IAA) are important regulators of abscission. IAA retards, whilst ethylene is a potent accelerator of the process. The general rule portrays that provided the flux of IAA to the abscission zone region is maintained; the cell separation is inhibited and abscission does not occur (Addicott, 1982; Brown, 1997). The antagonistic relationship between ethylene and auxin is evident in the general interaction between these two hormones. The auxin status of the abscission zone controls the sensitivity to ethylene; thus, any factor that affects the supply of auxin to the zone will also affect the sensitivity to ethylene. On the other hand, ethylene is a potent inhibitor of auxin transport and may elevate sensitivity of the zone to itself by enhancing destruction, increasing conjugation and interfering with transport of auxin (Beyer and Morgan, 1971; Beyer, 1973). Our previous study has shown that open flowers of *Dendrobium* cv Miss Teen inflorescences did not abscise whereas flower buds showed abscission in response to ethylene treatment (Bunya-atichart *et al.*, 2006). Our study was set up to test if endogenous auxin prevents normal abscission of open flowers of *Dendrobium* cv Miss Teen and decreases sensitivity of the abscission zone to ethylene.

We applied inhibitors of auxin action, CMPA [2-(4-chlorophenoxy)-2-methyl propionic acid], and auxin transport, TIBA (2,3,5-triiodobenzoic acid) as lanolin paste to open flowers prior to ethylene treatment. Ethylene treatment had no effect on abscission of open flowers without auxin inhibitors (Fig. 1 right). Open flowers treated with auxin inhibitors but without ethylene treatment showed 80–100% abscission (Fig. 1 left). This suggested inhibitors of auxin action and transport may increase sensitivity of open flowers to ethylene resulting in abscission. On the other hand, endogenous auxin may move from open flowers to the abscission zone at the base of pedicels and decrease sensitivity to ethylene resulting in no abscission. Fumigation with 1-methylcyclopropene (1-MCP), an inhibitor of ethylene action (Blankenship and Dole, 2003), delayed abscission of

A. Ramina et al. (eds.), Advances in Plant Ethylene Research: Proceedings of the 7th International Symposium on the Plant Hormone Ethylene, 297–299.

open flowers treated with TIBA and prevented abscission of open flowers treated with CMPA (Fig. 2 right). This suggested that auxin antagonists acted through ethylene action. Based on our results, we conclude that (a) endogenous auxin prevents flower abscission induced by ethylene, and (b) that the effect of anti-auxin compounds seems related to that of ethylene.

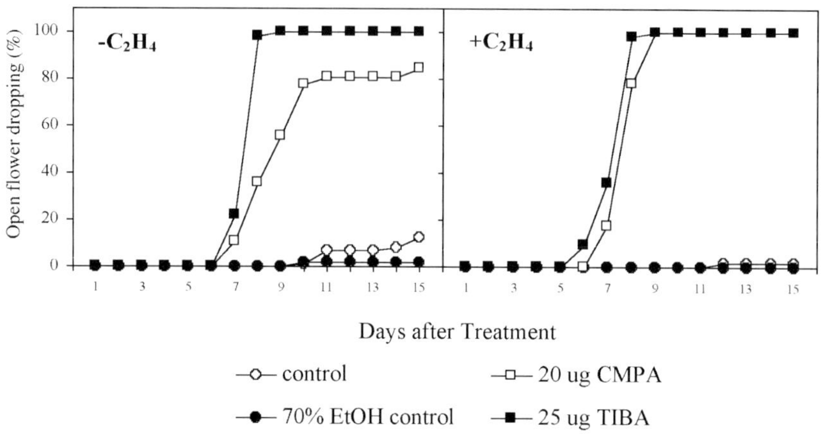

Fig. 1. Open flower dropping of *Dendrobium* cv Miss Teen inflorescences as affected by auxin inhibitors with and without ethylene treatment.

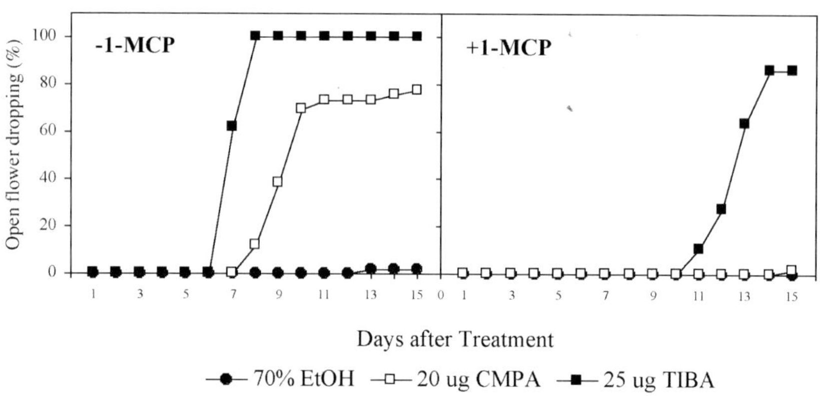

Fig. 2. Open flower dropping of *Dendrobium* cv Miss Teen inflorescences as affected auxin inhibitors with and without 1-MCP treatment.

Acknowledgements

The research was financially supported by the Thailand Research Fund.

References

Addicott FT (1982) Abscission. University of California Press, Berkeley

Beyer EM, Morgan PW (1971) Abscission: the role of ethylene modification of auxin transport. Plant Physiol 48: 208–212

Beyer EM (1973) Abscission: support for the role of ethylene modification of auxin transport. Plant Physiol 52: 1–5

Blankenship SM, Dole JM (2003) 1-Methylcyclopropene: a review. Postharvest Biol Technol 28: 1–25

Brown KM (1997) Ethylene and abscission. Physiol Plant 100: 567–576

Bunya-atichart K, Ketsa S, Van Doorn WG (2006) High floral bud abscission and lack of open flower abscission in *Dendrobium* cv Miss Teen: rapid reduction of ethylene sensitivity in the abscission zone. Funct Plant Biol 39: 539–546

Pre-harvest application of proprietary elicitor delays fruit senescence

Linden, J.C.[1] and Stoner, R.J.[2]

[1]Department of Chemical and Biological Engineering, Colorado State University, Fort Collins, CO 80523 USA. [2]AgriHouse Inc., Berthoud, CO 80513 USA. www.agrihouse.com. (Corresponding author: jlinden@colostate.edu)

The proprietary elicitor, YEA!, is an organic patented material derived from exoskeletons of crustaceans. YEA! appears to be responsible for actually reducing ethylene biosynthesis. This was tested using triple response assays on etiolated *Arabidopsis thaliana* seedlings, which is conducted using seeds germinated in the dark. Normally elongated hypocotyls (stems) are severely reduced in length by low concentrations of ethylene. When the seeds were germinated on agar medium containing 0.1 mg/mL YEA!, the results were the same as controls and indicated ethylene was not produced by the seedlings. Ethylene is a plant hormone that is associated with senescence. Sorenson compared YEA! with a water control by spraying the leaves once and two applications to the soil around orange trees several weeks preceding harvest. Fruit from treated trees was picked later than from untreated control trees, as abscission was delayed. In addition, post-harvest gassing procedure with 10 ppm ethylene gas for 4 days at 20°C was more severe than normally used on picked citrus to induce degreening.

Data in Fig. 1 represent citrus treated with YEA! in comparison with fruit picked from trees treated in an identical manner, except with water. Application of YEA! results in the delay of fruit sensecence, and gassed fruit does not exhibit expected signs of aging, as demonstrated by lower quantity of ethanol produced during a 3 month storage period. Because the fruit treated with YEA! requires more severe than normal gassing with ethylene to initiate post-harvest degreening, it may be concluded that ethylene biosynthesis is reduced by YEA! application in citrus.

A. Ramina et al. (eds.), Advances in Plant Ethylene Research: Proceedings of the 7th International Symposium on the Plant Hormone Ethylene, 301–302.

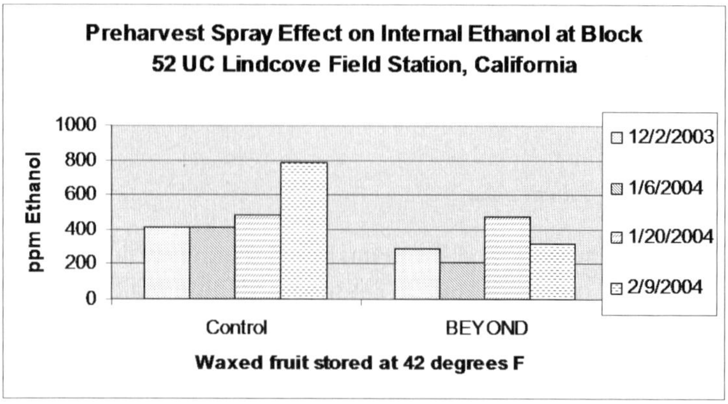

Fig. 1. Ethanol production internalized in citrus fruit at four sampling dates during three months of storage following pre-harvest treatment of trees with YEA! is lower than in fruit from water-treated control trees and indicates lower rates of fruit senescence and better fruit quality with YEA! treatments. Data kindly provided by David Sorenson: david.sorenson@fruitgrowers.com. The Use of Preharvest Plant Elicitors to Improve PostHarvest Quality of Fruit, FGS Packing Services for Sunkist, September 16, 2004.

The triple response measurements represented in Fig. 2 indicate that YEA! does not stimulate ethylene production.

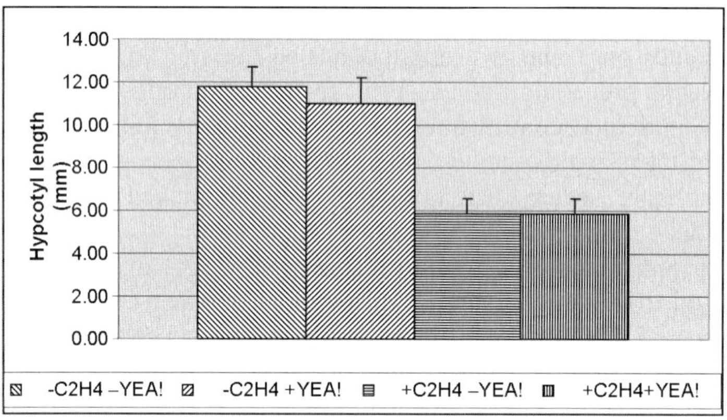

Fig. 2. Triple response hypocotyl elongations of etiolated seedlings of *Arabidopsis thaliana* exposed to zero or 0.1 mg/mL YEA! and zero or 1 μL/L C_2H_4 in four combinations from image analysis of scanned materials. Left: minus C_2H_4 minus YEA!; second: minus C_2H_4 plus YEA!; third: plus C_2H_4 minus YEA!; and right: plus C_2H_4 minus YEA!. The error bars represent standard deviations from two independent studies of each treatment consisting of 15–20 analyses each.

Calcium signaling in water stress-induced leaf abscission in citrus plants

Agustí, J., Conesa, A., Cercós, M., Talón, M. and Tadeo, F.R.*

Instituto Valenciano de Investigaciones Agrarias - Centro de Genómica. Apartado Oficial. 46113 Moncada (Valencia) Spain. (*Corresponding author: ftadeo@ivia.es)

Abscission is a widespread physiological process that allows the natural separation of organs from the parent plant. Abscission of plant organs takes place through a highly coordinated sequence of biochemical events that leads to wall breakdown in a discrete group of cells located in predictable positions in the plant known as abscission zones (AZs). Leaf abscission in citrus plants occurs during rewatering after a period of water stress in a process controlled by abscisic acid-directed ethylene production (Gómes-Cadenas *et al.*, 1996). Plants need calcium (Ca^{2+}) to strengthen cell walls and Ca^{2+} deficiencies have been associated with abscission (Addicott, 1982). On the other hand, treatments with Ca^{2+} have showed the ability of this cation to retard abscission (Poovaiah, 1973). Beside this, Ca^{2+} is a fundamental component of cellular signaling as well and it has been established that Ca^{2+}-triggered events are critical for both normal cellular activity and for adapted stress responses (Sanders *et al.*, 2002).

Abscission is a common response of plants under stress conditions and our goal in this study has been to detect differentially expressed genes encoding Ca^{2+}- and CaM-binding proteins in laminar-AZs of Clementine mandarin (*Citrus clementina*) plants subjected to a water stress/rehydration cycle. To this purpose, we have used a macroarray-based approach. A set of nylon filters was prepared with 381 cDNA clones, including 78 cDNAs encoding Ca^{2+}- and CaM-binding proteins, arrayed in quadruplicate. The cDNA probes spotted onto nylon filters were coming from two Clementine mandarin cDNA libraries (SSH and normalized full-length libraries, Forment *et al.*, 2005; Terol *et al.*, 2005). RNA samples for macroarray hybridization were prepared from petioles and laminar-AZ-enriched petioles harvested at the onset (0 h) and at the end of the water stress period (24 h) and after 1 and 6 h of rehydration. Hybridization data were normalized using previously detected housekeeping genes. Probes showing significant differential gene expression between samples were identified using the

A. Ramina et al. (eds.), Advances in Plant Ethylene Research: Proceedings of the 7th International Symposium on the Plant Hormone Ethylene, 303–304.

linear models in microarrays (LIMMA) library of the Bioconductor software package (Gentleman *et al.*, 2004). Differences in gene expression were considered to be significant when *P*-value was lower than 0.01 and induction or repression ratio was equal or higher than twofold. The macroarray results were verified by qRT-PCR using a set of six genes.

A 27% of the genes printed in the macroarray encoding Ca^{2+}- and CaM-binding proteins (21 out of 78 genes) appeared to be predominantly expressed in the laminar-AZ at the onset of the rehydration period. This set included genes encoding Ca^{2+} efflux carriers (2), calmodulin (CaM) and CaM-like proteins (3), Ca^{2+}-dependent protein kinases (CDPKs, 2), endoplasmic reticulum Ca^{2+}-binding proteins (3), Ca^{2+}-binding EF-hand-containing proteins (2) and CaM-binding proteins (9). The expression of Ca^{2+}-transporting ATPases (ACA8- and ECA1-like genes) and two CDPKs were transiently upregulated early in rehydration. Several genes encoding CaM-binding proteins were transiently or continuously upregulated up to 6 h after rehydration highlighting Ca^{2+}-dependent physiological processes occurring during citrus leaf abscission (DGK, phospholipid biosynthesis; NDK, pyrimidine nucleotide metabolism; GAD, GABA metabolism; RbohB, oxidative burst; five genes with unknown functions). These results suggest that a particular portion of the cellular Ca^{2+} signaling network would be active in citrus laminar-AZ during early stages of rehydration and therefore during the onset of leaf abscission.

References

Addicott FT. (1982) Abscission. University of California Press, Los Angeles

Forment J *et al.* (2005) Development of a citrus genome-wide EST collection and cDNA microarray as resources for genomic studies. Plant Mol Biol 57: 375–391

Gentleman RC *et al.* (2004) Bioconductor: open software development for computational biology and bioinformatics. Genome Biol 5: R80

Gómez-Cadenas A *et al.* (1996) Leaf abscission induced by ethylene in water-stressed intact seedlings of cleopatra mandarin requires previous abscisic acid accumulation in roots. Plant Physiol 112: 401–408

Poovaiah BW, Leopold AC. (1973) Inhibition of abscission by calcium. Plant Physiol 51: 848–851

Sanders D *et al.* (2002) Calcium at the crossroads of signaling. Plant Cell 14: S401–S417

Terol J *et al.* (2005) Analysis of citrus ESTs from standard and normalized full-length cDNA libraries. XIII Plant and Animal Genome Congress, San Diego

Control of pistillate flower abortion in 'Serr' walnuts in Chile by inhibiting ethylene biosynthesis with AVG

Lemus, G.[1], González, C.[1] and Retamales, J.[2, 3]*

[1]INIA-La Platina, Chile. Santa Rosa 11.610, Santiago, Chile; [2]Valent Bio-Sciences Corporation, Field R&D, Av. Kennedy 5735, Las Condes, Santiago, Chile; [3]Universidad de Chile, Facultad de Ciencias Agronómicas. (*Corresponding author: julio.retamales@valent.com)

Summary

Pistillate flower abortion (PFA), induced by excessive pollen load on female flowers and leading eventually to their abscission, affects severely fruit set and potential fruit yield in 'Serr' walnut variety. Such variety, owing to its nut quality, is extensively planted in Chile and in other walnut-producing areas in the world (California). During two seasons, trials were carried out to define possibilities of control of PFA by applying AVG (aminoethoxy-vinylglycine), an ethylene biosynthesis inhibitor. In the first season, AVG concentrations of 62.5, 125 and 250 mg/l were evaluated, using either 1000 or 2000 liters per hectare. Fruit set that season averaged 25% in the untreated controls. Application of AVG at 125 and 250 mg/l increased fruit set up to 80%, with AVG applied at 125 mg/l (i.e. 125 ppm of AVG) and with a spray coverage of 1000 liters per hectare being rated as the most appropriate treatment. During 2005/2006 season, AVG applications using 125 ppm of AVG were performed in several orchards. Natural fruit set in the untreated controls ranged from 35 to 83%. Despite less incidence of PFA in the latter season as compared with the previous one, in most of the orchards AVG application led to increased fruit set, particularly when PFA was high, resulting, thus, in significant improvements in yield potential in such cases. These results are indicating that AVG can be a powerful tool to overcome PFA and subsequent yield losses in 'Serr' walnut orchards under Chilean Central Valley conditions.

A. Ramina et al. (eds.), Advances in Plant Ethylene Research: Proceedings of the 7th International Symposium on the Plant Hormone Ethylene, 305–307.

1. Introduction and Methodology

'Serr' walnut variety presents pistillate flower abscission (PFA), induced by excessive pollen load. This problem is common to other walnut varieties, but less intense. It has been postulated that when pistillate flowers receive high pollen load, this results in high rate of ethylene biosynthesis inducing flower/fruitlet abscission. Therefore, it seemed relevant to assay effects of the ethylene biosynthesis inhibitor AVG (aminoethoxyvinylglycine) on PFA. Consequently, the commercial formulation ReTain (a.i. 15% AVG, Valent BioSciences Corp.) was studied as a tool to increase walnut fruit set from season 2004/2005 onwards. In 2004, rates of 62.5, 125 and 250 mg/l of AVG were applied at 20% female flower receptivity stage, using either 1000 or 2000 liters per hectare on three adult 'Serr' orchards in the Chilean Central Valley area. Fruit set and nut quality were measured on treated trees, being compared with untreated ones as control (UTC). Subsequently, a rate of 125 ppm of AVG applied with 1000 liters per hectare was used in season 2005/2006.

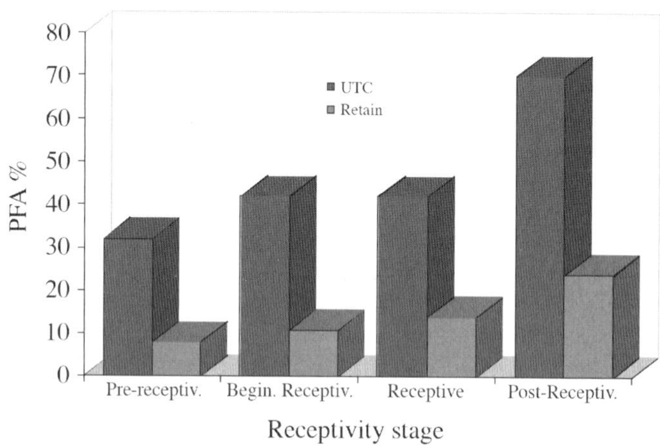

Fig. 1. Pistillate flower abortion (%) in different receptivity stages. Buin, Chile, 2005/2006. ReTain: applied with AVG, UTC (untreated control).

2. Results

In the season 2004/2005, AVG increased fruit set in average from 35 to 83%, according to the chemical's concentration. Although, control of PFA was obtained with AVG being applied in all flower stages (Fig. 1), owing that only a single treatment is desired, our data suggests that an early application is to be recommended. Further, AVG applications were effective in the entire tree canopy by decreasing PFA along all the positions on tagged branches, particularly on the apical and basal portions (data not shown). Moreover, AVG treatment was consistent in reducing PFA in orchards located in different places of the Chilean walnut production area (Central Valley), with 833 g/ha of the commercial product (i.e. 125 ppm of AVG) being suggested as an adequate concentration to improve fruit set on 'Serr'.

Effect of 1-MCP on quality and physiological changes in mandarins during degreening treatment

Salvador, A.*, Carvalho, C.P., Monterde, A., Navarro, P. and Martínez-Jávega, J.M.

Centro de Tecnología Postcosecha, Instituto Valenciano de Investigaciones Agrarias, Apartado Oficial, 46113 Moncada, Valencia, Spain.
(*Corresponding author: asalvado@ivia.es)

Ethylene-induced degreening is the postharvest technique used to modify the external colour of the early citrus fruit, stimulating an adequate development of its yellow–orange external colour. This process accelerates senescence by increasing the respiration rate, transpiration and dryness of button ends. The most common physiological disorders induced by degreening treatment are the rind aging as well as calyx browning and abscission. This effect could be minimized with the reduction of the exposure time to ethylene and avoiding doses of ethylene and temperatures higher to the recommended ones. Cronjé *et al.* (2005) states that the abscission process can be positively manipulated by application of 1-methylcyclopropene (1-MCP), an ethylene action inhibitor (Blankenship and Dole, 2003), but its efficacy depends on factors such as degreening, timing and concentration of 1-MCP application.

Mandarins cv 'Clemenules' and 'Oronules' were treated with 1-MCP at concentration of 0.5 $\mu L\ L^{-1}$ before or after degreening treatment with ethylene and then maintained at 20°C for up 8 days.

Application of 1-MCP before degreening treatment resulted in a desiccating effect of the rind close to stem and an important browning of the calyx was observed, although the calyxes were not abscised (Fig. 1). When the 1-MCP was applied after degreening treatment, a significant increase in calyx abscission was observed compared to control, this effect being lower when 1-MCP was applied later. An important increase in ethylene production was observed just after 1-MCP application (Fig. 2). 1-MCP also induced an increase in respiration rate (Data not shown). 1-MCP did not produce important changes in parameters such as colour index, maturity index and weight loss. The lower flavour values were presented by the fruit treated with 1-MCP when it was applied close to degreening treatment (Data not shown).

A. Ramina et al. (eds.), Advances in Plant Ethylene Research: Proceedings of the 7th International Symposium on the Plant Hormone Ethylene, 309–310.

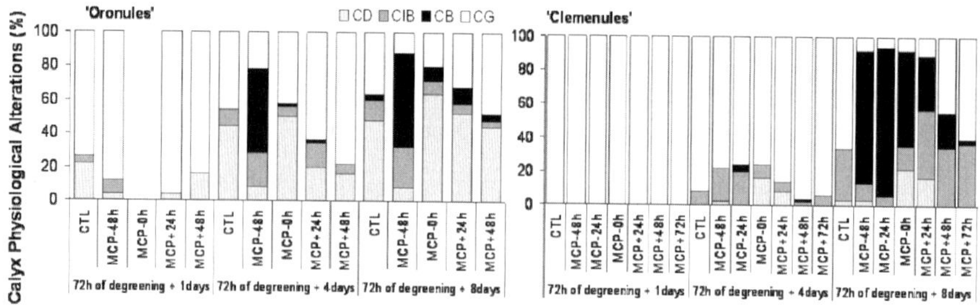

Fig. 1. Effect of 1-MCP on the incidence of calyx physiological alterations (CD – calyx drop; CIB – calyx initiating browning; CB – calyx browning; CG – calyx green) caused by degreening treatment in 'Oronules' and 'Clemenules' mandarins.

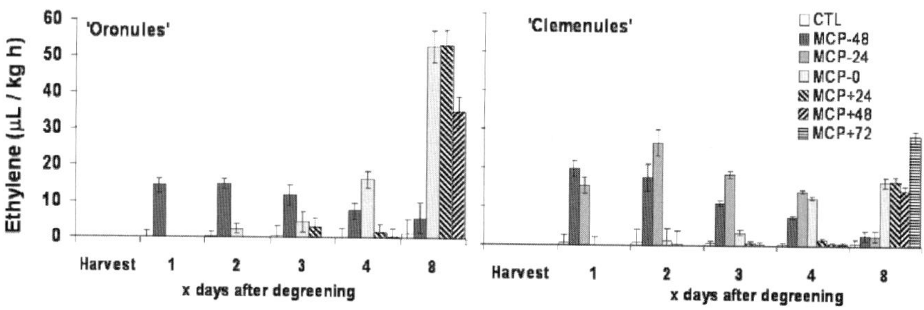

Fig. 2. Effect of 1-MCP treatment (before or after degreening treatment) on ethylene production in degreened 'Oronules' and 'Clemenules' mandarins. Vertical bars indicate LSD ($P < 0.05$).

References

Blankenship S; Dole JM (2003) 1-Methylcyclopropene: a review. Postharvest Biology and Technology, 28: 1–25.

Cronjé PJR; Crouch EM; Huysamer M (2005) Postharvest calyx retention of citrus fruit. Acta Horticulturae, 682, 1: 369–374.

Ethylene sensitivity and postharvest performance of *Azorina vidalii* (Campanulaceae)

Dalla Guda, C., Semeria, L. and Farina, E.*

CRA – Istituto Sperimentale per la Floricoltura – 18038 Sanremo (Italy).
(*Corresponding author: tecniche@istflori.it)

Azorina vidalii (Campanulaceae) is an ornamental shrub native to Azores that produces long stems bearing many pink waxy bells on the terminal raceme (Crook, 1951). This study explored the sensitivity to exogenous ethylene of *Azorina vidalii* flowers in relation to their developmental stage in conformity with an approach to select new crops for ornamental purposes (Vonk Noordegraaf, 2000).

Different exposure times (0–3–8–24 h) to the same initial ethylene level (1 p.p.m.) were provided on cut flowering stems maintained at 20°C during the hormone treatments and the subsequent vase life. Data on each flower in 12 stems were collected daily in order to determine flower development and/or ethylene-induced damages.

The exposure to ethylene modulates the vase life of the flowering stems of *Azorina vidalii*. Dramatic effects were already observed at 3–8 h exposure to 1 p.p.m. ethylene. In the control stems flower life of fully open flowers (O) was 11–12 days whereas the ethylene-treated ones collapsed quickly. The faster (1–6 days) was the senescence, the longer the exposure to ethylene (Fig. 1). The flowers at the first stage of anthesis when the lobes of the corolla start to open (HO) or the coloured buds (CB) showed the same trend. On the contrary, the green buds, in more immature developmental stage (GB), were injured in a less dramatic way or decayed in a longer period. In the control about 93% of the flower buds (HO–CB and GB stage) reached the fully opening stage while percentages of 28%, 44% or 62% were recorded in the treated ones after ethylene exposure for 24, 8 or 3 h, respectively. Stem vase life was reduced from 11–12 to 1–2 days by increasing the exposure time to the hormone (Fig. 2).

For a commonly accepted sensitivity test it should be necessary to consider *Azorina vidalii* as sensible to ethylene like other Campanulas (Kato *et al.*, 2002). The endogenous ethylene production is proved to be a process of no or little importance in the senescing cut stems of *Azorina vidalii*.

A. Ramina et al. (eds.), Advances in Plant Ethylene Research: Proceedings of the 7th International Symposium on the Plant Hormone Ethylene, 311–313.

From a practical point of view (Gibson *et al.*, 2000; Woodson *et al.*, 2003), it is necessary and sufficient to harvest *Azorina* stems in an early stage of flower opening or in a stage preliminary to bud opening and store them in an ethylene-free atmosphere.

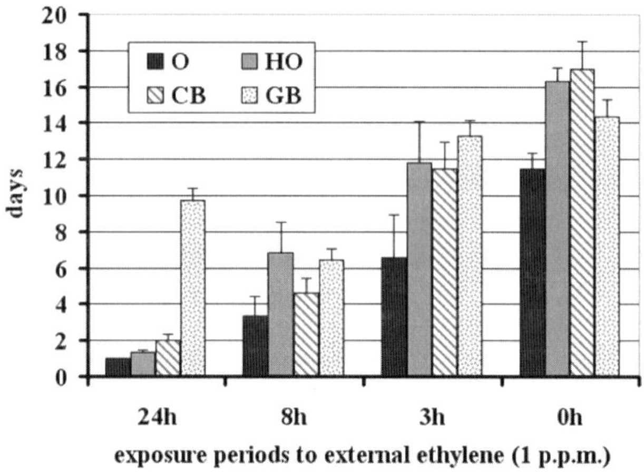

Fig. 1. *Azorina vidalii.* Flower damage appearance time (days) related to the exposure to 1 p.p.m. external ethylene in controlled environment (20°C). Initial flower stage: O = fully open flowers, HO = half-open flowers, CB = flower buds start to colour GB = immature green flower buds.

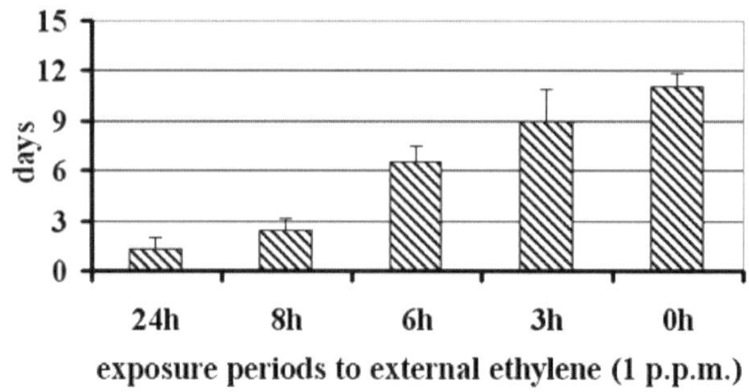

Fig. 2. Stem vase life of *Azorina vidalii* related to different exposure periods to external ethylene (1 p.p.m.).

References

Crook CH (1951) Campanulas. Country Life Ltd, London; Charles Scribner's Sons, New York

Gibson JL, Whipker BE, Blankenship S, Boyette M, Creswell T, Miles J, Peet, M (2000) Ethylene: Sources, Symptoms and Prevention for Greenhouse Crops. Horticulture Information Leaflet 530 – New 7:1–8

Kato M, Shimizu H, Onozaki T, Tanikawa N, Ikeda H, Hisamatsu T, Ichimura K (2002) Role of ethylene in senescence of pollinated and unpollinated *Campanula medium* flowers. J Jpn Soc Hortic Sci 71:385–387

Vonk Noordegraaf C (2000) An approach to select new ornamental crops. Acta Hortic 541:75–78

Woodson WR, Jones ML (2003) In search of eternal youth: the delay of postharvest senescence. Acta Hortic 624:305–314

Effects of exogenous ethylene on alstroemeria floral sticks

Berger, H.*, Galletti, L. and Villaseca, M.

Centro de Estudios Postcosecha (CEPOC). Facultad de Ciencias Agronómicas, Universidad de Chile, Santiago, Chile. (*Corresponding author: hberger@uchile.cl)

Alstroemeria belongs to Alstroemeriaceae family, and currently is extensively utilized as cut flower. Considering that it is an ethylene-sensitive flower, low concentration of this gas may deteriorate floral sticks, diminishing its commercial value. The effect of different ethylene concentrations on flower senescence of Alstroemeria var. Irena was evaluated. Flowers were harvested from a commercial glasshouse in three stages of development: II, III, and IV (Schiapacasse, 1988 modified) (Fig. 1), and exposed to different concentration of exogenous ethylene (Table 1). Respiratory rate was measured and floral sticks deterioration was evaluated until its senescence expressed as tepals abscission and leaves yellowing. The floral sticks monitored for respiration rate were placed in individual glass containers. Carbon dioxide was analyzed with gas chromatographs, Hewlett Packard 5890 Series II equipped with TCD detector. Daily, during 12 days leaves color (L*, a*, and b*) and respiration rate were measured on three floral sticks per treatment. The statistical design was based on a complete randomized factorial 3*3, with three replicates. Dates were analyzed by ANOVA and the differences were separated by Tukey's test. Exposure to concentrations of ethylene was directly related to initial respiratory rate for stage II, showing earlier and higher values of respiratory rate; in subsequent days, respiration increased significantly, reducing its vase life. In floral sticks harvested at stage IV, ethylene application accelerates the final respiratory rate decline. Basal leaves color changes observed in floral sticks were detected in all the stages, earlier and more intense in the first stages of development and with higher ethylene concentration. Ethylene had a negative effect on flower quality harvested in all stages of development, showing a rapid tepal abscission reducing its vase life. This effect was accentuated in advanced stages of development.

A. Ramina et al. (eds.), Advances in Plant Ethylene Research: Proceedings of the 7th International Symposium on the Plant Hormone Ethylene, 315–316.

Fig. 1. Stages of floral development.

The vase life was on average up to 8 days for the control, while the application of ethylene reduces this period to 4 and 5 days for the high and low concentrations, respectively.

Table 1. Treatments.

Floral sticks stage	C_2H_4 (ppm)		
	10.5	2.0	0
II	T1	T2	T3
III	T4	T5	T6
IV	T7	T8	T9

References

Berger, H., Galletti, L., and Villaseca, M. (2005) Determinación de la tasa respiratoria de Alstroemeria var. Irena. Avances en Horticultura Ornamental-Universidad Austral, Valdivia, Chile. 166 p.

Collier, D. (1997) Changes in respiration, protein and carbohydrates of tulip tepals and alstroemeria petals during development. J. Plant Physiol. 150: 446–451.

Halevy, A and Mayak, S. (1979) Senescence and postharvest physiology of cut flowers Part 1. Hortic. Rev. 1: 204–236.

Manzur, M. (1997) Bioactive agents from dryland biodiversity of Latin America "Bioprospección y Conservación de la Biodiversidad en Chile", Noticiero 5(2).

Schiapacasse, F. (1988) Conservación de Alstroemeria spp. Indices de madurez de cosecha y soluciones azucaradas. Tesis Ing. Agr. P. Universidad Católica de Chile. 83 p.

A massive approach to identify genes involved in ethylene induced abscission of apple fruitlets

De Franceschi, F.[1], Dal Cin, V.[1], Botton, A.[1], Caniato, E.[1], Barbaro, E.[2], Velasco, R.[2] and Ramina, A.[1]*

[1]Department of Environmental Agronomy and Crop Science, University of Padova, Viale dell'Università 16, I-35020 Legnaro (Padova), Italy. [2]Experimental Institute for Agriculture, Via Mach 2, San Michele all'Adige, I-38010 Trento, Italy. (*Corresponding author: angelo.ramina@unipd.it)

Abscission was studied in immature apple fruits (cv Golden Delicious) during the physiological drop. Fruitlet shedding is preceded by a stimulation of ethylene biosynthesis and a gain in sensitivity to the hormone. Experiments were performed on abscising fruitlet (AF) and non-abscising fruitlet (NAF) populations. AF were obtained from lateral fruitlets of trees sprayed with benzylaminopurine (BAP) at 200 ppm, 17 days after petal fall (APF) with a fruit cross diameter of about 10–12 mm. NAF were obtained from central flowers grown in clusters where all the lateral flowers had been removed at bloom. cDNA-AFLP with 53 primers combinations was used to identify putative genes differentially induced by ethylene and involved in abscission and senescence processes. 169, 70 and 39 differentially expressed sequences were isolated from cortex, peduncle and seeds, respectively. All sequences were classified as up- or down-regulated in relation to abscission; according to BLASTX, 85% of these sequences showed significant homology with known genes, 8% were unknown and 7% showed no homology. All clones were classified according to the Gene Ontology criteria by cellular component, process and function. Expression analyses of genes putatively involved in signal transduction and sugar metabolism were carried out by semiquantitative RT-PCR on agarose gel.

Through the gene ontology classification, we identified those cellular compartments, functional groups and processes that were more abundant and so, putatively, involved in abscission. More clones were up- than down-regulated, except for peduncles where an opposite behaviour was found. In seeds the difference between clones up- and down-regulated was much more relevant with the previous ones being much more frequent, even if the number of genes analysed was not very large. From our analy-

A. Ramina et al. (eds.), Advances in Plant Ethylene Research: Proceedings of the 7th International Symposium on the Plant Hormone Ethylene, 317–318.

sis the involvement of plastids, endomembranes and mitochondria seemed very important.

With functional analysis we found several ESTs involved in enzymatic activities in different tissues: in particular genes involved in hydrolytic, oxydoreductasic and transferasic pathways were found. Among genes associated with hydrolytic activity, a high number of ESTs encode enzymes with proteolytic action putatively involved in proteolysis (such as ubiquitination), while enzymes with transferasic action are mainly kinases. Moreover, many ESTs encoding hormone, sugar and ion transporters especially in seed were found. In peduncles, up-regulated genes encoding regulation factors involved in transcription are present.

The regulation of protein metabolism appears to be the most significant process affected in all tissues during abscission. Signal transduction mechanisms mediated by auxin, ethylene or jasmonates are equally affected. This is confirmed by the fact that several enzymes with kinase activity responsible for phosphorylative events are involved.

The cDNA-AFLP approach and the annotation of genes according to the gene ontology criteria were used to identify metabolic activities and biological processes putatively involved in apple fruitlets abscission. At present, we are carrying out expression analysis of genes involved in sugar metabolism and transport and in signal transduction by semiquantitative RT-PCR.

Is ethylene directly involved in floral abscission in *Cucurbita pepo*?

Rosales, R.[1], García-Ayllón, S.[1], Jamilena, M.[2], Gómez, P.[3] and Garrido, D.[1]*

[1]Department of Plant Physiology, University of Granada, Fuentenueva S/N, 18071, Granada (Spain). [2]Department of Applied Biology (Genetic area). University of Almería. Escuela Politécnica Superior, 04120 Almería (Spain). [3]C.I.D.H. (Centro de Investigación y Desarrollo Hortícola) Autovía del Mediterráneo S/N, La Mojonera, Almería (Spain). (*Corresponding author: dgarrido@ugr.es)

1. Introduction

In zucchini (*Cucurbita pepo*), different cultivars present a distinctive floral abscission capacity. In greenhouses in the south-east of Spain, the summer conditions (high temperatures and long days) promote the production of bisexual flowers with an inhibition in the floral abscission during fruit formation ('sticky flower' syndrome) (Gòmez *et al.*, 2004). For the farmers, the abnormal abscission generates severe economical losses, since the flower has to be removed manually to avoid fruit rot. Ethylene is known to be involved in abscission mainly by increasing the synthesis of hydrolytic enzymes such as endo-1,4-β-glucanases (cellulases) and polygalacturonases (del Campillo, 1999; Brown, 1997). To study the implication of ethylene in floral organ abscission in zucchini, we have used as a plant material two cultivars with different floral abscission behaviour, Cavili (high percentage of 'sticky flower') and Cora (almost normal abscission). In these two cultivars we have measured ethylene production in female, bisexual and male flowers, during summer and winter conditions, moreover, we have analysed the polygalacturonase (PG) and cellulase activity accordingly with the viscosity method (Rejòn-Palomares *et al.*, 1996) in the abscission zone (AZ) of female and bisexual flowers, and in female flowers treated with ethylene.

A. Ramina et al. (eds.), Advances in Plant Ethylene Research: Proceedings of the 7th International Symposium on the Plant Hormone Ethylene, 319–321.

2. Results and Discussion

In all the cases, ethylene release (Fig. 1) was higher in Cora than in Cavili and higher in the female flowers than in the bisexual ones (with an abscission arrest). With respect to the enzymes involved in abscission, their activity was always higher in Cora female flowers. In female flowers, the levels of PG and cellulase activities always surpassed those of bisexual flowers with abnormal abscission (Table 1). Treatments with ethrel were able to trigger PG and cellulase activities in the AZ of zucchini female flowers in both cultivars (Fig. 2).

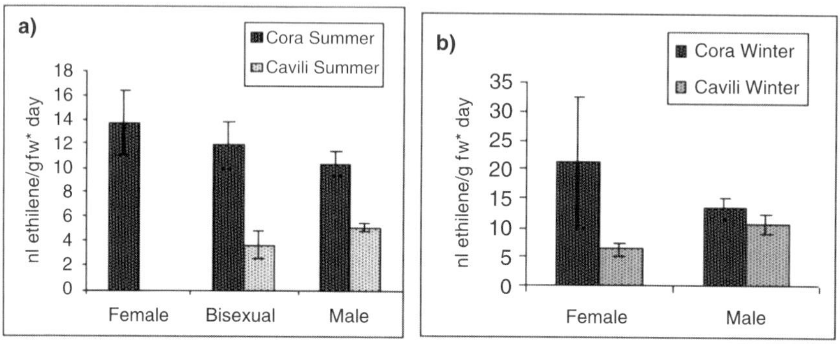

Fig. 1. Ethylene released in summer **(a)** and winter **(b)** conditions in Cora and Cavili flowers at the stage of anthesis.

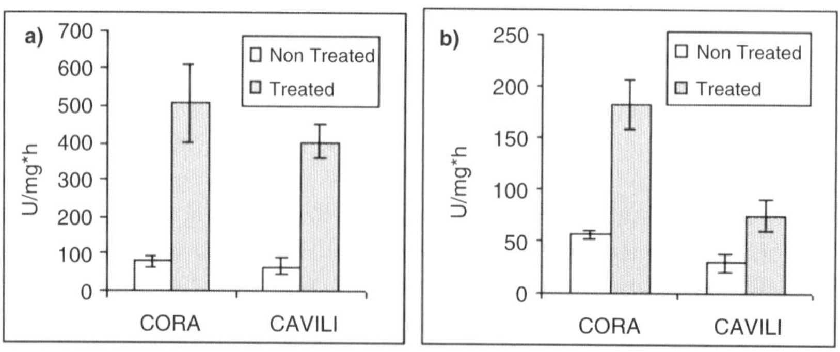

Fig. 2. Polygalacturonase **(a)** and cellulase **(b)** activities in the AZ of Cora and Cavili female flowers (2 days before anthesis) before and after ethrel treatment (0.12%).

Actually, we are performing expression analysis with the cDNA clones encoding for ACCoxidase, ACCsynthase, polygalacturonase and cellulase to confirm the data presented in this work.

3. Conclusion

Our results suggest that ethylene is necessary for normal flower abscission. The syndrome 'sticky flower' is related to a lower production of ethylene, along with an arrest of the hydrolytic enzymes activation.

Table 1. Polygalacturonase and cellulase activities.

	Polygalacturonase activity U/mg•h		**Cellulase activity U/mg•h**	
	AZ of mature female flowers	AZ of bisexual flowers. Sticky flower	AZ of mature female flowers	AZ of bisexual flowers. Sticky flower
CORA	377.94 ± 18	218.29 ± 40	152.91 ± 10.8	17.62 ± 1.9
CAVILI	116.89 ± 25.4	54.68 ± 71	19.75 ± 7.5	28.71 ± 10.2

Acknowledgements

This work was supported by a FPU grant and project AGL2005-0667700-CO2 (Ministerio de Educación y Ciencia, Spain).

References

Brown D.M. 1997. *Physiologia Plantarum* 100, 567–576.

del Campillo E. 1999. *Current Topics in Developmental Biology* 46, 39–61.

Gómez P. *et al.* 2004. Evaluation of flower abscission and sex expression in different cultivars of zucchini squash (*Cucurbita pepo*). A. Lebeda and H.S. Paris (Eds). Progress in Cucurbit Genetics and Breding Research.

Rejón-Palomares A, García-Garrido JM, Ocampo JA, García-Romera I. 1996. *Simbiosis* 21, 249–261.

6. ETHYLENE INVOLVEMENT IN BIOTIC AND ABIOTIC STRESSES

The role of ethylene in rhizobacteria-induced systemic resistance (ISR)

Pieterse, C.M.J.*, Van der Ent, S., Van Pelt, J.A. and Van Loon, L.C.

Section Phytopathology, Institute of Environmental Biology, Faculty of Science, Utrecht University, P.O. Box 800.84, 3508 TB Utrecht, The Netherlands. Phone +31 30 253 6887; Fax +31 30 251 8366. (*Corresponding author: C.M.J.Pieterse@bio.uu.nl)

1. Introduction

To protect themselves from disease, plants have evolved sophisticated defense mechanisms in which the signal molecules salicylic acid (SA), jasmonic acid (JA) and ethylene (ET) often play crucial role (Pieterse and Van Loon, 1999). Elucidation of signaling pathways controlling disease resistance is a major objective in research on plant–pathogen interactions. The capacity of a plant to develop a broad-spectrum, systemic acquired resistance (SAR) after primary infection with a necrotizing pathogen is well known and its signal transduction pathway extensively studied (Durrant and Dong, 2004). Plants of which the roots have been colonized by specific strains of nonpathogenic fluorescent *Pseudomonas* spp. develop a phenotypically similar form of protection that is called rhizobacteria-mediated induced systemic resistance (ISR) (Van Loon *et al.*, 1998). In contrast to pathogen-induced SAR, which is regulated by SA, rhizobacteria-mediated ISR is controlled by a signaling pathway in which ET and JA play key roles (Pieterse *et al.*, 1998). In the past decade, the model plant species *Arabidopsis thaliana* was explored to study the molecular basis of rhizobacteria-mediated ISR (Pieterse *et al.*, 2002). Here we review the current knowledge of the signal transduction steps involved in the ISR pathway that leads from recognition of the rhizobacteria in the roots to systemic expression of broad-spectrum disease resistance in above-ground foliar tissues.

A. Ramina et al. (eds.), Advances in Plant Ethylene Research: Proceedings of the 7th International Symposium on the Plant Hormone Ethylene, 325–331.

2. Rhizobacteria-Induced Systemic Resistance (ISR)

To study the signal transduction pathway of ISR, an Arabidopsis-based model system was developed. In this model system, the nonpathogenic rhizobacterial strain *Pseudomonas fluorescens* WCS417r is used as the inducing agent (Pieterse *et al.*, 1996). Colonization of Arabidopsis roots by ISR-inducing WCS417r bacteria protects the plants against different types of pathogens, including the bacterial pathogens *Pseudomonas syringae* pv. *tomato* (*Pst*) DC3000, *Xanthomonas campestris* pv. *armoraciae*, and *Erwinia carotovora* pv. *carotovora*, the fungal root pathogen *Fusarium oxysporum* f.sp. *raphani*, the fungal leaf pathogens *Alternaria brassicicola* and *Botrytis cinerea*, and the oomycete pathogen *Hyaloperonospora parasitica* (Pieterse *et al.*, 1996; Ton *et al.*, 2002a).

Research on the molecular mechanism of ISR was initially focused on the role of pathogenesis-related (PR)-proteins, as the accumulation of these proteins was considered to be strictly correlated with induced disease resistance (Van Loon *et al.*, 2006). However, Arabidopsis plants expressing WCS417r-mediated ISR showed enhanced resistance against *F. oxysporum* and *Pst* DC3000, but this did not coincide with the activation of the SAR marker genes *PR-1*, *PR-2*, and *PR-5* (Pieterse *et al.*, 1996). Determination of SA levels in ISR-expressing Arabidopsis plants revealed that ISR is not associated with increased accumulation of SA (Pieterse *et al.*, 2000). Moreover, WCS417r-mediated ISR was expressed normally in SA-nonaccumulating Arabidopsis NahG plants (Pieterse *et al.*, 1996), and in the SA biosynthesis mutants *eds5/sid1* and *sid2* (Pieterse *et al.*, 2002). This led to the conclusion that WCS417r-mediated ISR is a SA-independent resistance mechanism, and that WCS417r-mediated ISR and pathogen-induced SAR are regulated by distinct signaling pathways.

3. Genetic Dissection of the ISR Pathway in Arabidopsis

Since SA was not involved in WCS417r-elicited ISR, the Arabidopsis JA-response mutants *jar1*, *coi1*, and *eds8*, and the ET-response mutant *etr1* were tested for their ability to express ISR. None of these mutants were unable to mount resistance against *Pst* DC3000 after colonization of the roots by WCS417r (Pieterse *et al.*, 1998; Ton *et al.*, 2002b; unpublished data), indicating that ISR requires responsiveness to both JA and ET (Fig. 1). To further elucidate the role of ET in the ISR signaling pathway, a large set of well-characterized ET-signaling mutants was analyzed. None of these mutants showed an ISR response against *Pst* DC3000 after colonization of the roots by WCS417r (Knoester *et al.*, 1999). These results confirmed that an intact ET-signaling pathway is required for the establishment of ISR. Particularly interesting was the analysis of the *eir1* mutant, which is ET-insensitive in the roots, but not in the shoot. This *eir1* mutant was incapable of showing ISR after root colonization by WCS417r. In contrast, after leaf infiltration with WCS417r it did show ISR, indicating that responsiveness to ET is required at the site of rhizobacterial induction (Knoester *et al.*, 1999). The observation that the ET-responsive Arabidopsis *AtTLP1* gene, encoding a thaumatin-like protein, is activated in the roots upon colonization by WCS417r, confirms that ET signaling is initiated in the roots by ISR-inducing rhizobacteria (Léon-Kloosterziel *et al.*, 2005).

Further evidence for the involvement of the ET-response pathway came from the identification of the Arabidopsis *ISR1* locus (Ton *et al.*, 1999). Genetic analysis of the progeny of a cross between the WCS417r-responsive ecotype Col-0 and the ISR-impaired ecotype RLD1 revealed a single locus, designated *ISR1*, to be important in the expression of ISR against several different pathogens (Ton *et al.*, 2002c). Accessions with the recessive *isr1* allele have reduced sensitivity to ET and enhanced susceptibility to *Pst* DC3000 (Ton *et al.*, 2001). These results strongly indicate that the Arabidopsis *ISR1* locus encodes a novel component in the ET-signal transduction pathway that is important for both basal resistance and ISR in Arabidopsis.

4. Dual Role for NPR1 in SAR and ISR

To investigate a possible involvement of the SAR regulatory protein NPR1 in ISR signaling, the *Arabidopsis npr1* mutant was tested in the ISR bioassay. Surprisingly, the *npr1* mutant was incapable of showing WCS417r-mediated ISR (Pieterse *et al.*, 1998; Van Wees *et al.*, 2000) (Fig. 1). This result clearly showed that WCS417r-mediated ISR, like SA-dependent SAR, is an NPR1-dependent defense response. Further analysis of the ISR signal-transduction pathway revealed that NPR1 acts downstream of the JA- and ET-dependent steps (Pieterse *et al.*, 1998). Because SAR is associated with NPR1-dependent *PR*-gene expression, and ISR is not, the action of NPR1 in ISR must be different from that in SAR. These different activities are not mutually exclusive because simultaneous activation of ISR and SAR can lead to an enhanced defensive activity compared to that observed with either type of induced resistance alone (Van Wees *et al.*, 2000). These results suggest that the NPR1 protein is important in regulating and intertwining different hormone-dependent defense pathways.

5. ISR Is Associated with Priming for Enhanced Defense

In Arabidopsis, both JA and ET activate specific sets of defense-related genes (Schenk *et al.*, 2000) but, when applied exogenously, each of both can induce resistance (Pieterse *et al.*, 1998; Van Wees *et al.*, 1999). To investigate how far ISR is associated with these changes in JA/ET-responsive gene expression, Van Wees *et al.* (1999) monitored the expression of a set of well-characterized JA- and/or ET-responsive, defense-related genes in Arabidopsis plants expressing WCS417r-mediated ISR. None of these genes was up-regulated in induced plants, neither locally in the roots, nor systemically in the leaves. This suggested that the resistance attained was not associated with major increases in the levels of either JA or ET. Indeed, analysis of JA and ET levels in leaves of ISR-expressing plants revealed no changes in the production of these signal molecules (Pieterse *et al.*, 2000; Hase *et al.*, 2003). Therefore, it had to be assumed that the JA and ET dependency of ISR is based on an enhanced sensitivity to these hormones, rather than on an increase in their production.

To identify ISR-related genes, the transcriptional response of over 8000 Arabidopsis genes was monitored during WCS417r-mediated ISR (Verhagen *et al.*, 2004). However, systemically in the leaves, none of the ~8000 genes tested showed a consistent change in expression in response to effective colonization of the roots by WCS417r, indicating that the onset of ISR in the leaves is not associated with detectable changes in gene expression. However, after challenge inoculation of WCS417r-induced plants with *Pst* DC3000, 81 genes showed an augmented expression pattern in ISR-expressing leaves compared to inoculated control leaves, suggesting that ISR-expressing plants are primed to respond faster and/or more strongly upon pathogen attack. The majority of the primed genes were predicted to be regulated by JA and/or ET signaling, confirming earlier findings that colonization of the roots by WCS417r primed Arabidopsis plants for augmented expression of the JA- and/or ET-responsive genes *AtVSP2, PDF1.2* and *HEL* (Van Wees *et al.*, 1999; Hase *et al.*, 2003).

Priming is a phenomenon that is associated with different types of induced resistance (Conrath *et al.*, 2002; Conrath *et al.*, 2006). It provides the plant with an enhanced capacity for rapid and effective activation of cellular defense responses once a pathogen is contacted, and it allows the plant to react more effectively to any invader encountered by boosting the defenses that are activated in the host. This mechanism could also explain the broad-spectrum action of induced resistance. Priming for defense may combine advantages of enhanced disease protection with low metabolic costs. Recently, Van Hulten *et al.* (2006) examined the costs and benefits of priming in comparison to activated defense in Arabidopsis. The study revealed that the benefits of priming-mediated resistance outweigh the costs under conditions of pathogen pressure, suggesting an evolutionary advantage of this mechanism of induced resistance over constitutive activation of defense responses. Figure 1 provides a schematic representation of the ISR signaling pathway in Arabidopsis.

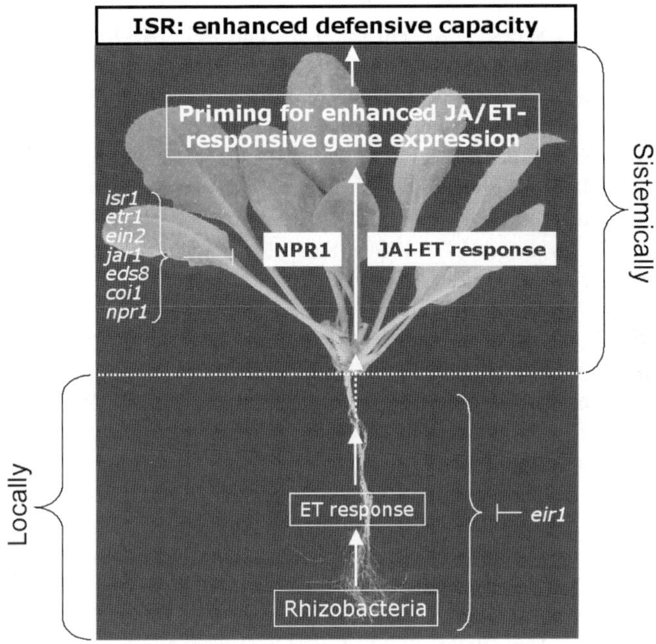

Fig. 1. Schematic representation of the *Pseudomonas fluorescens* WCS417r-mediated ISR signaling pathway in Arabidopsis.

References

Conrath U, Pieterse CMJ, Mauch-Mani B (2002) Priming in plant-pathogen interactions. Trends Plant Sci. 7:210–216

Conrath U, Beckers GJM, Flors V, García-Agustín P, Jakab G, Mauch F, Newman M-A, Pieterse CMJ, Poinssot B, Pozo MJ, Pugin A, Schaffrath U, Ton J, Wendehenne W, Zimmerli L, Mauch-Mani B (2006) Priming: getting ready for battle. Mol. Plant-Microbe Interact. 19(10):1062–1071

Durrant WE, Dong X (2004) Systemic acquired resistance. Annu. Rev. Phytopathol. 42:185–209

Hase S, Van Pelt JA, Van Loon LC, Pieterse CMJ (2003) Colonization of Arabidopsis roots by *Pseudomonas fluorescens* primes the plant to produce higher levels of ethylene upon pathogen infection. Physiol. Mol. Plant Pathol. 62:219–226

Knoester M, Pieterse CMJ, Bol JF, Van Loon LC (1999) Systemic resistance in Arabidopsis induced by rhizobacteria requires ethylene-dependent signaling at the site of application. Mol. Plant-Microbe Interact. 12:720–727

Léon-Kloosterziel KM, Verhagen BWM, Keurentjes JJB, Van Pelt JA, Rep M, Van Loon LC, Pieterse CMJ (2005) Colonization of the Arabidopsis rhizosphere by fluorescent *Pseudomonas* spp. activates a root-specific, ethylene-responsive *PR-5* gene in the vascular bundle. Plant Mol. Biol. 57:731–748

Pieterse CMJ, Van Wees SCM, Hoffland E, Van Pelt JA, Van Loon LC (1996) Systemic resistance in Arabidopsis induced by biocontrol bacteria is independent of salicylic acid accumulation and pathogenesis-related gene expression. Plant Cell 8:1225–1237

Pieterse CMJ, Van Wees SCM, Van Pelt JA, Knoester M, Laan R, Gerrits H, Weisbeek PJ, Van Loon LC (1998) A novel signaling pathway controlling induced systemic resistance in Arabidopsis. Plant Cell 10:1571–1580

Pieterse CMJ, Van Loon LC (1999) Salicylic acid-independent plant defence pathways. Trends Plant Sci. 4:52–58

Pieterse CMJ, Van Pelt JA, Ton J, Parchmann S, Mueller MJ, Buchala AJ, Métraux J-P, Van Loon LC (2000) Rhizobacteria-mediated induced systemic resistance (ISR) in *Arabidopsis* requires sensitivity to jasmonate and ethylene but is not accompanied by an increase in their production. Physiol. Mol. Plant Pathol. 57:123–134

Pieterse CMJ, Van Wees SCM, Ton J, Van Pelt JA, Van Loon LC (2002) Signalling in rhizobacteria-induced systemic resistance in *Arabidopsis thaliana*. Plant Biol. 4:535–544

Schenk PM, Kazan K, Wilson I, Anderson JP, Richmond T, Somerville SC, Manners JM (2000) Coordinated plant defense responses in Arabidopsis revealed by microarray analysis. Proc. Natl. Acad. Sci. USA 97:11655–11660

Ton J, Davison S, Van Wees SCM, Van Loon LC, Pieterse CMJ (2001) The Arabidopsis *ISR1* locus controlling rhizobacteria-mediated induced systemic resistance is involved in ethylene signaling. Plant Physiol. 125:652–661

Ton J, Pieterse CMJ, Van Loon LC (1999) Identification of a locus in Arabidopsis controlling both the expression of rhizobacteria-mediated induced systemic resistance (ISR) and basal resistance against *Pseudomonas syringae* pv. *tomato*. Mol. Plant-Microbe Interact. 12:911–918

Ton J, De Vos M, Robben C, Buchala A, Métraux J-P, Van Loon LC, Pieterse CMJ (2002a) Characterization of Arabidopsis enhanced disease susceptibility mutants that are affected in systemically induced resistance. Plant J. 29:11–21

Ton J, Van Pelt JA, Van Loon LC, Pieterse CMJ (2002b) Differential effectiveness of salicylate-dependent and jasmonate/ethylene-dependent induced resistance in Arabidopsis. Mol. Plant-Microbe Interact. 15:27–34

Ton J, Van Pelt JA, Van Loon LC, Pieterse CMJ (2002c) The Arabidopsis *ISR1* locus is required for rhizobacteria-mediated induced systemic resistance against different pathogens. Plant Biol. 4:224–227

Van Hulten M, Pelser M, Van Loon LC, Pieterse CMJ, Ton J (2006) Costs and benefits of priming for defense in Arabidopsis. Proc. Natl. Acad. Sci. USA 103:5602–5607

Van Loon LC, Bakker PAHM, Pieterse CMJ (1998) Systemic resistance induced by rhizosphere bacteria. Annu. Rev. Phytopathol. 36:453–483

Van Loon LC, Rep M, Pieterse CMJ (2006) Significance of inducible defense-related proteins in infected plants. Annu. Rev. Phytopathol. 44:135–162

Van Wees SCM, Luijendijk M, Smoorenburg I, Van Loon LC, Pieterse CMJ (1999) Rhizobacteria-mediated induced systemic resistance (ISR) in Arabidopsis is not associated with a direct effect on expression of known defense-related genes but stimulates the expression of the jasmonate-inducible gene *Atvsp* upon challenge. Plant Mol. Biol. 41:537–549

Van Wees SCM, De Swart EAM, Van Pelt JA, Van Loon LC, Pieterse CMJ (2000) Enhancement of induced disease resistance by simultaneous activation of salicylate- and jasmonate-dependent defense pathways in *Arabidopsis thaliana*. Proc. Natl. Acad. Sci. USA 97:8711–8716

Verhagen BWM, Glazebrook J, Zhu T, Chang H-S, Van Loon LC, Pieterse CMJ (2004) The transcriptome of rhizobacteria-induced systemic resistance in *Arabidopsis*. Mol. Plant-Microbe Interact. 17:895–908

Ethylene receptor signaling and plant salt-stress responses

Cao, W.H., Liu, J., Chen, T., Cao, Y.R., He, X.J., Mu, R.L., Zhou, H.L., Xie, C., Chen, S.Y.* and Zhang, J.S.*

National Key Lab of Plant Genomic, Institute of Genetics and Developmental Biology, Chinese Academy of Sciences, Beijing 100101, China; Telephone: 8610-64886859; Fax: 8610-64873428. (*Corresponding authors: jszhang@genetics.ac.cn, sychen@genetics.ac.cn)

1. Introduction

Ethylene has long been regarded as a stress hormone. However, the roles of the ethylene signaling in abiotic stress responses remain an open question. Previously, we have cloned tobacco ethylene receptor genes *NTHK1* and *NTHK2*, and studied their expression in response to different abiotic stresses. We found that both genes were induced by wounding and dehydration (Zhang *et al.*, 2001a; Zhang *et al.*, 2001b). *In situ* mRNA hybridization and immunohistochemistry analysis of the NTHK1 protein revealed that both mRNA and protein appeared first in the palisade cell layer upon cutting and then gradually spread to other sponge cells of the leaf (Zhang *et al.*, 2001a, Xie *et al.*, 2002). When exposed to salt stress, the *NTHK1* gene expression was induced by salt treatment whereas the *NTHK2* expression was not affected (Zhang *et al.*, 2001a; Zhang *et al.*, 2001b). This differential expression pattern indicated different roles of various receptors in multiple stress responses, and also implied a possible specific function for NTHK1 in plant salt-stress responses. The biochemical property of NTHK1 and NTHK2 has been investigated and it is found that NTHK1 had Ser/Thr kinase activity whereas NTHK2 had serine/threonine and histidine kinase activity in the presence of Mn^{2+} and Ca^{2+} respectively (Xie *et al.*, 2003; Zhang *et al.*, 2004). NTHK1 reduced expression of a salt-induced *NAC* gene (He *et al.*, 2005).

In this study, we generated transgenic plants overexpressing the tobacco ethylene receptor homolog NTHK1 and investigated the plant responses to salt stress. These studies have significance in elucidating the role of ethylene signaling in plant salt-stress response.

A. Ramina et al. (eds.), Advances in Plant Ethylene Research: Proceedings of the 7th International Symposium on the Plant Hormone Ethylene, 333–339.

2. Results and Discussion

2.1 Phenotype and ethylene sensitivity of the NTHK1-transgenic plants

To investigate the function of tobacco ethylene receptor homolog gene *NTHK1* in plant, we transformed the *NTHK1* gene containing a 50 bp 5′-untranslated region (UTR), the open reading frame (ORF) and the 90 bp 3′-UTR sequences, driven by the 35S promoter, into *Arabidopsis* plants. It is assumed that similar downstream components existed in *Arabidopsis* as in tobacco plants since the *Arabidopsis* ethylene receptor gain-of-function mutant gene *etr1-1* functioned in a tobacco background. Totally, 40 individual transgenic lines were obtained and 90% of these lines showed large rosette, and late flowering as exemplified by line S10 (Figs. 1A and B). Homozygous T3 lines were recovered and two lines (S1 and S10, single insertion) were selected for further analysis. The large rosette was most likely due to the enlargement of the epidermal cells on leaf surface (Fig. 1C) whereas the late flowering probably resulted from the late development of the reproductive apex in the *NTHK1*-transgenic plants (Figs. 1C and D). It is not known if the mesophyll cells were also enlarged. The NTHK1 protein expression was confirmed in young leaves of the transgenic plants by immunohistochemical method (Fig. 1E).

NTHK1 is an ethylene receptor homolog gene and introduction of this gene may alter plant ethylene sensitivity. We then examined the ethylene sensitivity of the two *NTHK1*-transgenic lines S1 and S10. Upon ACC (precursor of the ethylene biosynthesis) treatment, the etiolated seedlings of the transgenic Arabidopsis showed reduced triple response when compared with the wild-type control (Fig. 1F), consistent with our previous observation in *NTHK1* transgenic tobacco seedlings (Xie *et al.*, 2002). The expression of an ethylene-inducible gene *chitinase B* was examined in both wild type and the transgenic plants. The results in Fig. 1G showed that this gene was inducible in wild-type plants upon ACC treatments whereas it is not significantly affected in the transgenic line S10. All these results indicate that the *NTHK1*-overexpressing transgenic plants are less sensitive to ethylene.

2.2 NTHK1 increases salt sensitivity of the transgenic plants and ACC can suppress this sensitivity

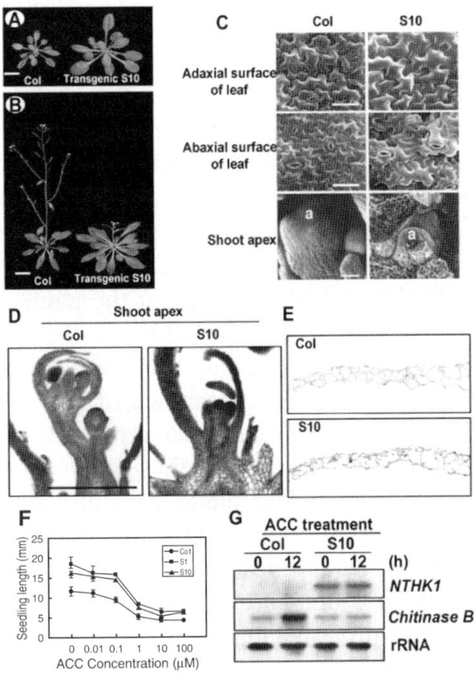

Fig. 1. (Color figure in the Annex, p.459). Phenotype and ethylene sensitivity of the *NTHK1*-transgenic plants. **(A)** Phenotype of the wild-type Col and the *NTHK1* transgenic line S10. Plants have grown in pot for 4 weeks and bar represents 1 cm. **(B)** Phenotype of the Col and the transgenic line S10. Plants have grown in pot for 6 weeks and bar represents 1 cm. **(C)** Scanning electron micrograph of the leaf surface and shoot apex from Col and the transgenic line S10. **a**, apex. Bars in the left panel are suitable for the right panel, with the first two bars representing 50 μm and the last bar 30 μm. **(D)** Longitudinal section of the shoot apex of the Col and the transgenic line S10. The bar represents 1000 μm and is suitable for both panels. **(E)** NTHK1 protein distribution revealed by immunohistochemical analysis on cross-section of a developing leaf from 12-day-old seedlings of Col or the transgenic line S10. Red color indicates positive signal. **(F)** ACC dose response curves of the seedling length for Col and the two transgenic lines S1 and S10. Values represent means of 25 measurements. **(G)** ACC effect on *Chitinase B* gene expression in Col and the transgenic line S10. Twelve-day-old seedlings were treated with 100 μM ACC and then subjected to RNA analysis with labeled *NTHK1* and *Chitinase B* cDNA probes.

Because the *NTHK1* mRNA accumulation was observed in response to salt stress in tobacco seedlings (Zhang *et al.*, 2001a), this gene may participate in salt-stress responses in plants. We tested the responses of the transgenic *Arabidopsis* under salt-stress conditions. Five-day-old seedlings were transferred onto the agar media containing various concentrations of NaCl and maintained for 7 days. It can be seen in Fig. 2A that under normal condition, the transgenic plants appeared to be similar to wild type. However, in 50 mM NaCl, the transgenic lines exhibited slight epinasty phenotype, i.e., backward growth of the leaf blade and the petiole. This epinasty phenotype was much severe in the transgenic lines at 100 mM NaCl treatment. The petioles of the leaves in the transgenic lines were very short and the leaf blade grew backward close to base of the plant. The whole plant appeared to be very compact. More than 90% of the salt-stressed transgenic lines have this phenotypic change. The epinasty phenotype was in contrast with the relatively normal appearance of the wild-type plants in the same plate of NaCl medium. It was also apparently different from the ethylene-caused symptom in Arabidopsis that has upward-erected petiole of light-green color, and small leaf blade with curvature (Fig. 2B). At higher NaCl concentrations (150 and 200 mM), both wild type and the transgenic lines showed apparent growth inhibition. These results indicate that *NTHK1*-overexpressing lines are more sensitive to salt stress when compared with the wild-type Col plants during this developmental stage.

Ethylene has been proposed to negatively regulate its receptor activity. It would suppress the salt-induced epinasty if this phenotype were the result of the NTHK1 function. To test whether the epinasty phenotype in *NTHK1*-transgenic *Arabidopsis* can be altered by ethylene, the plants with the salt-induced epinasty were transferred to the salt medium with or without ethylene precursor ACC, and the phenotypic change was examined. ACC can be easily absorbed by plants and converted to ethylene. As shown in Fig. 2C, when the S1 and S10 transgenic plants with the severe epinasty phenotype were still transferred onto the salt medium, the phenotype remained unchanged. However, when the transgenic plants with the same phenotype were transferred onto the salt medium plus ACC, they were all rescued rapidly from the second or the third day of the treatment and exhibited the phenotype resembling that of wild-type plants. Other plant hormones cannot rescue the salt-induced epinasty phenotype (data not shown). These results indicated that the transgenic plants needed more ethylene to counteract the NTHK1 function that leading to the epinasty under salt stress and ethylene negatively regulated the NTHK1 action.

2.3 Comparison of the phenotype of Arabidopsis ethylene response mutants under salt stress

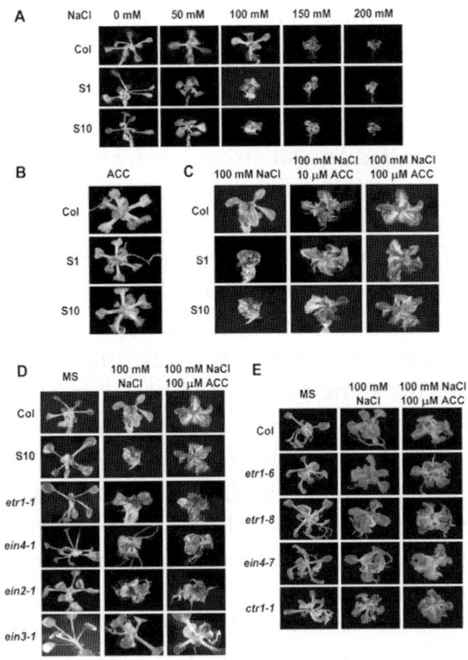

Fig. 2. (Color figure in the Annex, p.460). Phenotype of the *NTHK1*-transgenic plants and the ethylene-response mutants under salt stress. **(A)** Comparison of the phenotype of the Col and the transgenic lines S1 and S10 upon treatment with various concentrations of NaCl. Five-day-old seedlings of the Col and the transgenic plants were transferred onto salt agar plates and maintained for 7 days to observe the phenotypic change. **(B)** Phenotype of ACC-treated Col and the two transgenic lines S1 and S10. **(C)** ACC suppression of the epinasty phenotype. The *NTHK1*-transgenic lines (S1 and S10) with the salt-induced epinasty phenotype were transferred onto 100 mM NaCl plus 10 μM or 100 μM ACC to observe the recovery of the epinasty phenotype. **(D)** Comparison of the phenotype of Col, the transgenic plants S10, and the ethylene-response mutants *etr1-1*, *ein4-1*, *ein2-1*, and *ein3-1* under NaCl or NaCl + ACC. **(E)** Comparison of the phenotype of the ethylene receptor loss-of-function mutants *etr1-6*, *etr1-8*, and *ein4-7* and ethylene-constitutive response mutant *ctr1-1* under NaCl or NaCl plus ACC treatment.

Overexpression of the ethylene receptor homolog NTHK1 in transgenic plants represents gain-of-function. Arabidopsis gain-of-function mutants of ethylene receptors, together with other ethylene-insensitive mutants in the ethylene-signaling pathway, were also tested for their responses in salt medium with or without ACC. The result in Fig. 2D showed that all of the *etr1-1* and *ein4-1* gain-of-function mutant plants exhibited similar epinasty phenotype as the *NTHK1*-transgenic plants S10 had, and *ein2-1*plants, with a mutation in the membrane-localized EIN2 protein of the ethylene signaling pathway, also showed similar phenotype. However, ACC cannot rescue this phenotype in the three mutants, due to the ethylene binding mutation (gain-of- function) in the *etr1-1*and *ein4-1*genes, and the loss-of-function mutation in the *ein2-1* gene. Under salt stress, *ein3-1* mutant, with a mutation in the transcription factor EIN3 of the ethylene signaling pathway, did not have the epinasty phenotype but showed wild type like phenotype (Fig. 2D).

We also tested if loss-of-function mutants of Arabidopsis ethylene receptors have any phenotypic change under salt stress. The results in Fig. 2E showed that none of the tested loss-of-function mutants *etr1-6*, *etr1-8*, and *ein4-7* had any epinasty phenotype upon salt stress as compared with the control plants under the same condition. These results further indicate that active receptor signaling or receptor activity led to the epinasty observed under salt stress. ACC treatment did not significantly change the phenotype of the salt-treated loss-of-function mutants (Fig. 2E). The ethylene-constitutive response mutant *ctr1-1* did not show significant phenotypic alteration either under both salt and ACC treatments in comparison with the Col plants except that the *ctr1-1* mutants appeared to be slightly smaller than the Col plants (Fig. 2E; data not shown). The NTHK1-transgenic tobacco plants were also sensitive to salt stress (Cao *et al.*, 2006) and NTHK1 different domains may have specific roles in regulation of plant growth and stress responses (Zhou *et al.*, 2006).

References

Cao, WH, Liu J, Zhou QY, Cao YR, Zheng SF, Du BX, Zhang JS, Chen SY (2006) Expression of tobacco ethylene receptor NTHK1 alters plant responses to salt stress. Plant Cell Environ 29: 1210–1219

He XJ, Mu RL, Cao WH, Zhang ZG, Zhang JS, Chen SY (2005) AtNAC2, a transcription factor downstream of ethylene and auxin signaling pathways, is involved in salt stress response and lateral root development. Plant J 44: 903–916

Xie C, Zhang ZG, Zhang JS, He XJ, Cao WH, He SJ, Chen SY (2002) Spatial expression and characterization of a putative ethylene receptor protein NTHK1 in tobacco. Plant Cell Physiol 43: 810–815

Xie C, Zhang JS, Zhou HL, Li J, Zhang ZG, Wang DW, Chen SY (2003) Serine/threonine kinase activity in the putative histidine kinase-like ethylene receptor NTHK1 from tobacco. Plant J 33: 385–393

Zhang JS, Xie C, Shen YG, Chen SY (2001a) A two-component gene (NTHK1) encoding a putative ethylene-receptor homolog is both developmentally and stress-regulated in tobacco. Theor Appl Genet 102: 815–824

Zhang JS, Xie C, Du BX, Wu XL, Chen SY (2001b) Tobacco two-component gene NTHK2. Chin Sci Bull 46: 574–577

Zhang ZG, Zhou HL, Chen T, Gong Y, Cao WH, Wang YJ, Zhang JS, Chen SY (2004) Evidence for serine/threonine and histidine kinase activity in the tobacco ethylene receptor protein NTHK2. Plant Physiol 136: 2971–2981

Zhou HL, Cao WH, Cao YR, Liu J, Hao YJ, Zhang JS, Chen SY (2006) Roles of ethylene receptor NTHK1 domains in plant growth, stress response and protein phosphorylation. FEBS Lett 580: 1239-1250.

Chemical-induced programmed cell death in tomato suspension cells is mediated through ethylene and lipid signalling

Woltering, E.J.[1]*, Yakimova, E.T.[2], Michaeli, R.[1] and Kapchina-Toteva, V.M.[3]

[1]Wageningen University and Research Centre, Agrotechnology and Food Sciences Group (AFSG). P.O. Box 17, 6700 AA Wageningen, The Netherlands. [2]Regional Center for Research and Extension Service in Floriculture and Agriculture (RCNPO), 1222 Sofia, Bulgaria. [3]Department of Plant Physiology, Faculty of Biology, University of Sofia, 8 Dragan Tsankov Blvd., 1164 Sofia, Bulgaria. (*Corresponding author: Ernst.Woltering@wur.nl).

Abstract

Chemical-induced cell death was studied in suspension-cultured tomato (*Lycopersicon esculentum* Mill.) cells (line MsK8) treated with the anti-cancer drug camptothecin (CPT) and the heavy metal $CdSO_4$. Within 24 h, chemical treatment induced cell death in a concentration-dependent manner. Chemical-induced cell death was alleviated by the addition of sub μM concentrations of peptide inhibitors specific to human caspases indicating that cell death proceeds through a mechanism with similarities to animal programmed cell death. Addition of antioxidants greatly reduced chemical-induced cell death indicating that reactive oxygen species (ROS) are instrumental in cell death. Inhibitors of phospholipase C (PLC) and phospholipase D (PLD) signalling pathway intermediates greatly reduced chemical-induced cell death. Ethylene, while not inducing cell death when applied alone, stimulated chemical-induced cell death. Application of the ethylene biosynthesis inhibitor aminoethoxy vinylglycine (AVG) reduced chemical-induced cell death. Together, the results show that toxic chemicals induce programmed cell death exhibiting apoptotic-like features. The cell death process requires increased ROS production and activation of PLC, PLD and ethylene signalling pathways.

Keywords: Antioxidants; cadmium; camptothecin, ethylene; *Lycopersicon esculentum* Mill; phospholipids; programmed cell death; tomato cell culture.

A. Ramina et al. (eds.), Advances in Plant Ethylene Research: Proceedings of the 7th International Symposium on the Plant Hormone Ethylene, 341–351.

Abbreviations: AA, ascorbic acid; AVG, aminoethoxy vinylglycine; CPT, camptothecin; FDA, fluorescein diacetate; IMP, inositolmonophosphatase; IP_3, phosphatidylinositol-triphosphate; PA, phosphatidic acid; L-PEA, Lysophosphatidylethanolamine; MeOSuc-AAPV-CMK, methoxysuccinyl-Ala-Ala-Pro-Val-chloromethylketone; PCD, programmed cell death; PLC, phospholipase C; PLD, phospholipase D; ROS, reactive oxygen species; U-73122, 1-[6-([17β)-3-metoxyestra-1,3,5(10)-trien-17-yl]amino)hexyl]-1H-pyrrole-2,5,-dione; VPE, vacuolar processing enzyme; YVAD-CMK, Tyr-Val-ala-Asp-chloromethyl-ketone; Z-Asp-CH_2-DCB, benzyoxycarbonyl-Asp-2,6-dichlorobenzoyloxymethylketone.

1. Introduction

Programmed cell death (PCD) is a functional term used to describe cell death aimed at eliminating redundant or harmful cells during the life cycle of multicellular organisms. PCD plays an essential role during development and morphogenesis by removing unwanted or misplaced cells in specific structures and organs. In addition, PCD is used in defensive mechanisms against infected or mutated cells. Deregulation of PCD is implicated in various human diseases ranging from cancer and autoimmune disorders to neurodegenerative diseases and AIDS. In animal cells, PCD is often associated with the occurrence of a specific set of cellular morphological features such as condensation of the nucleus and the cytoplasm, nuclear and DNA fragmentation (DNA laddering) and the formation of cellular debris-containing vesicles called apoptotic bodies (Steller, 1995). The cell death process showing such morphological features is called apoptosis. A small family of regulatory aspartate-specific cysteine proteases called caspases is responsible for at least some of these specific features of apoptotic cells (Hengartner, 2000).

Also in plants, PCD is an essential process during growth and development. Cellular suicide is involved in for example xylogenesis, aerenchyma formation, plant reproductive events, leaf senescence, petal senescence and endosperm degradation. Furthermore, PCD plays an important role during cell death in response to a variety of pathogens and during responses to various abiotic stresses such as heat shock, toxic chemicals, ozone exposure, UV radiation and hypoxia (Overmeyer *et al.*, 2003; Van Doorn and Woltering, 2005).

In a number of plant systems, PCD is accompanied by chromatin condensation and nuclear and DNA fragmentation and by formation of apoptotic-like bodies (De Jong *et al.*, 2000; Wang *et al.*, 1996). These morphological similarities between dying animal and plant cells suggest the existence of an evolutionary conserved cell death mechanism.

Caspases belong to a class of specific cysteine proteases that show a high degree of specificity with an absolute requirement for cleavage adjacent an aspartic acid residue and a recognition sequence of at least four amino acids *N*-terminal to this cleavage site. Generally, apoptotic cell death in mammalian cells involves a sequence of caspase activation events in which initiator caspases activate down-stream executioner caspases that process a variety of target proteins eventually leading to the apoptotic phenotype. An increasing number of reports show that inhibitors to various mammalian caspases markedly suppress plant cell death as well (Woltering *et al.*, 2002). In tomato suspension cells, treatment with low concentrations of e.g. topoisomerase 1 inhibitor camptothecin (CPT) or fumonisin-B1 resulted in cell death exhibiting typical apoptotic features such as cytoplasmic shrinkage and nuclear and DNA fragmentation. Cell death induced by these chemicals was effectively inhibited by the human caspase-1 and caspase-3 inhibitors as well as by the non-specific caspase inhibitor Z-asp-CH_2-DCB (De Jong *et al.*, 2000; De Jong *et al.*, 2002). These inhibitors proved to be effective at very low concentrations, with 50% inhibition of chemical-induced cell death at concentrations of 1–10 nM (Woltering et *al.*, 2002). These results suggest that also in plants, caspase-like proteases are involved in cell death. Apart from the possible involvement of caspase-like proteases in plant cell death also other key components of the animal apoptotic pathway (e.g. increased production of reactive oxygen species (ROS), the involvement of mitochondrial factors, and the existence of anti-apoptotic pathways) may have counterparts in plant cells indicating that PCD in plants may proceed through a similar mechanism as in animal cells (for review see Hoeberichts and Woltering, 2003).

Plant cell cultures represent an effective model system to study the effect of stimulators and inhibitors of metabolic pathways involved in cell death (De Jong *et al.*, 2000; De Jong *et al.*, 2002). We have used a pharmacological approach to elucidate some of the pathways involved in chemical-induced PCD.

2. Materials and Methods

The experiments were conducted with tomato (*Lycopersicon esculentum* Mill.) suspension cells, (line Msk8) grown in a liquid Murashige–Skoog medium, supplemented with 5 μM α-naphtalene acetic acid, 1 μM 6-benzyladenine, 3% (w/v) sucrose and vitamins (De Jong *et al.*, 2000). The cells were subcultured every 7 days by 1:4 dilution with fresh medium and kept on a rotary shaker at 25°C. For treatments, cells were used 5 days after subculture. Cell death inducers and inhibitors were added simultaneously to 5 ml of suspension culture in 30 ml flasks with a gas-tight screw-cap. Most of the chemicals for the treatments were taken from concentrated stocks allowing μL volumes to be added to the cell cultures. YVAD-CMK, Z-Asp-CH_2-DCB and lysophosphatidylethanolamine (L-PEA) were dissolved in DMSO (final solvent concentration in treated cell culture 0.1% v/v), the other chemicals were dissolved in water. DMSO was tested alone and no effect on cell viability was detected.

Cell death was determined 24 h after the treatments. 250 μL from the cell suspension was taken and diluted with 4 ml tap water in small Petri dishes. 0.002% fluorescein diacetate (FDA) was used to stain the living cells and the cell counting was executed by fluorescence microscopy (Axiovert, Zeiss) at 100x magnification. For each sample, three different non-overlapping microscope fields were randomly chosen. The cell death was calculated as a percentage of dead cells to the total number of cells. Data were processed statistically, presented as average values from at least three independent experiments and compared by SD.

3. Results and Discussion

The applied cell death-inducing chemicals (cadmium and CPT) caused cell death in a concentration-dependent manner and dead cells showed a typical apoptotic-like morphology exhibiting severely shrunken cytoplasm and condensed nuclei (data not shown). This type of PCD, showing some, but not all features of animal apoptosis was recently classified as non-lysosomal PCD (Van Doorn and Woltering, 2005).

By administration of specific peptide inhibitors, it was previously shown that chemical-induced cell death in animal cell lines is apoptotic and that it involves participation of caspases. The application of YVAD-CMK, Asp-Glu-Val-Asp-aldehyde (DEVD-CHO) or z-Val-Ala-Asp-floromethylketone (zVAD-FMK) to Rat-1 fibroblasts incubated in 10 μM $CdCl_2$ effectively blocked the cell death and prevented PARP cleavage (Kim *et al.*, 2000). Also, in fish (trout) liver cells DEVD-dependent protease activation, internucleosomal DNA cleavage and enhanced ROS formation have been found to actively participate in cell death execution (De Faverney *et al.*, 2001).

Here we show that peptide inhibitors specific to human caspases (caspase1 inhibitor YVAD-CMK and the broad-range caspase inhibitor Z-Asp- CH_2-DCB) are potent inhibitors of chemical-induced cell death in plant cells (Fig. 1). As a negative control, cells were treated with a caspase-unrelated peptide inhibitor methoxysuccinyl-Ala-Ala-Pro-Val-chloromethylketone (MeOSuc-AAPV-CMK) bearing a similar chloromethylketone (CMK) active moiety as the caspase 1 inhibitor. This peptide inhibitor did not affect chemical-induced cell death.

Recently some of the targets of these caspase inhibitors have been identified in plants. In oat leaves, a specific class of subtilisin-like serine proteases, involved in victorin-induced cell death, was found to exhibit caspase-like activity and could be blocked by human caspase inhibitors (Coffeen and Wolpert, 2004). In addition, caspase-related plant proteases, vacuolar processing enzymes (VPEs), were reported to exhibit caspase-like activity and VPE was shown to be involved in cell death in tobacco induced by TMV (Hatsugai *et al.*, 2004), and in bacteria, fungus and virus-induced cell death in *Arabidopsis thaliana* (Rojo *et al.*, 2004). The effect of the caspase inhibitors on chemical-induced cell death in tomato suspension cells is further evidence that chemicals induce PCD and implies that caspase-like proteases such as subtilisin-like serine porteases or VPEs may be involved.

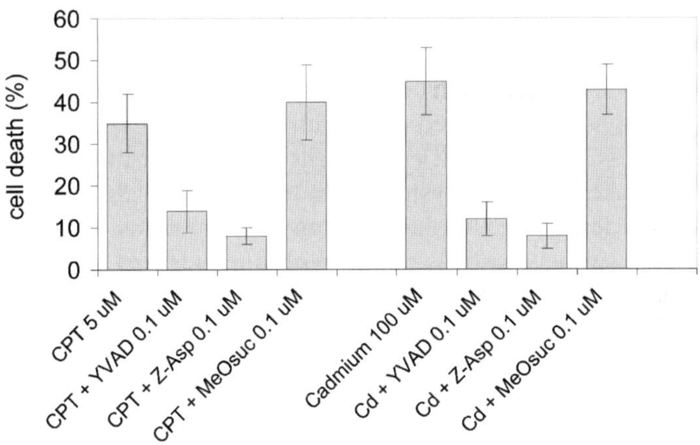

Fig. 1. Effect of caspase inhibitors on chemical-induced cell death. MeOsuc was added as a negative control. Non-treated cells contained approximately 5% dead cells. Vertical bars represent +/– SD.

Treatment of the cells with antioxidants (catalase, ascorbic acid and quercetin) effectively blocked chemical-induced cell death, indicating that the increase in ROS is causatively related to the cell death (Fig. 2). This is in line with the finding that H_2O_2 scavengers prevented H_2O_2 accumulation and reduced the symptoms of cadmium toxicity in pea leaves (Romero-Puertas *et al.*, 2004). Similarly, cell death inhibition by catalase has been reported in animal cells exposed to cadmium. Co-incubation of rat glioma cells with catalase and cadmium strongly inhibited cadmium-induced

DNA ladder formation, indicating that H_2O_2 is at least partially involved in apoptotic activity of cadmium in animal systems (Wätjen and Beyersmann, 2004).

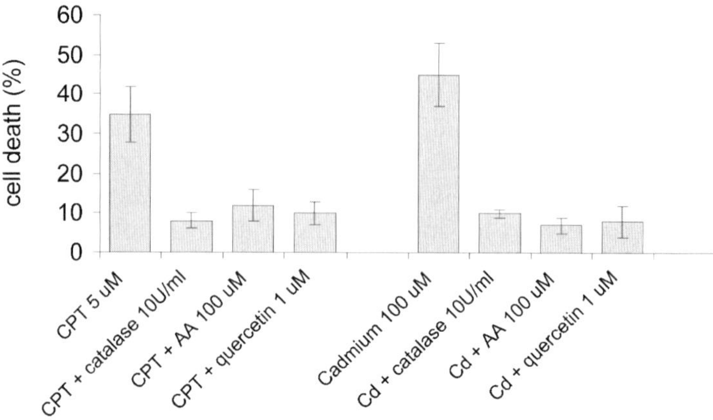

Fig. 2. Effect of antioxidants on chemical-induced cell death. Non-treated cells contained approximately 5% dead cells. Vertical bars: +/– SD.

Suppressing ethylene production with AVG suppressed cell death and addition of ethylene stimulated cell death (Fig. 3). This shows that ethylene signalling is an important component of chemical-induced cell death. Also in a number of other systems ethylene was closely associated with increased cell death (e.g. in plant–pathogen interactions, following ozone application, in suspension cells treated with fungal toxins and in a diversity of developmental processes) (Overmeyer *et al.*, 2003; Woltering *et al.*, 2003). The specific role of ethylene in stimulating cell death in most of these systems has, however, not yet been elucidated.

The stimulating effect of ethylene on cell death in CPT-treated tomato cells was related to an increased production of CPT-induced hydrogen peroxide (De Jong *et al.*, 2002). Similarly, in carrot cells exposed to carbon starvation ethylene has been shown to play a role in activation of NADPH oxidase and the addition of ethylene inhibitors reduced the cell death (Chae and Lee, 2001). Given the ROS dependence of cell death induced by chemicals, a similar mechanism may underlie the stimulating effect of ethylene in chemical-induced cell death.

To investigate the role of lipid signalling in chemical-induced cell death we have tested the effect of a number of inhibitors and intermediates associated with PLC and PLD signalling. Some of these chemicals have earlier been shown to suppress cell death in e.g. suspension-cultured carrot cells and maize roots (He *et al.*, 1996; Koch *et al.*, 1998). Collectively, our data show that lipid signalling and calcium play an important role in chemical-induced cell death.

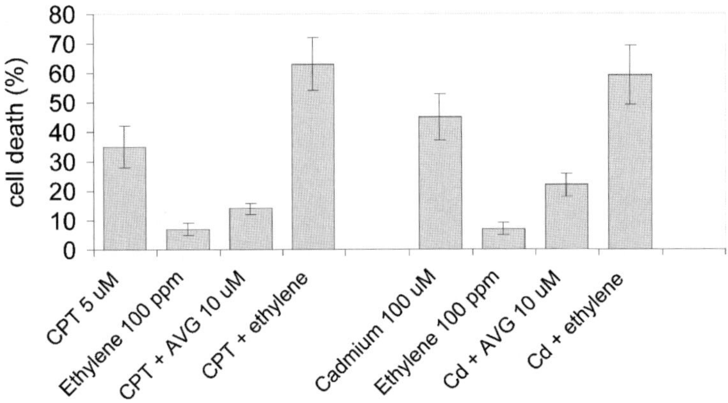

Fig. 3. Effect of ethylene and AVG on chemical-induced cell death. Non-treated cells contained approximately 5% dead cells. Vertical bars represent +/– SD.

Inhibition of IMP by LiCl may regulate the level of free inositol thus affecting processes in inositide cascade. Plant IMP has been isolated and cloned from young tomato fruits and the three IMPs have been found sensitive to lithium inhibition (Gillaspy *et al.*, 1995). The participation of IMP in the hypersensitive response in *Citrus limon* innoculated with *Alternaria alternata* has been shown by lithium inhibition (Ortega and Perez, 2001). In our experiments, LiCl inhibits chemical-induced cell death (Fig. 4), which indicates that chemicals may stimulate PLC and may initiate the further signalling through increased levels of phosphatidylinositol-triphosphate (IP_3), cytosolic calcium and PA.

In line with this hypothesis we show that several inhibitors of either PLC (U73122, neomycin) or PLD (L-PEA, 1-butanol) inhibit the effect of chemicals on cell death (Figs. 4, 5). The involvement of PLC, PLD and their products, especially PA, in the generation of ROS following elicitation with *N*-acetylchitoologosaccharide has recently been demonstrated in cultured rice cells. Elicitor-enhanced accumulation of H_2O_2 was prevented by application of U73122 and 1-butanol (Yamaguchi *et al.*, 2003). This indicates that PLD and PLC signalling are instrumental in the production of ROS and, subsequently the induction or execution of cell death.

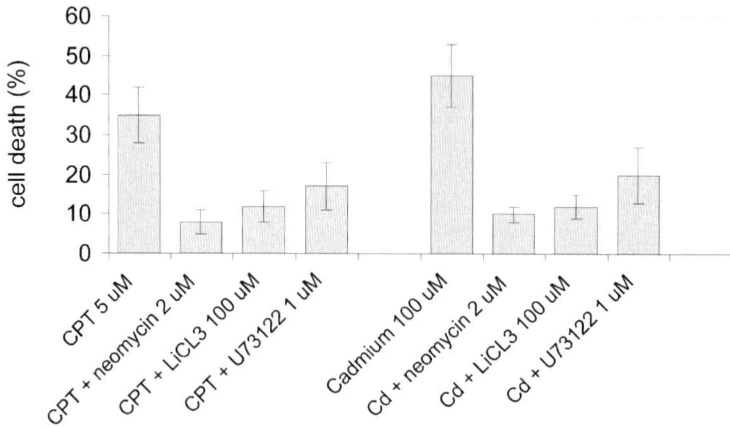

Fig. 4. Effect of PLC inhibitors on chemical-induced cell death. Non-treated cells contained approximately 5% dead cells. Vertical bars: +/– SD.

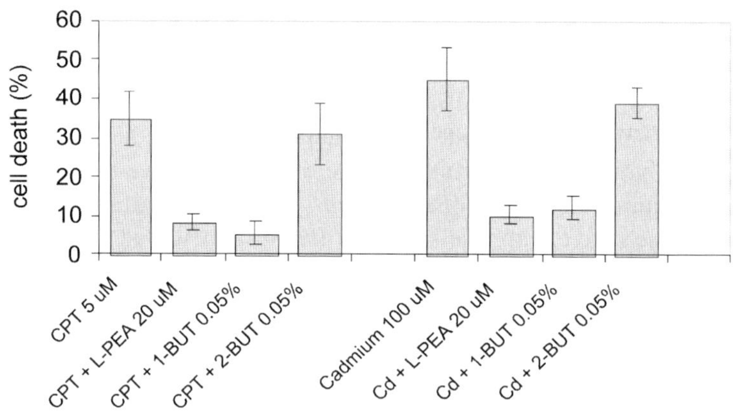

Fig. 5. Effect of PLD inhibitors on chemical-induced cell death. 2-butanol was added as a negative control. Non-treated cells contained approximately 5% dead cells. Vertical bars: +/– SD.

In conclusion, we showed that chemical-induced cell death is mediated by proteases with caspase-like activity and associated with increased ROS production. We showed that ethylene, PLC and PLD signalling pathways are instrumentally involved in cell death. A challenge for future research is the elucidation of the exact sequence of events and possible cross-talk between the different signalling routes involved in ROS formation and cell death. A model can be drawn based on our present results and recent findings by other authors (Fig. 6). In this view, low concentrations of toxic chemicals lead, through PLC/PLD signalling and possibly activation of calcium-dependent protein kinases and/or MAP kinases to increased oxidative stress, which is instrumental in subsequent PCD mediated by caspase-like proteolytic activity. Ethylene apparently mediates the severity of oxidative stress through a yet unidentified mechanism.

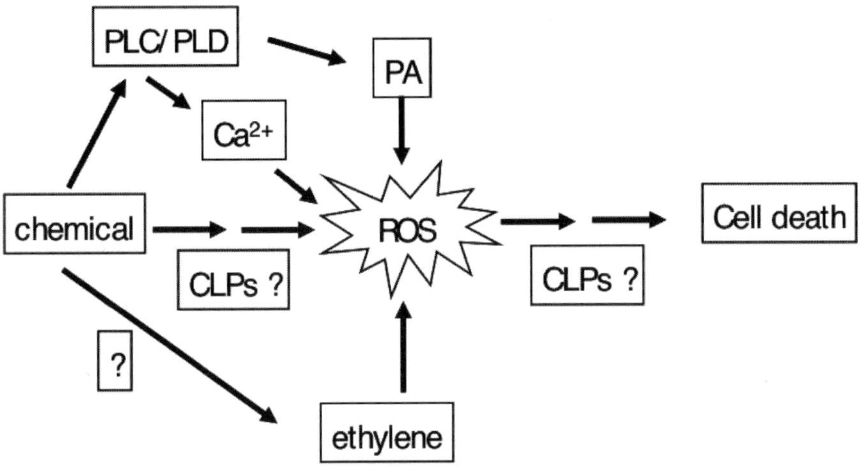

Fig. 6. Hypothetical model for chemical-induced cell death in tomato suspension cells. CLPs = caspase-like proteases.

Acknowledgements

Elena Yakimova and Veneta Kapchina were supported by grants from EU FP5 contract HPRI-CT-1999-00029, EU COST Action 844 and the Ministry of Education and Science, National Science Fund of Bulgaria project B-1502/05. The authors are grateful to Teun Munnik and Ana Laxalt for stimulating discussions on the role of PLD and PLC in cell death.

References

Chae SH, Lee WS (2001) Ethylene- and enzyme-mediated superoxide production and cell death in carrot cells grown under carbon starvation. Plant Cell Rept 20:256–261

Coffeen WC, Wolpert TJ (2004) Purification and characterisation of serine proteases that exhibit caspase-like activity and are associated with programmed cell death in *Avena sativa*. Plant Cell 16:857–873

De Jong AJ, Hoeberichts FA, Yakimova ET, Maximova E, Woltering EJ (2000) Chemical-induced apoptotic cell death in tomato cells: Involvement of caspase-like proteases. Planta 211:656–662

De Jong AJ, Yakimova ET, Kapchina VM, Woltering EJ (2002) A critical role of ethylene in hydrogen peroxide release during programmed cell death in tomato suspension cells. Planta 214:537–545

De Faverney CR, Devaux A, Lafaurie M, Girard JP, Bailly B, Rahmani R (2001) Cadmium induces apoptosis and genotoxicity in rainbow trout hepatocytes through generation of reactive oxygen species. Aquat Toxicol 53:65–76

Gillaspy GE, Keddie JS, Oda K, Gruissem W (1995) Plant inositol monophosphatase is a lithium-sensitive enzyme encoded by multigene family. Plant Cell 7:2175–2185

Hatsugai N, Kuroyanagi M, Yamada K, Meshi T, Tsuda S, Kondo M, Nishimura M, Hara-Nishimura I (2004) A plant vacuolar protease, VPE, mediates virus-induced hypersensitive cell death. Science 305:855–858

He Ch-J, Morgan PW, Drew MC (1996) Transduction of an ethylene signal is required for cell death and lysis in the root cortex of maize during aerenchyma formation induced by hypoxia. Plant Physiol 112:463–472

Hengartner MO (2000) The biochemistry of apoptosis. Nature 407:770–776

Hoeberichts FA, Woltering EJ (2003) Multiple mediators of plant programmed cell death: interplay of conserved cell death mechanisms and plant-specific regulators. BioEssays 25:47–57

Kim MS, Kim BJ, Woo HN, Kim KW, Kim KB, Kim IK, Jung YK (2000) Cadmium induces caspase-mediated cell death: suppression by Bcl-2. Toxicology 145:27–37

Koch W, Wagner C, Seitz HU (1998) Elicitor-induced cell death and phytoalexin synthesis in *Daucus carota* L. Planta 206:523–532

Ortega X, Perez LM (2001) Participation of the phosphoinositide metabolism in the hypersensitive response of Citrus limon against *Alternaria alternata*. Biol Res 34:43–50

Overmeyer K, Brosche M, Kangasjarvi J (2003) Reactive oxygen species and hormonal control of cell death. Trends Plant Sci 8:335–342

Rojo E, Martin R, Carter C, Zouhar J, Pan S, Plotnikova J, Jin H, Paneque M, Sánchez-Serano JJ, Baker B, Ausubel FM, Raikhel NV (2004) VPEγ exhibits a caspase-like activity that contributes to defense against pathogens. Curr Biol 14:1897–1906

Romero-Puertas MC, Rodrigues-Serrano M, Corpas FJ, Gomez M, del Rio LA, Sandalio LM (2004) Cadmium-induced subcellular accumulation of O_2 and H_2O_2 in pea leaves. Plant Cell Environ 27:1122–1134

Steller H (1995) Mechanisms and genes of cellular suicide. Science 267:1445–1449

Van Doorn WG, Woltering EJ (2005) Many ways to exit? Cell death categories in plants. Trends Plant Sci 10:117–122

Wang H, Juan LI, Bostock RM, Gilchrist DG (1996) Apoptosis: a functional paradigm of programmed cell death induced by a host-selective phytotoxin and invoked during development. Plant Cell 8:375–391

Wätjen W, Beyersmann D (2004) Cadmium-induced apoptosis in C6 glioma cells: Influence of oxidative stress. BioMetals 17:65–78

Woltering EJ, van der Bent A, Hoeberichts FA (2002) Do plant caspases exist? Plant Physiol 130:1764–1769
Woltering EJ, de Jong AJ, Iakimova E, Kapchina V, Hoeberichts FA (2003) Ethylene: mediator of oxidative stress and programmed cell death in plants. In: Vendrell M, Klee H, Pech JC, Romojaro F (eds) Biology and Biotechnology of the Plant Hormone Ethylene III. IOS Press, Amsterdam, The Netherlands, pp 315–323
Yamaguchi T, Minami E, Shibuya N (2003) Activation of phospholipases by *N*-acetylchitoologosaccharide elicitor in suspension-cultured rice cells mediates reactive oxygen generation. Physiol Plant 118:361–370

Terpenoid biosynthesis and signaling in legume plants in response to herbivorous damage

Arimura, G.[1]*, Volpe, V.[1,2], Kunert, M.[1], Maffei, M.[2] and Boland, W.[1]

[1]Department of Bioorganic Chemistry, Max Planck Institute for Chemical Ecology, Hans-Knöll-Strasse 8, D-07745 Jena, Germany, Tel. 49 3641 571213; Fax. 49 3641 571202. [2]Department of Plant Biology and Centre of Excellence CEBIOVEM, University of Turin, Viale P.A. Mattioli 25, I-10125 Turin, Italy, Tel. 390 11 6705967; Fax. 390 11 2365967. (*Corresponding author: garimura@ice.mpg.de)

1. Introduction

Herbivore-induced plant volatiles (HIPVs) contribute to attracting the natural enemies of the herbivores and, hence, benefit the plant indirectly. Volatile terpenoids are the major products among the HIPVs from legumes, e.g., lima bean (Phaseolus lunatus) and Medicago truncatula (Leitner et *al.*, 2005). Up to now, only little has been known about the early events following tissue damage and the cross talk between the signaling networks that are up-regulated by the feeding herbivore. The interaction of various pathways in the networks is assumed to result in an integrated overall response that initiates the emission of a characteristic volatile pattern (Fig. 1). A subset of oxylipin compounds (jasmonic acid [JA], its precursors, and related compounds) very likely act as a master switch for herbivore-stimulated plant responses that controls the activation of distinct sets of defense genes leading to terpenoid formation (Arimura *et al.*, 2005). In addition, antagonistic or synergistic cross-reactions with other regulators such as ethylene control and coordinate the formation of a characteristic blend of volatiles (Schmelz *et al.*, 2003; Mithöfer *et al.*, 2005).

In this paper, we demonstrate the results obtained from the analysis using MecWorm, a mechanical caterpillar which has been designed to mimic herbivore-caused tissue damage. It has previously been reported that MecWorm-wounded lima bean leaves emit a blend of HIPVs that resembles that induced by feeding lepidopteran larvae (Mithöfer *et al.*, 2005). This discovery is novel and very important, since is generally believed that insect oral factors play a major role over physical wounding in the induction of HIPVs (Paré and Tumlinson, 1999). In fact, continuous damage is

A. Ramina et al. (eds.), Advances in Plant Ethylene Research: Proceedings of the 7th International Symposium on the Plant Hormone Ethylene, 353–358.

sufficient to induce HIPVs without any assistance from insect oral factors. Here, in order to further support this finding, we collected headspace volatiles from lima bean plants following different types of leaf wounding. Furthermore, we demonstrate the mode of calcium-dependent ethylene production in *M. truncatula*.

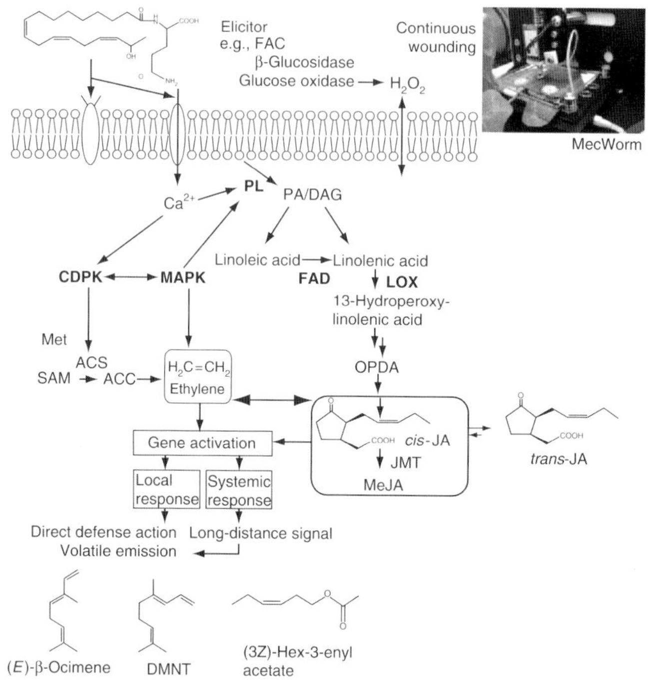

Fig. 1. Schematic representation of the signaling pathways required for herbivore-induced responses in plants. Elements in bold represent enzymes. Abbreviations: ACC, 1-aminocyclopropane-1-carboxylic acid; ACS, ACC synthase; CDPK, calcium-dependent protein kinase; DAG, diacylglycerol acid; DMNT, (*E*)-4,8-dimethyl-1,3,7-nonatriene; FAC, fatty acid-amino acid conjugate; FAD, ω-3 fatty acid desaturase; JA, jasmonic acid; JMT, JA carboxyl methyltransferase; LOX, lipoxygenase; MAPK, mitogen-activated protein kinase; MeJA, methyl JA; OPDA, 12-oxophytodienoic acid; PL, phospholipase; PA, phosphatidic; SAM, *S*-adenosyl methionine.

2. Materials and Methods

2.1 Plant treatment

For infestation of lima bean (*Phaseolus lunatus*; 4 weeks old), 7–8 third-instar *Spodoptera littoralis* larvae were placed on leaves of a lima bean plant. Manual wounding was performed on lima bean leaves by punching 6 mm diameter of 18 holes and 4–6 holes into each primary leaf and other small leaves, respectively. The MecWorm device and its operation (Raupe1.3) were described in our earlier paper (Mithöfer *et al.*, 2005). A single punch was programmed to obtain three circle sectors every 4 s and to obtain the next three every 18 s.

2.2 Analysis of plant volatiles with SuperQ adsorbent

For headspace analysis, lima bean leaves were subjected to caterpillar or single mechanical wounding treatment, or controls, were enclosed in a glass container (2.5 l). When plants were exposed to MecWorm wounding, a primary leaf of plants was enclosed in a Plexiglas cabinet (approximately 500 ml) and punched for the time indicated. The emitted volatiles were trapped onto SuperQ adsorbent (10 mg, Alltech Associates, Deerfield, IL, USA) using the closed loop dripping system. The collected volatiles were eluted with dichloromethane (3 × 20 μl) containing *n*-bromodecane (200 ng μl^{-1}) as an internal standard. Samples were analyzed on a Thermo-Quest/Finnigan TRACE GC 2000 with a TRACE MS (Manchester, UK) equipped with an EC™-5 capillary column (0.25 mm i.d. × 15 m with 0.25 mm film, Alltech Associates).

2.3 Ethylene analysis

1,2-Bis-(2-aminophenoxy)ethane-*N*,*N*,*N'*,*N'*-tetra acetic acid (BAPTA) was applied by placing the roots of freshly unearthed *M. truncatula* plants in water containing the Ca^{2+} chelator (0.2 mM BAPTA aq. solution). Controls were placed in water. Six *Spodoptera exigua* larvae were immediately allowed to feed on each plant. Ethylene production was measured in real time with a photoacoustic laser spectrometer consisting of a fine-tunable CO_2 laser and two resonant photoacoustic cells (Beßler *et al.*, 1998).

3. Results and Discussion

3.1 Monoterpene emitted from damaged leaves

The monoterpene (*E*)-β-ocimene was the most significant volatile in lima bean plants following herbivore and MecWorm wounding (Mithöfer *et al.*, 2005). When lima bean leaves were damaged by feeding *Spodoptera littoralis* or MecWorm wounding, the emission of (*E*)-β-Ocimene was significantly induced 1 day and 2 days after exposing the leaves to the damage (Fig. 2).

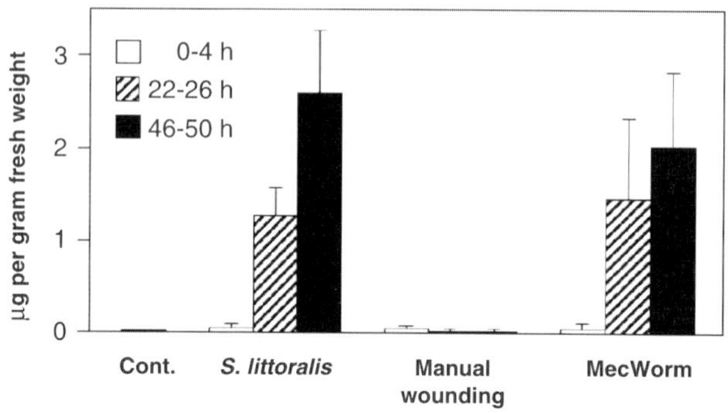

Fig. 2. (*E*)-β-Ocimene emissions from lima bean leaves following *Spodoptera littoralis* feeding, manual, or MecWorm wounding. Data represent the mean + standard error ($n = 4$).

By contrast, nonwounded plants and plants in which the leaves were manually wounded by punching once at the onset of the experiments did not induce the emission of (*E*)-β-ocimene up to 2 days. These results indicate that mechanical damage by a feeding larva is sufficient to induce HIPVs even without any assistance from oral factors.

3.2 Calcium-dependent ethylene production

Membrane depolarization and Ca^{2+} influx at the very edge of the damaged area belong to the early events after tissue damage by herbivores (Maffei *et al.*, 2004). To test whether or not this Ca^{2+} influx is also involved in the herbivory-induced ethylene production, the *M. truncatula* plants were treated with the Ca2+-chelator BAPTA following by Spodoptera *exigua* treatment. In addition, we studied whether and how herbivory-promoted Ca^{2+} signaling might impact ethylene production (Fig. 3). The level of diurnal ethylene emitted from uninfested plants of *M. truncatula* (10–20 nl/h) was ca. tenfold higher than that reported for lima bean (Arimura *et al.*,

2002). Irrespective of this high emission level, the ethylene emission from herbivore-damaged plants increased further after the onset of the second light period.

Infested plants, treated with BAPTA, displayed an emission rate similar to that of control plants during the first 27 h. Treatment of uninfested plants with BAPTA strongly reduced the ethylene emission, indicating that Ca^{2+} is required to maintain a consistent level of ethylene. The involvement of the ethylene signaling in the herbivore- or MecWorm-induced HIPV formation is a promising field of future research.

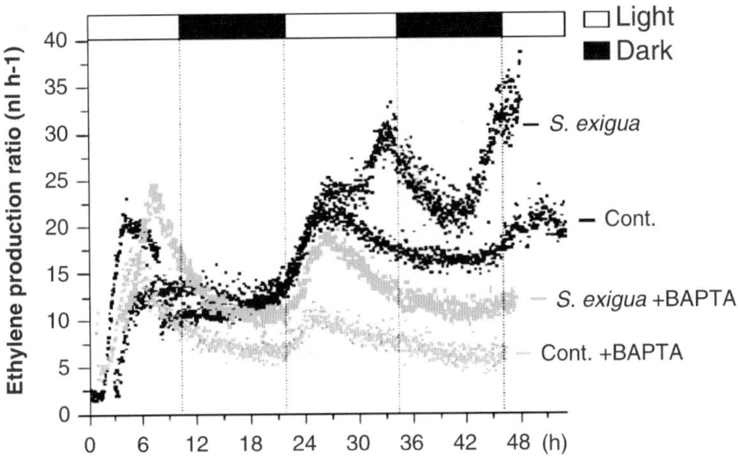

Fig. 3. Calcium signaling may contribute to regulating ethylene production in *M. truncatula* and ethylene production in a herbivore (*S. exigua*)-infested and/or BAPTA-exposed plant. The unexposed plant served as a control. An independent experiment was repeated twice, with similar results each time. A single set of the data is represented here.

References

Arimura G, Ozawa R, Nishioka T, Boland W, Koch T, Kühnemann F, Takabayashi J (2002) Herbivore-induced volatiles induce the emission of ethylene in neighboring lima bean plants. Plant J 29:87–98

Arimura G, Kost C, Boland W (2005) Herbivore-induced, indirect plant defences. Biochim Biophys Acta 1734:91–111

Beßler B, Schmitgen S, Kühnemann F, Gäbler R, Urban W (1998) Light-dependent production of ethylene in *Tillandsia usneoides* L. Planta 205:140–144

Horiuchi J, Arimura G, Ozawa R, Shimoda T, Takabayashi J, Nishioka T (2001) Exogenous ACC enhances volatiles production mediated by jasmonic acid in lima bean leaves. FEBS Lett 509:332–336

Leitner M, Boland W, Mithöfer A (2005) Direct and indirect defences induced by piercing-sucking and chewing herbivores in *Medicago truncatula*. New Phytol 167:597–606

Maffei M, Bossi S, Spiteller D, Mithöfer A, Boland W (2004) Effects of feeding *Spodoptera littoralis* on lima bean leaves. I. Membrane potentials, intracellular calcium variations, oral secretions, and regurgitate components. Plant Physiol 134:1752–1762

Mithöfer A, Wanner G, Boland W (2005) Effects of feeding *Spodoptera littoralis* on lima bean leaves. II. Continuous mechanical wounding resembling insect feeding is sufficient to elicit herbivory-related volatile emission. Plant Physiol 137:1160–1168

Paré PW, Tumlinson JH (1999) Plant volatiles as a defense against insect herbivores. Plant Physiol 121:325–332

Schmelz EA, Alborn HT, Tumlinson JH (2003) Synergistic interactions between volicitin, jasmonic acid and ethylene mediate insect-induced volatile emission in *Zea mays*. Physiol Plant 117:403–412

Metabolomic approaches to understand ethylene mediated defenses in *Arabidopsis thaliana* against *Botrytis cinerea*

Lloyd, A.J.*, Smith, A.R. and Mur, L.A.J.

Institute of Biological Sciences, University of Wales, Aberystwyth, Edward Llwyd Building, Ceredigion, Wales SY23 3DA, UK. (*Corresponding author: ajl01@aber.ac.uk)

1. Introduction

Botrytis cinerea (teleomorph, *Botryotinia fuckeliana*), more commonly known as the cause of grey mould, has a wide host range of over 200 plant species, including many economically important crops such as tomato, grape and potato. This necrotrophic fungus secretes cell wall degrading enzymes which kill host plant tissue in advance of its own growth, metabolising any nutrients that are released (Govrin and Levine, 2000). Resistance against necrotrophs is dependent upon the synergistic action of jasmonic acid (JA) and Ethylene signaling cascades (Thomma *et al.*, 1999a; Penninckx *et al.*, 1996; Knoester *et al.*, 1998; Penninckx *et al.*, 1998; Thomma *et al.*, 1998). However, much of the role of ethylene in plant interactions with necrotrophic fungi remains to be elucidated.

The model plant *Arabidopsis thaliana* has been shown to be susceptible to *Botrytis cinerea* and offers a wide range of well-characterized mutants with altered ethylene perception and transduction (Guzman and Ecker, 1990). These mutants may exhibit enhanced or compromised defence against necrotrophs. The use of these mutants will enable snapshots of the metabolome to be obtained during metabolic reprogramming of the host due to attack (Kieber, 1997). The 'metabolome' is defined as the quantitative complement of all the low molecular weight molecules present under defined conditions (Oliver *et al.*, 1998). Mutated genes will cause qualitative changes in metabolite pools compared to wild type. These changes can be revealed by producing biochemical fingerprints of samples without focusing on the identification of individual compounds (Goodacre *et al.*, 2003; Johnson *et al.*, 2003; Kaderbhai *et al.*, 2003; Allwood *et al.*, 2006). Fourier Transform Infrared Spectroscopy (FT-IR) is a rapid screening tool for such metabolite fingerprinting (Goodacre *et al.*, 1996). FT-IR uses the

A. Ramina et al. (eds.), Advances in Plant Ethylene Research: Proceedings of the 7th International Symposium on the Plant Hormone Ethylene, 359–368.

vibrational characteristics of chemical bonds within molecules within the samples to produce a biochemical 'fingerprint' (Bauer and Richter, 1996). FT-IR has been used to differentiate mutants by their metabolic complements (Oliver *et al.*, 1998; Winder *et al.*, 2004; Goodacre *et al.*, 2000) and to classify bacterial samples (Winson and Kell, 1997), and is increasingly being applied to plant biology (Johnson *et al.*, 2003; Johnson *et al.*, 2004).

Metabolite fingerprinting can provide evidence of biochemical discrimination and provide biochemical targets for further profiling analysis. More targeted metabolic analysis can be preformed using electrospray mass spectroscopy (ESI-MS) which has been used to identify chemical compounds in plant samples (Goodacre *et al.*, 2003; Allwood *et al.*, 2006).

Very few metabolomic analyses have been preformed to investigate the effects of two interacting organisms. Recently, Allwood *et al.* (2006) adopted a metabolomic approach to elucidate key discriminatory nonpolar metabolites in response to challenge of *Brachypodium distachyon* with *Magnaporthe grisea.*

The objective of this study was to apply metabolomic fingerprinting and profiling, supported by mathematical and statistical models to analyze the *B.cinerea–A.thaliana* interaction focusing on the role of ethylene. Mutants studied included the ethylene insensitive *etr1-1,* which has a dominant mutation in the gene that encodes the ethylene receptor (Bleecker *et al.*, 1988) and the constitutive ethylene response mutant, *ctr1-1* (Kieber *et al.*, 1993) that is mutated in the MAP3 kinase CTR1 gene. This study demonstrates the development of highly reproducible, robust and representative datasets showing distinct genotypic and response-specific biochemical profiles illustrating the potential to identify targets and novel hypotheses for further investigation.

2. Materials and Methods

2.1 Challenge of A.thaliana *with* B.cinerea

All experimental plant material was grown in a Fotron 600-H growth cabinet (Fisons Environmental Equipment) set at 24°C with an 8 h photoperiod (80 $\mu mol^{-2} s^{-1}$ at shelf height, with 50–80% relative humidity)

B. cinerea was cultured on potato dextrose agar plates for 4 weeks (PDA; Oxoid) at 20°C with a 12 h photoperiod in a Gallenkamp illuminated cooled incubator (Sanyo Biomedical Europe BV). Conidia were harvested from the surface of the plates by flooding with sterile potato dextrose broth (PDB; Oxoid) and dislodging conidia with an L-shaped glass rod, filtering and then diluted to 1×10^5 spores ml^{-1} in PDB. Plants were inoculated by spraying using a mini airbrush (Amtech) attached to a compressor (35/20 Bambi air Ltd.), subsequently covered with propagator hoods and placed in Fotron 600-H growth cabinets set as described above.

2.2 Histochemical staining methods

The first staining technique was using trypan blue (Koske and Gemma, 1989) where leaves were placed in Petri dishes containing 3 mm filter paper (Whatman) soaked in clearing solution (5 M chloral hydrate in 2.4:1:1 of 95% ethanol, 90% lactic acid, chloroform). The leaves were treated until they were fully cleared of chlorophyll then stained in a solution of 0.1% trypan blue, in lactoglycerol (1:1:1 85% lactic acid: 90% glycerol:ddH_2O) for 48 h.

The second staining method was the KOH-aniline blue fluorescence technique (Hood and Shew, 1996) where leaves were autoclaved at 121°C in 5 ml of 1 M KOH for 2 min, then rinsed in sterile distilled water and stained with 0.05% aniline blue dye (No. 12642 George T. Gurr Ltd. in 0.067 M K_2HPO_4 pH9).

Leaves were mounted on glass slides in lactoglycerol, with the exception of leaves stained for fluorescent aniline blue which were mounted in the stain solution. A cover slip was placed over the top and observed under a Zeiss microscope (Axioplan) with white light or UV illumination (excitation 355 nm, emission 450 nm).

2.3 Fourier Transform Infrared analysis

Plant samples were ground up with a pestle and mortar and 10% of ddH_2O was added. Aliquots of 5 μl plant material were loaded into a 400 well aluminium plate and oven-dried (50°C, 50 min) then placed onto the motorized stage of a reflectance thin layer chromatography (TLC) accessory attached to a Brucker IFS28 FT-IR Spectrometer (Bruker Spectrospin Ltd.) equipped with a liquid nitrogen-cooled MCT (mercury-cadmium-telluride) detector. OPUS 2.1 software was used running under the IBM OS/2 Warp operating system at a resolution of 3.84 cm^{-1}, collecting spectra over the wavenumbers 4000 cm^{-1}–600 cm^{-1} at an acquired at a rate of 20 s^{-1}.

2.4 Electrospray injection mass spectroscopy

Samples were ground using a ball mill (MM200, Retsch) and 1 ml of chloroform: methanol: sterile dH_2O (1:2.5:1) was added. The samples were then mixed using a shaking table (Janke and Kunkel, VX 2E) in a cold room at 3°C for 15 min. The aqueous layer was removed and 0.5 ml of sterile ultra pure dH_2O was added. The extracts were mixed, centrifuged at 3°C at 14000 rpm (Hettich, EBA 12R) for 3 min and then dried down in a Savant AES2000 (Thermo Electron Corp) automatic environmental speed vacuum concentrator. Metabolomic profiling of the extracts was carried out using FI-ESI-MS on a micromass LCT mass spectrometer. All extracts were introduced into the electrospray source via a Waters 2695 Alliance Separations Module (Waters Ltd.) in both ES −ve and ES +ve ionisation mode. Extracts were reconstituted in 100 μl 80% [v/v] methanol and data were acquired over the *m/z* range 100–1400 Th.

2.5 Cluster analysis

Datasets obtained were imported into MATLAB version 6.5 (The MathWorks Inc) running under Microsoft Windows NT. The data were first analyzed using unsupervised Principal Component Analysis (PCA; Causton, 1987; Jolliffe, 1986) which models on the natural variability of the dataset and in this case was used to reduce the dimensionality of the data for supervised Discriminant Function Analysis (DFA; Manly, 1994). PC-DFA discriminates between groups based on the retained PCs and the *a priori* knowledge of the class structure of the data set. DFA maximizes between-class variance whilst minimizes within-class variance. Validation of the PC-DFA cluster analysis was achieved by constructing PC-DFA on biological replicates, the 'training set', and projecting the remaining two biological replicates 'test set' into the PC-DFA ordinate space (Jarvis and Goodacre, 2004).

3. Results and Discussion

Generalized macroscopic symptoms on wild-type *A.thaliana* Columbia-0 (Col-0) included grey/brown lesions after 72 h post-infection (hpi) along with water soaking (Fig. 1a). Grey/brown mould growing on the surface of affected areas was observed after 96 hpi. *Etr1-1* showed premature symptoms with lesions appearing at 48 hpi, progressing to full chlorotic leaf collapse by 120 hpi (Fig. 1a). In contrast, whole leaf collapse occurred later in Col-0. These observations suggested that the ethylene insensitive mutant was more susceptible to *Botrytis* and thus ethylene may be an important regulator of plant defenses against this necrotroph (Thomma *et al.*, 1999b). On the other hand, ctr1-1 showed delayed symptom responses and increased resistance to the necrotroph with minimal lesion development but increased whole leaf collapse at 96/120 hpi (Fig. 1a).

Microscopic analysis using aniline blue staining (and UV light) and vital staining using trypan blue showed that fungal germination and hyphal development was independent of the plant genotype (data not shown). By 24 hpi a fungal burst of vesicle-like bodies occurred which may be involved in tissue maceration, followed by colonization and formation of asexual conidiophores by 72 hpi (Fig. 1b).

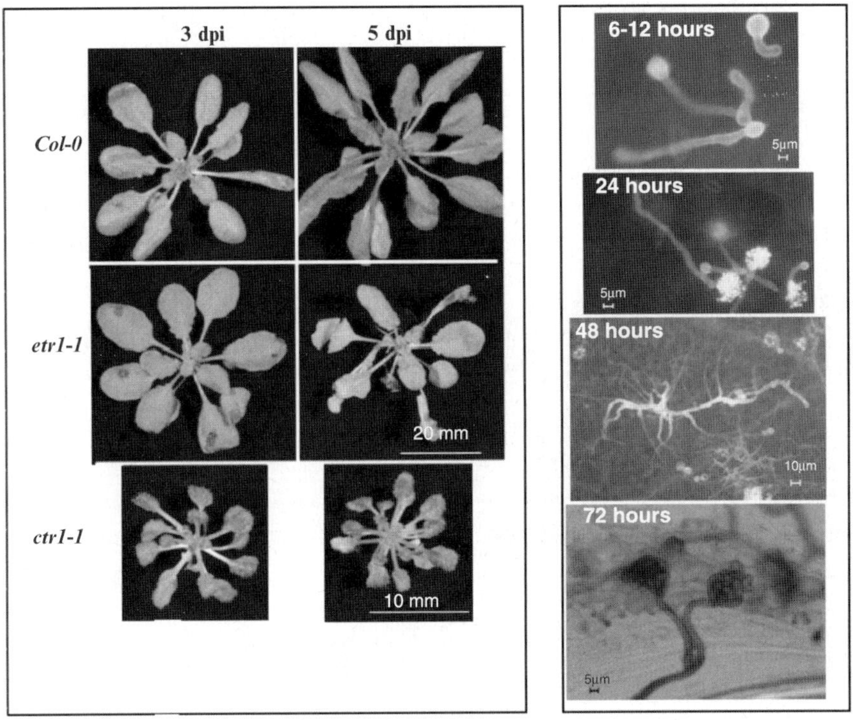

Fig. 1. (Color figure in the Annex, p.461). (**a**) Infection sites of *B. cinerea* on *A. thaliana* at 72 and 120 hpi, showing heightened lesion development on *etr1-1* compared to the wild-type Col-0. Increased chlorosis/necrosis is shown by *ctr1-1* at 120 hpi compared to Col-0. (**b**) *B.cinerea* development using Aniline Blue (and UV light) and vital staining using Trypan Blue. Between 6 and 12 hpi asynchronized germination occurs followed by the fungal burst. Colonization and conidiophore formation occur by 72 hpi. These events were independent of genotype.

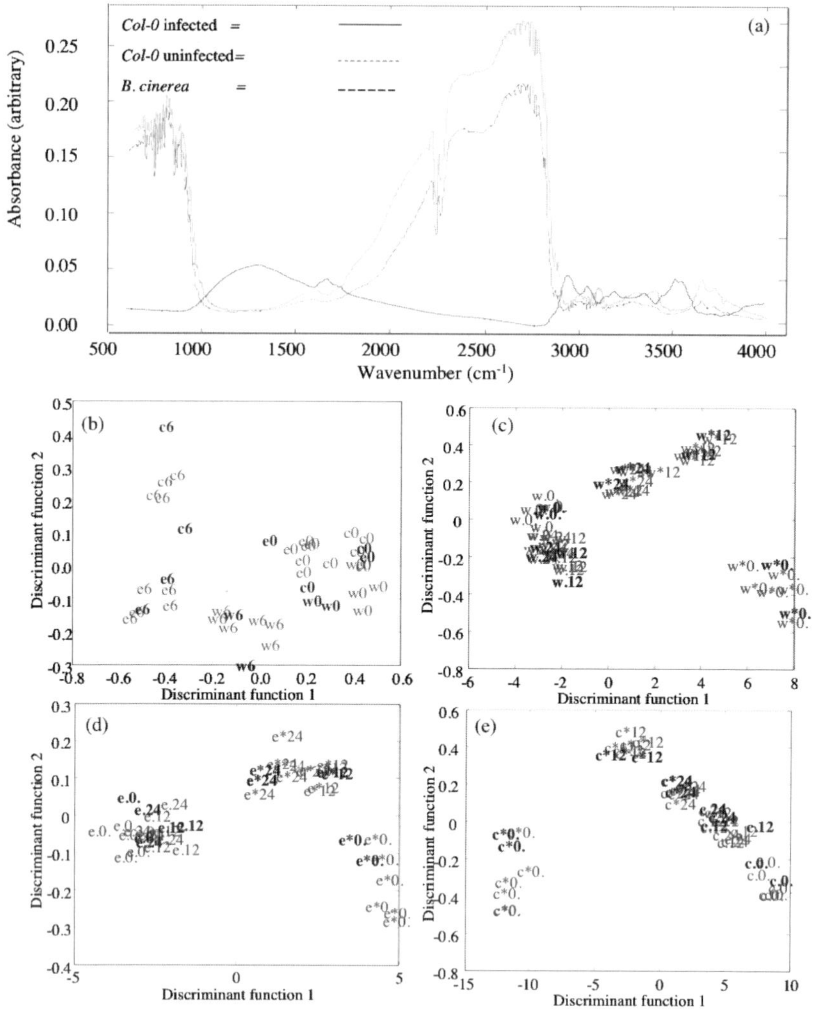

Fig. 2. (Color figure in the Annex, p.462). Fourier-transform infrared (FT-IR) spectra with discriminant function analysis (DFA) of the *B.cinerea-A thaliana* interaction (**a**) FT-IR spectra of *B.cinerea* alone (large dashed line), uninfected *A.thaliana* (solid line) and infected *A.thaliana* (Col-0) (small dashed line) (**b**) DFA of Col-0, *etr1-1* and *ctr1-1* (w, e, and c respectively) 0 and 6 hpi (0, 6 respectively) based upon the first 12 PCs accounting for <99.95% of variance. Validation involved training sets based on six replicates on which the remaining two replicates have been projected (bold). (**c–f**) validated DFA models (as before) of *B.cinerea* inoculated (*) and controls over 24 h (0, 12, 24) with (**c**) Col-0 (**d**) *etr1-1* (**e**) *ctr1-1*, explaining <99.95% of total explained variance. The effects of pathogen inoculation are indicated with arrows.

FT-IR was employed to access the unique biochemical fingerprint events that occur prior to necrosis following challenge of *A. thaliana* wild-type (Col-0) and ethylene mutants *ctr1*- and *etr1-1* with *B. cinerea*. FT-IR spectra of pure *B. cinerea* samples and challenged plant material were highly different, which was to be expected (Fig. 2a). However, the spectra for infected and non-infected plant material showed very little variation and could not be interpreted by eye (Fig. 2a). Therefore, chemometrics were employed to de-convolute these spectra allowing the assessment of whether the spectra contained sufficient biochemical information to allow differentiation between challenged plant material and controls.

The unsupervised principle component analysis, PCA, could not separate the challenged material from the controls due to the highly heterogeneous plant samples. However, discriminant function analysis (DFA), a supervised technique requiring *a priori* knowledge of the class structure of the dataset was performed on the principle components (PCs). Projection validation by the superimposition of the 'test set' on the corresponding 'training set' suggested highly reproducible experimental conditions. The validated DFA plots for the three genotypes, Col-0, *etr1-1* and *ctr1-1*, over 24 h are shown in Figs. 2b–e respectively.

The infection response differed with genotype, and thus it was likely that FT-IR was detecting metabolite changes associated with the enhanced/compromised defense due to the specific ethylene mutant. These biochemical changes could be detected with FT-IR as early as 6 hpi (Fig. 2b) suggesting different responses to the initial germination and hyphal development of the fungus.

In terms of identifying discriminatory metabolites, FT-IR does not have sufficient resolution so therefore the more targeted analysis of electrospray injection mass spectroscopy (ESI-MS) was employed. Again, as with the FT-IR datasets, PC-DFA was used to differentiate infected and non-challenged ESI-MS spectra for each genotype and validated suggesting highly reproducible models (Fig. 3). Examination of the DF loading vectors identified ions used to create the PC-DFA models (Figs. 4a and b).

Mass ions that discriminated between plant–pathogen interactions were tentatively identified (data not shown) but confirmation of these key metabolites is the next logical step using electrospray injection tandem mass spectroscopy (ESI-MS-MS).

Therefore, to conclude, metabolic changes can be detected in response to pathogenic challenge before visible localized symptoms (i.e., at 6 h post inoculation) producing robust and reproducible mathematical models irrespective of sources of variation. This study has demonstrated the potential of applying metabolomics to generate novel hypotheses and in this case, to enable the classification of control and challenged *A.thaliana* ethylene mutants, which will ultimately aid in the identification of key discriminatory metabolites and pathways involved in plant defense.

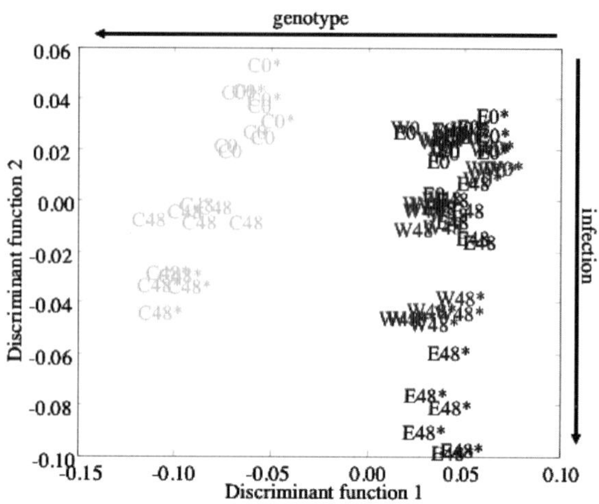

Fig. 3. Electrospray injection mass spectroscopy (ESI-MS) dataset with discriminant function analysis (DFA) of the *B.cinerea* interaction (*) and controls (just PDB; no symbol) with *Col-0, etr1-1* and *ctr1-1* (W, E, and C respectively) at 0 and 48 hpi (0, 48 respectively) based upon the first 12 PCs accounting for <99.95% of total explained variance.

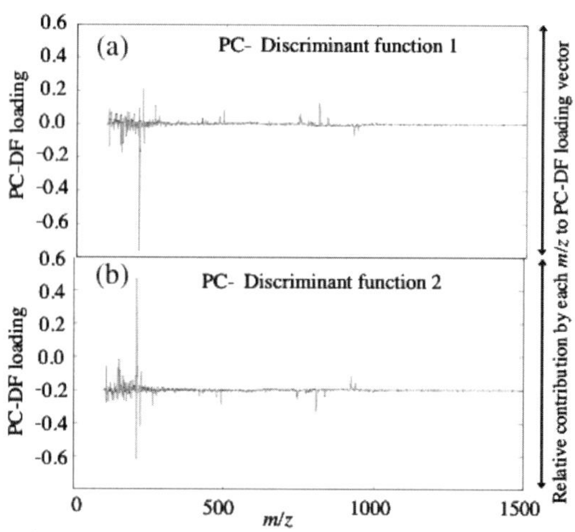

Fig. 4. PC-DFA loading vectors plotted against wavenumber (cm^{-1}) for (**a**) PC-DF1 and (**b**) PC-DF2.

References

Allwood JA, Ellis DI, Heald JK, Goodacre R, Mur LAJ (2006) Metabolomic approaches reveal that phosphatidic and phosphatidyl glycerol phospholipids are major discriminatory non-polar metabolites in responses by *Brachypodium distachyon* to challenge by *Magnaporthe grisea*. Plant Journal 46: 351–369

Bauer G, Richter W (1996) Optical characterization of epitaxial semiconductor layers, Springer-Verlag, Berlin pp 225–232

Bleecker A, Estelle M, Somerville C, Kende H (1988) Insensitivity to ethylene conferred by a dominant mutation in *Arabidopsis thaliana*. Science 241: 1086–1089

Causton DR (1987). A Biologist's Advanced Mathematics. London: Allen & Unwin

Goodacre R, Timmins EM, Rooney PJ, Rowland JJ, Kell DB (1996) Rapid identification of *Streptococcus* and *Enterococcus* species using diffuse reflectance-absorbance Fourier transform infrared spectroscopy and artificial neural networks. FEMS Microbiology Letters 140(2–3): 233–239

Goodacre R, Shann B, Gilbert RJ, Timmins ÉM, Mc-Govern AC, Alsberg BK, Kell DB, Logan NA (2000) The detection of the dipicolinic acid biomarker in *Bacillus* spores using Curie-point pyrolysis mass spectrometry and Fourier-transform infrared spectroscopy. Analytical Chemistry 72: 119–127

Goodacre R, York EV, Heald JK, Scott, IM (2003) Chemometric discrimination of unfractionated plant extracts analyzed by electrospray mass spectrometry. Phytochemistry 62 (6): 859–863

Govrin EM, Levine A (2000) The hypersensitive reaction facilitates plant infection by the necrotrophic fungus *Botrytis cinerea*. Current Biology 10: 751–757

Guzman P, Ecker JR (1990) Exploiting the triple response of *Arabidopsis* to identify ethylene-related mutants. Plant Cell 2: 513–523

Hood ME, Shew D (1996) Application of KOH-aniline blue fluorescence in the study of plant-fungal interactions. Phytopathology 86: 704–708

Jarvis RM, Goodacre R (2004) Rapid discrimination of bacteria using surface enhanced Raman spectroscopy. Analytical Chemistry 76: 40–47

Johnson HE, Broadhurst D, Goodacre R, Smith AR (2003) Metabolic fingerprinting of salt-stressed tomatoes. Phytochemistry 62(6): 919–28

Johnson HE, Broadhurst D, Kell DB, Theodorou M, Merry RJ, Griffith GW (2004) High-throughput metabolic fingerprinting of legume silage fermentations via Fourier transform infrared spectroscopy and chemometrics. Applied Environmental Microbiology 70(3):1583–1592

Jolliffe IT (1986) Principal Component Analysis. New York: Springer-Verlag

Kaderbhai NN, Broadhurst DI, Ellis DI, Goodacre R, Kell DB (2003) Functional genomics via metabolic footprinting: monitoring metabolite secretion by *Escherichia coli* tryptophan metabolism mutants using FT–IR and direct injection electrospray mass spectrometry. Comparative and Functional Genomics 4: 376–391

Kieber JJ (1997) The ethylene response pathway in *Arabidopsis*. Annual Review of Plant Physiology and Plant Molecular Biology 48: 277–296

Kieber JJ, Rothenberg M, Roman G, Feldmann KA, Ecker JR (1993) *CTR1*, a negative regulator of the ethylene response pathway in *Arabidopsis*, encodes a member of the raf family of protein kinases. Cell 72: 427–441

Knoester M, Van Loon LC, Van Den Heuvel J, Hennig J. Bol JF, Linthorst HJM (1998) Ethylene-insensitive tobacco lacks nonhost resistance against soil-borne fungi. Proceedings of the National Academy of Sciences, USA 95: 1933–1937

Koske RE, Gemma JN (1989) A modified procedure for staining roots to detect VA mycorrhizas. Mycological Research 92: 486–505

Manly BFJ (1994) Multivariate Statistical Methods: A Primer. London: Chapman & Hall

Oliver SG, Winson MK, Kell DB, Baganz F (1998) Systematic functional analysis of the yeast genome. Trends in Biotechnology 16: 373–378

Penninckx IAMA, Eggermont K, Terras FRG, Thomma BPHJ, De Samblanx GW, Buchala A, Metraux J-P, Manners JM, Broekaert WF (1996) Pathogen-induced systemic activation of a plant defensin gene in *Arabidopsis* follows a salicylic acid-independent pathway. Plant Cell 8: 2309–2323

Penninckx IAMA, Thomma BPHJ, Buchala A, Métraux J-P, Broekaert W (1998) Cooperative activation of jasmonate and ethylene response pathways in parallel is required for induction of a plant defensin gene in Arabidopsis. Plant Cell 10: 2103–2114

Thomma BPHJ, Eggermont K, Penninckx IAMA, Mauch-Mani B, Vogelsang R, Cammue, BPA, Broekaert WF (1998) Separate jasmonate-dependent and salicylated dependent defense-response pathways in *Arabidopsis* are essential for resistance to distinct microbial pathogens. Proceedings of the National Academy of Sciences, USA 95: 15107–15111

Thomma BPHJ, Eggermont K, Tierens KFM-J, Broekaert WF (1999a) Requirement of Functional *Ethylene-Insensitive 2* Gene for Efficient Resistance of Arabidopsis to Infection by *Botrytis cinerea* Plant Physiology 121: 1093–1101

Thomma BPHJ, Nelissen I, Eggermont K, Broekaert WF (1999b) Deficiency in phytoalexin production causes enhanced susceptibility of *Arabidopsis thaliana* to the fungus *Alternaria brassicicola*. Plant Journal 19: 163–171

Winder CL, Carr E, Goodacre R, Seviour R (2004) The rapid identification of *Acinetobacter* species using Fourier transform infrared spectroscopy. Journal of Applied Microbiology 96: 328–339

Winson MK, Kell DB (1997) If you've got it, flaunt it—rapid screening for microbial biocatalysts. Trends in Biotechnology 15(4): 120–122

Characterization of tomato *Sl-MBF1* transcriptional coactivator gene family

Sanchez-Ballesta, M.T., Hommel, M., Jaimes-Miranda, F., Tournier, B., Zegzouti, H., Mila, I., Latché, A., Pech, J.C., Bouzayen, M. and Regad, F.*

Laboratoire de Génomique et Biotechnologie des Fruits INRA/INP-ENSAT UMR990, Castanet-Tolosan, F31326, France. (*Corresponding author: regad@ensat.fr)

1. Introduction

The regulation of eukaryotic gene activity at the transcriptional level involves multiprotein complexes and requires the concerted action of transcription factors and coactivator proteins. Transcription factors bind to DNA in a sequence-specific manner and essentially mark a gene for activation or repression through the recruitment of coactivator or corepressor proteins (Spiegelman and Heinrich, 2004). Coactivators are proteins that connect a transacting factor with a component of the basal transcription machinery allowing transcriptional activation to proceed (Roeder, 1991; Li *et al*., 1994). Initially, multiprotein bridging factor 1 (MBF1) genes were identified in *Bombix mori* as coactivators (Roeder, 1991) and thereafter shown to contribute with other proteins to the building of TAF complexes (TATA box protein associated factors) that are essential for transcriptional initiation. MBF1 proteins recruit the basal transcription machinery to specific transcription factors bound to their target promoters (Naar *et al*., 2001). It is quite clear now that transcriptional regulation at the level of coactivators is a substantial part of the mechanisms underlying the control of gene expression and that coactivators allow functional integration of transcription factors from different types (Spiegelman and Heinrich, 2004).

We have cloned the first plant MBF1-like gene (named ER24) on the basis of its regulation by the plant hormone ethylene in tomato fruit. In addition, this gene was shown to be induced during fruit ripening (Zegzouti *et al*., 1999). Three stress-regulated MBF1 genes were subsequently identified in *Arabidopsis thaliana* and their encoded proteins were able to functionally complement the MBF1 yeast mutant and to bridge TBP and the yeast transcriptional activator involved in amino acids and purines biosynthesis (GCN4) *in vitro* (Tsuda *et al*., 2004). It was also reported that

A. Ramina et al. (eds.), Advances in Plant Ethylene Research: Proceedings of the 7th International Symposium on the Plant Hormone Ethylene, 369–375.

ER24-like genes were up-regulated by heat-shock treatment in *Retama raetam* (Pnueli *et al.*, 2002) and by drought and heat shock in tobacco and Arabidopsis (Rizhsky *et al.*, 2002; Rizhsky *et al.*, 2004). Direct evidence of the involvement of MBF1 in plant responses to environmental stresses came from ectopic expression of the MBF1c gene which resulted in enhancing tolerance of the transgenic Arabidopsis lines to heat and osmotic stresses (Suzuki *et al.*, 2005). These data suggest that MBF1-like genes may represent primary targets of physiological signals associated with a variety of plant developmental processes.

We describe in here the isolation and characterization of three new members of the MBF1 gene family in the tomato. Like Sl-ER24, the remaining tomato MBF1 genes, encode functional transcriptional coactivators as demonstrated by complementation assays of the MBF1-disruptant *S. cerevisiae* strain (Δmbf1). Detailed expression studies revealed a distinct pattern of expression for Sl-ER24 compared to other MBF1-like genes, suggesting a specific role for ER24 in ethylene responses and in fruit ripening.

2. Materials and Methods

2.1 Plant material and treatments

Solanum lycopersicum (Ailsa Craig) plants were grown in soil under standard greenhouse conditions. Late immature green fruits, unable to produce ripening-related ethylene, were harvested and treated either with ethylene (50 $\mu l\ l^{-1}$) for 15 min, 1 h and 6 h, or with air. Samples corresponding to treated and untreated fruit, flowers, roots, leaves and stem were harvested for RNA extraction.

2.2 Isolation of tomato genomic clones

Genomic DNA was extracted from tomato leaves by a CTAB method. Genomic clones were isolated by PCR amplification and the amplified fragments were cloned into pGEM-T easy vector (Promega) and fully sequenced. Comparative analysis between the genomic clone and the corresponding cDNA sequences allowed introns and exons delimitation.

2.3 Functional complementation of yeast mutant strains

Sl-MBF1a-c and Sl-ER24 ORFs were cloned into the pYES TOPO vector and expressed in the Δmbf1 mutant strain.

2.4 Expression analyses by Quantitative RT-PCR

Real-time quantitative PCR was performed using SYBR GREEN PCR Master Mix. To determine relative fold differences for each sample in each experiment, the Ct value for Sl-ER24, Sl-MBF1a, Sl-MBF1b and Sl-MBF1c genes was normalized to the Ct value for Sl-Actin-51 a constitutive reference gene and was calculated relative to a calibrator using the formula 2-ΔΔCt.

3. Results and Discussion

3.1 Tomato MBF1 are encoded by a small multigene family

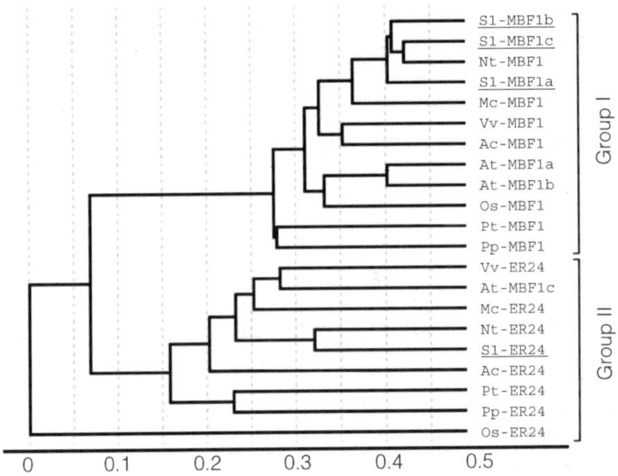

Fig. 1. Phylogenetic tree of plant MBF1s. Group I: Sl-MBF1a, Sl-MBF1b, Sl-MBF1c (S. lycopersicum), Nt-MBF1 (N. tabacum), Mc-MBF1 (M. crystallinum), Vv-MBF1 (*V. vinifera*), Ac-MBF1 (*A. cepa*), At-MBF1a, At-MBF1b (*A. thaliana*), Pt-MBF1 (*P. taeda*), Pp-MBF1 (*P. patens*). Group II: Vv- ER24, At-MBF1c, Mc-ER24, Nt-ER24, Sl-ER24, Ac-ER24, Pt-ER24, Pp-ER24. Members of the tomato MBF1 family are underlined.

Sl-ER24, Sl-MBF1a, Sl-MBF1b and Sl-MBF1c full-length cDNA clones encode putative proteins of 16.1 kDa, 15.3 kDa, 15.3 kDa and 15.4 kDa, respectively. High degree of sequence conservation was found between Sl-MBF1a, Sl-MBF1b and Sl-MBF1c deduced proteins ranging from 92.1 to 94.3% homology and 96.4 to 98.6% similarity. By contrast, Sl-ER24 derived protein displays no more than 44.2% homology and 62.6% similarity with the remaining tomato MBF1 proteins. Phylogenetic analyses indicate that tomato MBF1 genes are divided in two distinct groups. Sl-MBF1a, b and c belong to the same group while Sl-ER24 lies in a separate clade (Fig. 1).

3.2 Structure analysis of tomato MBF1 genomic clones

The genomic structure of MBF1 genes is remarkably conserved between Arabidopsis and tomato except for the size of the introns that are significantly larger in the tomato genes. Sl-ER24 and its putative Arabidopsis ortholog (At-MBF1c) are highly divergent from the remaining members of the multigene family and exhibit an intronless genomic structure (Fig. 2).

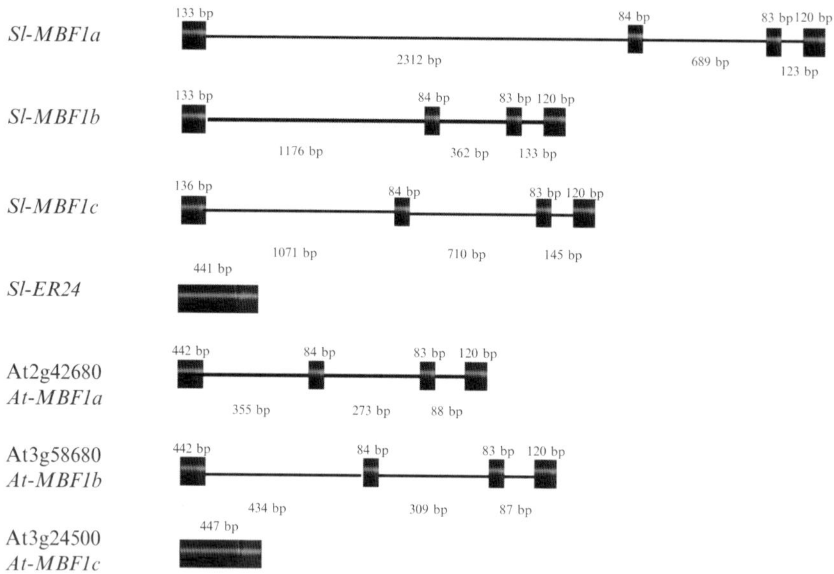

Fig. 2. Comparison of exon/intron structure of tomato and Arabidopsis MBF1 genes. Black boxes represent exons and lines connecting boxes denote introns. Numbers indicate intron and exon sizes in bp.

3.3 Tomato MBF1 gene family encodes functional transcriptional coactivators

A complementation test of yeast MBF1 mutation was used to assess the ability of the tomato MBF1 genes to function as transcriptional co-activators. Yeast Δmbfl strain was transformed with each of the tomato MBF1 cDNAs (Takemaru *et al.*, 1998). The ability of the tomato MBF1 genes to complement the yeast MBF1-disrupted strain reveals that all four tomato MBF1 genes encode functional transcriptional coactivator and indicates that the mechanism by which MBF1 proteins regulate transcription is well conserved among eukaryotic organisms from yeast to higher plants (data not shown).

3.4 Members of the tomato MBF1 gene family show distinct expression patterns

While Sl-ER24, Sl-MBF1a, Sl-MBF1b and Sl-MBF1c transcripts are detected in all tissues tested, Sl-ER24 and Sl-MBF1a show the most contrasted tissue-specific variation (Fig. 3). Sl-MBF1b transcripts are the most abundant in all tissues tested except in ripening fruit where Sl-ER24 transcripts display slightly higher accumulation levels. Sl-ER24 transcript accumulation is highest in ripening fruit and lowest in the stem whereas Sl-MBF1a transcripts accumulate preferentially in roots. Sl-ER24 is preferentially expressed in fruit and shows the weakest expression in leaves. The specific expression patterns of tomato MBF1 genes is indicative of distinct roles for the encoded proteins.

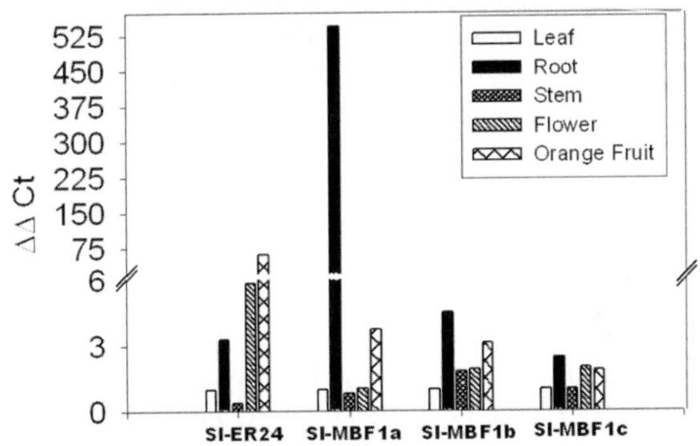

Fig. 3. Tissue-specific expression of tomato MBF1 genes. The levels of Sl-ER24, Sl-MBF1a, Sl-MBF1b and Sl-MBF1c transcripts were assessed by real time quantitative PCR using total RNAs isolated from leaves, root, stem, flower and orange fruit (65 dpa). ΔΔCt refers to the fold difference in each Sl-MBF1 expression relative to the leaf taken as a reference sample.

3.5 Sl-ER24 is highly induced during fruit ripening

While *Sl-MBF1a–c* show either weak or no regulation during fruit ripening S1-ER24 displays a clear ripening-associated regulation that is strongly impaired in the tomato ripening mutants *nor*, *rin* and *Nr*. Moreover, the expression of *Sl-ER24* was found to be transiently up-regulated upon ethylene treatment in immature green fruit where endogenous ethylene production is very low (data not shown). These data studies strongly suggest that Sl-ER24 could be involved in fruit development and ripening.

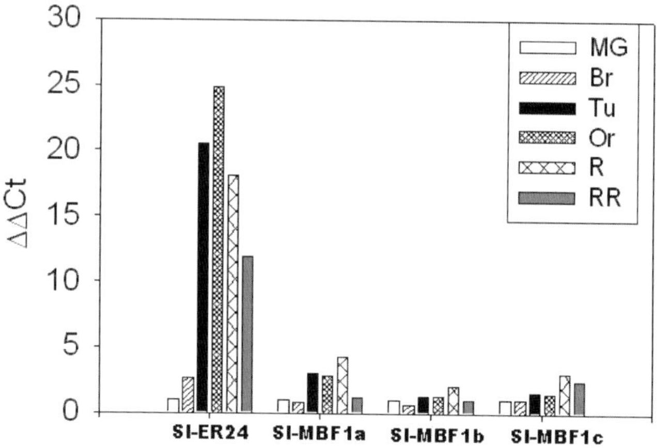

Fig. 4. Ripening-associated expression of tomato MBF1 genes. **A**: Transcript accumulation of Sl-ER24, Sl-MBF1a, Sl-MBF1b and Sl-MBF1c genes during tomato fruit ripening. Mature-green fruit (MG, 43 dpa); breaker fruit (Br, 53 dpa), turning fruit (Tu, 60 dpa); orange fruit (Or, 65 dpa); red fruit (R, 69 dpa) and red-ripe fruit (RR, 76 dpa).

References

Li, F.Q., Ueda, H. and Hirose, S. 1994. Mediators of activation of fushi tarazu gene transcription by BmFTZ-F1. Mol. Cell Biol. 14: 3013–3021

Naar, A.M., Lemon, B.D. and Tjian, R. 2001. Transcriptional co-activator complexes. Annu. Rev. Biochem. 70: 475–501.

Pnueli, L., Hallak-Herr, E., Rozenberg, M., Cohen, M., Goloubinoff, P., Kaplan, A. and Mittler, R. 2002. Molecular and biochemical mechanisms associated with dormancy and drought tolerance in the desert legume *Retama raetam*. Plant J. 31: 319–30.

Rizhsky, L., Liang, H. and Mittler, R. 2002. The combined effect of drought stress and heat-shock on gene expression in tobacco. Plant Physiol. 130: 1143–1151

Rizhsky, L., Liang, H., Shuman, J., Shulaev, V., Davletova, S. and Mittler, R. 2004. When defense pathways collide. The response of Arabidopsis to a combination of drought and heat stress. Plant Physiol. 134: 1683–1696.

Roeder, R.G. 1991. The complexities of eukaryotic transcription initiation: regulation of preinitiation complex assembly. Trends Biochem. Sci. 16: 402–408.

Spiegelman, B.M. and Heinrich, R. 2004. Biological control through regulated transcriptional co-activators. Cell 119: 157–167.

Suzuki, N., Rizhsky, L., Liang, H., Shuman, J., Shulaev, V. and Mittler, R. 2005. Enhanced tolerance to environmental stress in transgenic plants expressing the transcriptional co-activator multiprotein bridging factor 1c. Plant Physiol. 139: 1313–1322.

Takemaru, K., Harashima, S., Ueda, H. and Hirose, S. 1998. Yeast co-activator MBF1 mediates GCN4-dependent transcriptional activation. Mol. Cell Biol. 18: 4971–4976. Abel, S., Theologis, A. 1994. Transient transformation of Arabidopsis leaf protoplasts: a versatile experimental system to study gene expression. Plant J. 5: 421–427.

Tsuda, K., Tsuji, T., Hirose, S. and Yamazaki, K. 2004. Three Arabidopsis MBF1 homologs with distinct expression profiles play roles as transcriptional co-activators. Plant Cell Physiol. 45: 225–231.

Zegzouti, H., Jones, B., Frasse, P., Marty, C., Maitre, B., Latché, A., Pech, J.C. and Bouzayen, M. 1999. Ethylene-regulated gene expression in tomato fruit: characterization of novel ethylene-responsive and ripening-related genes isolated by differential display. Plant J. 18: 589–600.

Response of canola plants at the transcriptional level to expression of a bacterial ACC deaminase in the roots

Czarny, J.C.[1]*, Shah, S.[2] and Glick, B.R.[1]

[1]Department of Biology, University of Waterloo, Waterloo, Ontario, Canada. [2]Alberta Research Council, Vegreville, Alberta, Canada. (*Corresponding author: jczarny@sciborg.uwaterloo.ca)

1. Introduction

A model was previously proposed to explain how ACC deaminase-containing plant growth-promoting bacteria can lower plant ethylene levels and in turn stimulate plant growth (Glick *et al.*, 1998), especially under stress conditions. In this model, some of a plant's ACC will be exuded along with other small molecules and may be taken up by the rhizosphere bacteria and utilized as a food source. ACC is cleaved by ACC deaminase to form ammonia and α-ketobutyrate, compounds that are further metabolized by the bacteria. As a result of acting as a sink for ACC and lowering its level within the plant, the amount of ethylene that is produced by the plant is also reduced. Thus, the inhibition of plant growth by ethylene (especially during periods of stress) is decreased and these plants generally have longer roots and shoots, and greater biomass.

Stress ethylene is produced by plants when challenged with a variety of stresses including chilling, wounding, drought, some types of pathogens and excess heavy metals (Hyodo, 1991). One model, which explains the effects of stress ethylene on plants, emphasizes the fact that in stressed plant tissues there is an initial very small peak of ethylene close in time to the onset of stress and then a second much larger peak some time later (Robinson *et al.*, 2001). The first peak, which consumes the existing pool of ACC within plant tissues, is thought to initiate a protective response by the plant. ACC synthase genes are transcribed and more ACC begins to accumulate to fuel the second peak of ethylene which is so large that processes such as cell death, senescence and chlorosis are initiated.

Among the metals that are most toxic to plants are those that displace essential metal ions in biological processes, including cadmium, zinc, mercury, copper, lead and nickel (Prasad and Strazalka, 2000). High metal concentrations in the soil can cause increased ethylene production, (Goren

A. Ramina et al. (eds.), Advances in Plant Ethylene Research: Proceedings of the 7th International Symposium on the Plant Hormone Ethylene, 377–382.

and Siegel, 1976; Burd *et al.*, 1998), inhibit root and shoot development, reduce CO_2 fixation and limit sugar translocation (Prasad and Strazalka, 2000). Reduction in plant ethylene levels using either transgenic plants containing the ACC deaminase gene, under the control of the root specific *rolD* promoter or ACC deaminase-containing plant growth-promoting bacteria has been shown to enhance survival when heavy metals are present. (Burd *et al.*, 1998; Grichko *et al.*, 2000) Presented in this paper are the preliminary results of expression changes that occur in transgenic *Brassica napus* (canola) plants, possessing a bacterial ACC deaminase gene, when grown in the presence of nickel.

2. Methods

2.1 Plant growth conditions

This work was performed using transgenic canola (*Brassica napus* var. Westar) described in (Stearns *et al.*, 2006). Seeds of transformed and non-transformed canola were surface sterilized then grown in Promix BX greenhouse mix, watered as needed with tap water, and fertilized beginning on day 10 once a week with a fertilizer containing 20% nitrogen, 20% phosphorous and 20% potassium. Nickel-spiked soil was prepared by addition of a solution of $NiSO_4$ to the greenhouse mix to a concentration of 175 ppm Ni.

2.2 Microarrays

Arabidopsis II 60-mer oligonucleotide microarrays were purchased from Agilent Technologies along with all hybridization and washing solutions. These arrays contain 21575 probes designed to bind within the coding region and representing 80% of *Arabidopsis thaliana* genes. Total RNA was extracted from canola shoot tissues using a cesium chloride/guanidine isothiocyanate (GIT) method (Davis *et al.*, 1986), further purified using RNeasy Midi columns (Qiagen). RNA from three individual plants was isolated, cleaned, then pooled prior to labeling with Cy-3 or Cy-5 using the fluorescent direct labeling kit (Agilent Technologies) by following the protocol for 22 K arrays. Paired samples were hybridized washed and dried according to the hybridization protocol then scanned using an Axon 4000A scanner with GenePix 1.4 software (molecular devices). LOWESS normalization and expression analysis was performed using GeneSpring GX 7.3 (Agilent Technologies). Each sample comparison was done in duplicate.

3. Results and Discussion

3.1 Biomass

When grown in nickel, non-transformed canola seedlings displayed effects of metal stress that increased in severity as the concentration of the metal was increased. At high nickel concentrations, cotyledon leaves appeared yellow and had a curled phenotype, while roots stopped developing shortly after germination. On the other hand, seedlings of transgenic plants expressing a bacterial ACC deaminase gene developed faster and had more lateral roots than non-transformed plants both in the presence and absence of nickel. They also had longer roots and shoots, a less curled phenotype and developed a green colour even at high nickel concentrations.

Transgenic and non-transformed plants were very similar in appearance when grown for three weeks in un-spiked soil. These same plants, however, displayed a remarkable difference in appearance, when nickel was added, an effect which resulted in a fivefold difference in biomass between non-transformed and transgenic plants (Fig. 1).

Fig. 1. Three week-old canola plants. (**A**) non-transformed in control soil; (**B**) transgenic in control soil; (**C**) non-transformed in soil spiked with 175 ppm nickel; (**D**) transgenic in soil spiked with 175 ppm nickel.

3.2 Expression changes

Transcript profiles from shoots of transgenic and non-transformed plants treated with nickel show between 216 and 275 differentially expressed genes when compared to plants grown without added nickel. These profiles differ significantly in the types of genes that are up- or down-regulated (Table 1). For example, when compared to shoot tissue from control plants, the non-transformed plants show up-regulation of many genes involved in oxidative stress such as oxidoreductases, peroxidases, glutathione S-transferase and superoxide dismutase. They also show severe down-regulation of transcripts for chlorophyll binding and moderate down-regulation of genes involved in carbohydrate metabolism such as sucrose synthase and starch phosphorylase. In contrast to what was observed with non-transformed plants, transgenic plants show no change in the expression of transcripts for chlorophyll binding proteins and even slight down-regulation of transcripts for oxidative stress proteins. Transgenic plants do, however, show down-regulation of carbohydrate metabolism and respiration genes, suggesting that although both non-transformed and transgenic plants suffer nutritional stress from excess nickel, without high levels of stress ethylene, the severity of symptoms is reduced. It also suggests that in the non-transformed plants, ethylene is inducing a strong oxidative stress, which may be the main cause of plant damage.

Table 1. Description and ID of genes that were differentially expressed in canola plants in the presence of nickel.

Up-regulated in non-transformed	
AT2G46370, AT4G03400	Auxin-responsive GH3
AT1G51950, AT3G23030	Auxin-responsive IAA18, IAA2
AT2G33380	Calcium-binding
AT3G01500, AT5G14740	Carbonic anhydrase
AT5G13930	Chalcone synthase
AT4G39950	Cytochrome P450 79B2
AT5G36220	Cytochrome P450 81D1 (CYP81D1) (CYP91A1)
AT4G31500	Cytochrome P450 83B1
AT4G30530	Defense-related
AT1G76180, AT2G21490, AT4G38410	dehydrin
AT1G56220	Dormancy/auxin family protein similar to auxin-repressed
AT4G15700	Glutaredoxin
AT2G02930, AT4G02520, AT1G10370, AT2G30860, AT1G69920, AT2G30870	Glutathione S-transferase
AT5G27380	Glutathione synthetase (GSH2)
AT1G68010	Glycerate dehydrogenase
AT3G16440, AT3G16470, AT1G52050	Jacalin lectin family protein
AT2G37710	Lectin protein kinase
AT5G48380	Leucine-rich repeat family protein
AT1G51805, AT1G51820	Leucine-rich repeat protein kinase
AT2G31880, AT1G09970	Leucine-rich repeat transmembrane protein kinase
AT3G45140	Lipoxygenase (LOX2)
AT2G16660, AT5G14120, AT3G14770	Nodulin family protein
AT5G24530, AT5G05600	Oxidoreductase, 2OG-Fe(II)
AT2G14610, AT2G14580	Pathogenesis-related protein 1 (PR-1)
AT4G25100	Superoxide dismutase [Fe], chloroplast (SODB)
AT4G23810, AT4G31550.2	WRKY family transcription factor
Up-regulated in transgenic	
AT1G67430, AT2G39460, AT3G06680, AT1G36240, AT1G69620, AT3G09500, AT2G43460, AT3G02190, AT3G56020	60S ribosomal protein L17, L23A, L29, L30, L34, L35, L38, L39, L41
AT3G60770, AT5G64140, AT5G56670	40S ribosomal protein S13, S28, S30
AT1G14040, AT2G03260	EXS family protein/ERD1/XPR1/SYG1 family protein
AT1G06830	Glutaredoxin family protein
AT5G61130	Glycosyl hydrolase family protein 17
AT1G27970	Nuclear transport factor 2 (NTF2)

AT1G12090	Protease inhibitor/seed storage/lipid transfer protein
AT1G24530	Transducin family protein
AT3G52730	Ubiquinol-cytochrome C reductase
Down-regulated in non-transformed	
AT3G27690, AT3G08940, AT3G54890, AT1G61520, AT2G34430, AT5G54270, AT1G29920, AT3G47470, AT1G15820	Chlorophyll A-B binding protein
AT2G15970, AT2G42530	Cold-acclimation protein, putative (FL3-5A3)
AT2G42530	Cold-responsive protein/cold-regulated protein (cor15b)
AT2G31360	Delta 9-desaturase (ADS2)
AT3G61580, AT2G46210	Delta 8-sphingolipid desaturase (SLD1)
AT4G25420, AT5G07200, AT1G60980	Gibberellin 20-oxidase
AT3G29320	Glucan phosphorylase
AT1G65960	Glutamate decarboxylase
AT5G37600	Glutamine synthetase
AT1G62510, AT4G12510, AT4G12520	Protease inhibitor/seed storage/lipid transfer protein (LTP)
AT5G54190, AT4G27440	Protochlorophyllide reductase A and B (POR A and B)
AT3G46970, AT1G10760	Starch phosphorylase, starch excess protein (SEX1)
AT5G20830, AT3G43190	Sucrose synthase
Down-regulated in transgenic	
AT1G71520, AT1G24590, AT5G67180, AT1G01250	AP2 domain-containing transcription factor
AT3G07490, AT4G12860	Calcium binding protein
AT3G03410	Calmodulin-related protein
AT5G26742, AT3G22310, AT5G65900	DEAD box RNA helicase
AT1G20440	Dehydrin (COR47)
AT1G60620	DNA-directed RNA polymerase
AT4G25480	DRE-binding protein (DREB1A)
AT4G24260	Endo-1,4-beta-glucanase
AT1G72280	Endoplasmic reticulum oxidoreductin 1 (ERO1)
AT3G10280, AT2G46720	Fatty acid elongase 3-ketoacyl-CoA synthase
AT5G53460	Glutamate synthase [NADH]
AT1G32760	Glutaredoxin
AT1G08990, AT1G48100	Glycoside hydrolase
AT5G02490, AT1G54400, AT2G41690	Heat-shock (HSC70-2) (HSP70-2), 18.0 kDa class I, factor
AT1G55020, AT1G72520	Lipoxygenase (LOX1)
AT5G17800, AT5G67300, AT3G10580, AT5G59570	myb family transcription factor
AT4G08300, AT1G21140, AT3G25190	Nodulin
AT2G45630, AT3G22920	Oxidoreductase
AT1G24110	Peroxidase
AT2G47120, AT3G26770, AT3G03980	Short-chain dehydrogenase
AT3G50740, AT4G15260, AT5G66690	UDP-glucoronosyl/UDP-glucosyl transferase
AT2G34830, AT4G01250, AT3G56400	WRKY family transcription factor
AT2G33500, AT5G54470, AT5G04340, AT1G04360, AT3G14320, AT1G67340, AT1G05890, AT1G64620, AT5G60850	Zinc finger (B-box type, C2H2 type, C3HC4-type RING finger and MYND type) Dof type
Down-regulated in both	
AT5G02270	ABC transporter family protein
AT3G06450	Anion exchange family protein
AT3G09960	Calcineurin-like phosphoesterase family protein
AT4G12860	Calcium binding protein
AT4G23700	Cation/hydrogen exchanger
AT3G19450	Cinnamyl-alcohol dehydrogenase
AT3G19450	Cinnamyl-alcohol dehydrogenase (CAD)
AT2G17870	Cold-shock DNA-binding family protein
AT1G31930	Extra-large guanine nucleotide binding protein

AT5G02490	Heat-shock (HSC70-2) (HSP70-2
AT5G59570	myb family transcription factor
AT4G08300, AT3G25190	Nodulin
AT5G47810	Phosphofructokinase
AT3G50740, AT5G66690	UDP-glucoronosyl/UDP-glucosyl transferase
AT3G14320, AT1G67340, AT1G05890	Zinc finger

References

Burd GI, Dixon DG, Glick BR (1998) A Plant Growth-promoting Bacterium that Decreases Nickel Toxicity in Seedlings. Applied and Environmental Microbiology 64: 3663

Davis LG, Dibner MD, Battey JB (1986) Preparation and analysis of RNA from eukaryotic cells. *In* Basic Methods in Molecular Biology. Elsevier Science Publishing Co., Inc., New York

Glick BR, Penrose DM, Li J (1998) A Model For the Lowering of Plant Ethylene Concentrations by Plant Growth-promoting Bacteria. Journal of Theoretical Biology 190: 63–68

Goren R, Siegel SM (1976) Mercury-induced Ethylene Formation and Abscission in Citrus and Coleus Explants. Plant Physiology 57: 628–631

Grichko VP, Filby B, Glick BR (2000) Increased ability of transgenic plants ex-pressing the bacterial enzyme ACC deaminase to accumulate Cd, Co, Cu, Ni, Pb, and Zn. Journal of Biotechnology 81: 45

Hyodo H (1991) Stress/wound ethylene. *In* AK Mattoo, JC Suttle, eds, The Plant Hormone Ethylene. CRC Press, Boca Raton

Prasad MNV, Strazalka K (2000) Physiology and biochemistry of metal toxicity and tolerance in plants. Kluwer Academic Publishers, Boston

Robison MM, Griffith M, Pauls KP, Glick BR (2001) Dual role for ethylene in susceptibility of tomato to verticillium wilt. Journal of Phytopathology 149: 385

Stearns JC, Shah S, Glick BR (2006) Increasing plant tolerance to metals in the environment. *In* N Willey, ed, Phytoremediation: Methods and Reviews. Hu-mana Press, Totowa

Cross talk between ethylene, nitric oxide and salicylic acid in ozone fumigated tobacco plants

Pasqualini, S.[1]*, Morettini, R.[1], Borgogni, A.[1], Ghering, C.[2], Meier, S.[2] and Ederli, L.[1]

[1]Dipartimento di Biologia Vegetale e Biotecnologie Agroambientali e Zootecniche Borgo XX Giugno, 74 06121 Perugia (Italy); [2]Department of Biotechnology, University of Western Cape, Bellville, South Africa. (*Corresponding author: spas@unipg.it)

1. Introduction

Stratospheric ozone (O_3) protects life from detrimental ultraviolet-B radiation, but tropospheric O_3 is a serious world-wide pollutant. It is generally accepted that O_3 initiates an oxidative burst and an active production of reactive oxygen species (ROS). Several signal molecules like ethylene (ET), jasmonic acid (JA), salicylic acid (SA) and nitric oxide (NO) have been hypothesized to act as a second or third messengers for O_3-induced gene expression. Using a pharmacological approach, the aim of this work was to inspect the time course of the NO, SA and ET accumulation under ozone stress in the O_3-sensitive tobacco cv BelW3 and the cross talk between these signaling molecules.

2. Materials and Methods

The *Nicotiana tabacum* L. cv BelW3 plants were grown in a climatic chamber with 14 h photoperiod, PPFR of 120 μmol m^{-2} s^{-1}, day/night air temperature of 25°C/20°C, and relative humidity 60–75%. The fourth leaf from the apex of four treated and four untreated (controls) 12-week-old plants were used in all experiments. For ozone treatment, plants were exposed for 5 h to 150 nL L^{-1} O_3. SA quantification was performed as described in Pasqualini *et al.*, (2002). ET determination was carried out in a gas chromatography. NO was evaluated spectrophotometrically by quantitative oxidation of oxyhaemoglobin (HbO_2) to methaemoglobin (metHb) as reported in Murphy and Noack (1994). cGMP was quantified in dried extracts by ELISA method.

A. Ramina et al. (eds.), Advances in Plant Ethylene Research: Proceedings of the 7th International Symposium on the Plant Hormone Ethylene, 383–385.

3. Results

NO levels showed a significant increase at 1.5 h and 4 h during ozone treatment (Fig. 1A). In many systems, the action of NO is carried out through the activation of soluble guanylate cyclase and the production of cGMP. Ozone fumigation induced a transient cGMP accumulation at 2 h from starting fumigation, thus subsequently to NO emission (Fig. 1B). We also measured significant effects of ozone fumigation on ethylene production in BelW3 plants. ET evolution increased within 2 h of starting treatment, reached a maximum at 5 h, then decreased to near control values (Fig. 1C). To evaluate the role of NO on ethylene evolution, NO accumulation was suppressed by pre-treating tobacco plants with cPTIO + LNNA and ethylene evolution measured. When NO accumulation was blocked, there was a dramatic reduction in ethylene production (Fig. 1C) and ozone-induced visual damage on the leaves was greatly reduced (5%). The results of HPLC analysis show that O_3 exposure induces an accumulation of SA, with a peak at 7 h from the start of treatment (Fig. 1D). When the leaves were pre-treated with AVG (ET inhibitor) or cPTIO + LNNA (NO scavenger and NOS inhibitor, respectively), there was no increase of SA level indicating that SA accumulation required both ET and NO.

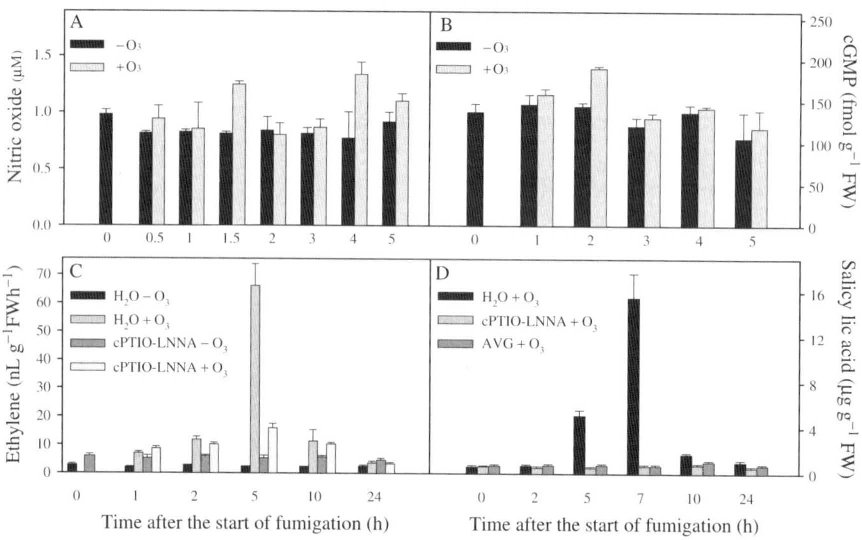

Fig. 1. NO (**A**), cGMP (**B**), ET (**C**) and SA (**D**) levels during and after an ozone fumigation (150 nL L^{-1} for 5 h) in leaves of the tobacco cv BelW3. * indicate the significativity at $p = 0.05$.

References

Murphy ME, Noack E (1994) Nitric-oxide assay using haemoglobin method. Oxygen Radicals in Biology Systems. Part C 233:240–250.

Pasqualini S, Della Torre G, Ferranti F, Ederli L, Piccioni C, Reale L, Antonielli M (2002) Salicylic acid modulates ozone-induced hypersensitive cell death in tobacco plants. Physiol Plant 115:204–212.

Role of ethylene in triggering ROS production in the tomato mutant *Nr* subjected to acute ozone treatment

Castagna, A.[1], Ederli, L.[2], Pasqualini, S.[2], Mensuali-Sodi, A.[3], Baldan, B.[4], Pardossi, A.[5] and Ranieri, A.[1]*

[1]DCBA University of Pisa, [2]DBVBAZ University of Perugia, [3]SSSUP "S. Anna" Pisa, [4]DB University of Padova, [5]DBPA University of Pisa. (*Corresponding author: aranieri@agr.unipi.it)

Tropospheric ozone (O_3) is believed to be the most important phytotoxic air pollutant in industrialised countries and one of the major anthropogenic stresses that threaten native and cultivated ecosystems. Although reactive oxygen species (ROS) generated following O_3 exposure are believed to act as key molecules that elicit plant response to O_3, the current information underlines the importance of the complex interactions between ROS and several phytohormones, among which ethylene (ET) (Kangasjärvi *et al.* 2005, Diara *et al.* 2005). Aimed to understand the role of the ET receptor *LeETR*3 (NR) in mediating the responses to O_3, the tomato mutant *Never ripe* (Nr), which carries a semi-dominant mutation in LeETR3 gene, and its isogenic wild type (cv Pearson) were exposed to 200 ppb of O_3 for 4 h.

The absence of major differences between *Nr* and Pearson in the appearance of O_3-induced lesions and the similar degree of plasma-membrane alteration detected by the vital non-permeable dye Evan's blue (Fig. 1A) made us hypothesize a similar O_3-sensitivity by the two genotypes. ET was undoubtedly involved in the progression of lesions also in *Nr*, as indicated by the almost complete prevention of leaf injury in both the genotypes by the inhibitor of ET synthesis AVG (data not shown). Both Pearson and *Nr* were found to increase ET evolution following O_3 exposure and exhibited the maximum ET emission after 2 h of fumigation (Fig. 1B), but in *Nr* ET emission was delayed and prolonged. The ET biosynthetic genes were induced in both O_3-treated Pearson and *Nr* plants, indicating that *Nr* was not impaired in any step of ET biosynthesis. However, similarly to what observed for ET evolution, in the mutant the response was delayed and long-lasting (Fig. 1B). In both the genotypes, O_3 activated a biphasic H_2O_2-dependent oxidative burst, which appeared to be ET-driven also in *Nr* leaves as indicated by the dramatic reduction in H_2O_2 accumulation observed in AVG-treated leaves (data not shown). Evidently the defective

A. Ramina et al. (eds.), Advances in Plant Ethylene Research: Proceedings of the 7th International Symposium on the Plant Hormone Ethylene, 387–388.

ET perception by *Nr* causes a delay in the achievement of the ET threshold needed to trigger ROS production. From the overall data the *Nr* mutant seems to retain a partial sensitivity to ET, suggesting a functional compensation by other ET receptors.

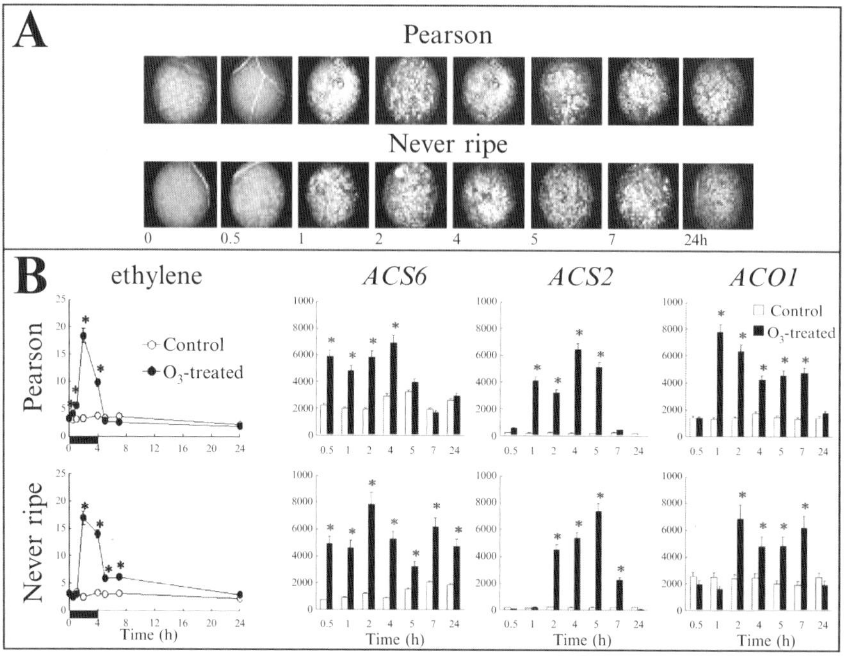

Fig. 1. (Color figure in the Annex, p.463). Evan's blue staining (**A**) and leaf ET emission and expression levels of *ACS6*, *ACS2* and *ACO1* mRNA (**B**) in Pearson and *Never ripe* tomato plants exposed to filtered air (control) or O_3 (200 ppb, 4 h; O_3-treated). Asterisk (*) indicates statistically significant differences between O_3-treated samples and controls at each time point (t-test; $P < 0.01$; ET $n = 3$; mRNA analyses $n = 4$).

References

Diara C, Castagna A, Baldan B, Mensuali Sodi A, Sahr T, Langebartels C, Sebastiani L, Ranieri A (2005) Differences in the kinetics and scale of signalling molecule production modulate the ozone sensitivity of hybrid poplar clones: the roles of H_2O_2, ethylene and salicylic acid. New Phytol 168:351–364

Kangasjärvi J, Jaspers P, Kollist H (2005) Signalling and cell death in ozone-exposed plants. Plant Cell Environ 28:1021–1036

Ozone as a tool for studying stress responses in tomato (*Solanum lycopersicum* L.). III. Ethylene, cyanide and the development of foliar symptoms in the autonecrotic mutant V20368

Antonelli, M.G.[1], Marabottini, R.[1], Santangelo, E.[1], Francini, A.[2], Pellegrini, E.[2], Di Baccio, D.[3], Ederli, L.[4], Nali, C.[2], Pasqualini, S.[4], Soressi, G.P.[1], Ciaffi, M.[1], Lorenzini, G.[2], Sebastiani, L.[3] and Badiani, M.[5]*.

[1]DAA, Università della Tuscia, Via S.C. De Lellis, I-01100 Viterbo, Italy; [2]Patologia Vegetale, Università di Pisa, Via di Marina Vecchia, 11, I-56100 Pisa, Italy; [3]Scuola Superiore Sant'Anna, Piazza Martiri della Libertà, 33, I-56127 Pisa, Italy; [4]BVBAZ, Università di Perugia, Borgo XX Giugno, 74, I-06121 Perugia, Italy; [5]BIOMAA, Università Mediterranea, I-89129, Reggio Calabria, Italy. (*Corresponding author: mbadiani@unirc.it)

Plants' exposure to ozone (O_3) activates a multitude of defense reactions – among which the launch of a cell death program (PCD) leading to the development of visible foliar symptoms – which are also deployed in response to other abiotic or biotic stressors. Ethylene (ET) plays a key role in these events.

During ET synthesis, stoichiometric amounts of cyanide (CN^-) are produced, which are normally detoxified to β-cyanoalanine through the action of β-cyanoalanine synthase (β-CAS). Cyanide induces the expression of the CN^--insensitive mitochondrial alternative oxidase (AOX), whose activation is ET-dependent. By providing a substitutive sink for the electron overflow arising from the CN^--dependent block of cytochrome *c* oxidase, AOX helps in decreasing the formation of reactive oxygen species (ROS), which trigger the launch of PCD.

The tomato (*Solanum lycopersicum* L.) mutant line V20368 develops severe foliar necrosis in response to the natural increase in air temperature and light intensity in the field. As this occurs together with a significant increase in foliar ET release, we wondered whether common molecular bases might underlie autonecrosis and O_3-induced PCD in V20368, and whether CN^- metabolism might play a role.

A. Ramina et al. (eds.), Advances in Plant Ethylene Research: Proceedings of the 7th International Symposium on the Plant Hormone Ethylene, 389–390.

V20368 and the non-autonecrotic control line MP-22 were exposed to a single pulse of 100 nL L^{-1} O_3 for 5 h, in the form of a square wave. Such O_3 dose was able to selectively induce O_3-dependent necrotic lesions, appearing 96 h after the end of O_3-fumigation, in the leaves of V20368, but not in those of MP-22 (not shown).

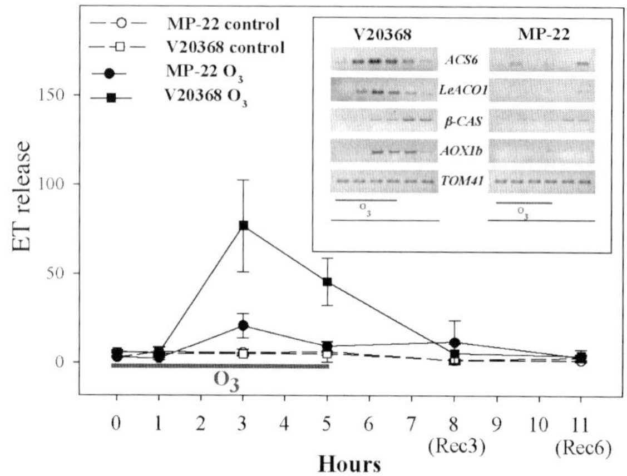

Fig. 1. Ethylene release and gene expression data in different tomato genotypes.

The foliar CN^- levels were increased by O_3 more in V20368 than in MP-22. This was a late event respect to the ET burst, as it peaked after 3 h of postexposure recovery (Rec3, data not shown).

The results presented here (Fig. 1) appear to support once more the link between an O_3-dependent ET burst and the development of foliar symptoms, and suggest that CN^- overproduction might take part in the launch of O_3-induced PCD. However, neither the ability for CN^- scavenging (*β-CAS* expression) nor the activation of CN^--insensitive respiration (*AOX* expression) were depressed in V20368, but they were instead potentiated. In principle, this would keep under control, at least in the short term, both CN^--overproduction and its promotive effects on ROS levels in the mitochondria, possibly contrasting the launching of ROS-dependent cell death.

So that, additional components and mechanisms involved in the development of spontaneous or O_3-induced foliar lesions need to be studied. Clearly, the V20368 tomato mutant appears to be a valuable tool for studying common molecular responses to different stressing agents.

Acknowledgment

This work contributes to a 2005 PRIN-MIUR Project.

The autonecrosis triggered by *Cf-2/Rcr3*^{esc} interaction in tomato (*Solanum lycopersicum* L.) involves an ethylene burst and the enzymatic machinery protecting the respiratory apparatus of mitochondria

Santangelo, E.[1], Marabottini, R.[1], Antonelli, M.[1], Badiani, M.[2], Pasqualini, S.[3] and Soressi, G.P.[1]*

[1]Dipartimento di Agrobiologia e Agrochimica, Università della Tuscia, Via S. C. De Lellis, I-01100 Viterbo, Italy. [2]Dipartimento BIOMAA, Università Mediterranea di Reggio Calabria – Loc. Feo di Vito, I-89129 Reggio Calabria, Italy. [3]Dipartimento di BVBAZ, Università di Perugia, Borgo XX Giugno, I-06121, Perugia, Italy. (*Corresponding author: santangelo@agrital.ccr.it)

High ethylene (ET) synthesis observed under stress conditions produces stoichiometric amounts of CN^- and CO_2. Normally, plant tissues have ample capacity to detoxify the cytosolic CN^- to β-cyanoalanine by the action of β-cyanoalanine synthase (β-CAS), mainly present in the mitochondrial fraction. Cyanide induces the expression of the CN^--insensitive mitochondrial alternative oxidase (AOX), which provides an alternative sink for the electron overflow arising from the CN^--dependent block of cytochrome *c* oxidase; this would keep low the formation and accumulation of partially reduced O_2 forms (ROS), which trigger the launch of PCD. This metabolic network was studied by adopting as a model system the V20368 autonecrotic tomato line, developing necrotic foliar lesions when exposed to increasing temperatures and light intensities; such a phenotype is due to the interaction between the *Cf-2* gene, from *S. pimpinellifolium*, conferring resistance to *Cladosporium fulvum* and the *Rcr3*^{esc} gene from *S. lycopersicum*, coding for a cysteine-proteinase.

Leaf samples from basal (B), central (C), and apical (A) plant sectors detached in the same date (September 10, 2005) from autonecrotic V20368 and MP22 (control) plants, were analyzed at four developmental stages (5–6th, 6–7th, 7–8th, 8–9th true leaves (Fig. 1)).

A strong *β-CAS* expression was observed at the apical, central, and basal leaves in the young plants (5–6th and 6–7th stages) of V20368, when both ethylene and HCN (not shown) levels began to increase (Fig. 2). At the

A. Ramina et al. (eds.), Advances in Plant Ethylene Research: Proceedings of the 7th International Symposium on the Plant Hormone Ethylene, 391–392.

maximum hormone level (central leaves of 7–8th stage) the *β-CAS* expression decreased, although always present in all the leaves of the oldest plants (8–9th and 7–8th stages) of V20368. Clear was the differential activation of the *AOX1b* (but not of the *AOX2*), in the central and basal leaves of the youngest plants and in the basal ones of the 6–7th stage. A weak expression of *AOX1b* was observed in the oldest plants where the cyanide level was the highest one.

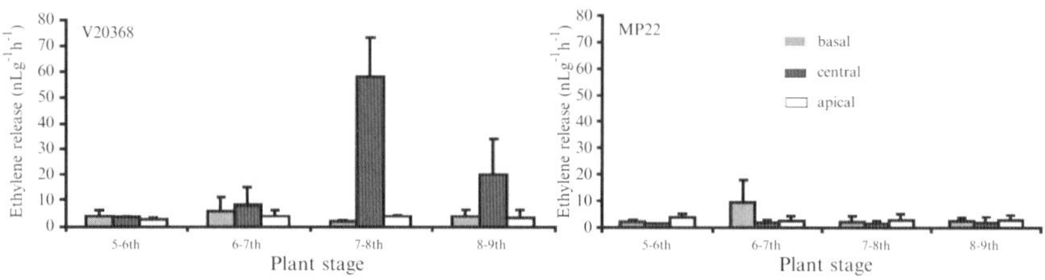

Fig. 1. Ethylene release of leaf samples from autonecrotic V20368 and MP22 (control) plants at four developmental stages (5–6th, 6–7th, 7–8th, 8–9th true leaves).

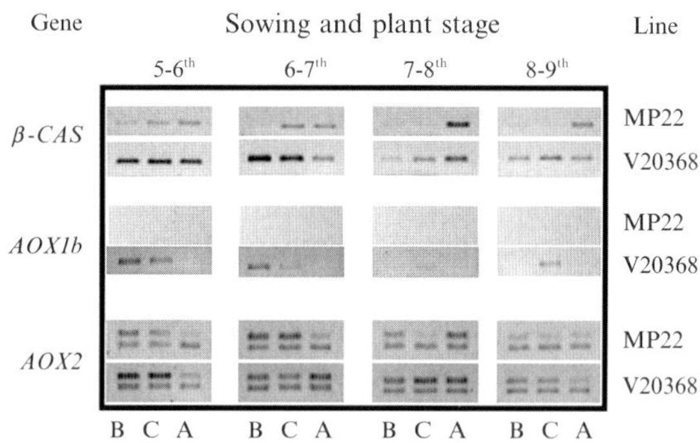

Fig. 2. Gene expression data.

Likely, the high HCN level in V20368 (remarkably higher than the control), associated to an ethylene burst, could overwhelm the detoxifying capacity of β-CAS and AOX1b, leading to an inhibition of mitochondria respiration, ROS overproduction, and cell death. In the studied case, the duration of the stress and the continuous evolution of both ET and HCN also played a crucial role.

Acknowledgments

This work contributes to a 2005 PRIN-MIUR Project.

Ethylene response of tomato fruits to the minimum dropping

Usuda, H.[1,2], Kitagawa, M.[3], Ito, Y.[1], Umehara, H.[1], Nakamura, N.[1], Roy, P.[1], Okadome, H.[1], Ishikawa, Y.[1], Satake, T.[2] and Shiina, T.[1]*

[1]National Food Research Institute, 2-1-12, Kannondai, Tsukuba Ibaraki 305-8642, Japan; [2]Graduate school of Life and Environment Sciences, University of Tsukuba, 1-1-1, Tennodai, Tsukuba, Ibaraki 305-8572, Japan; [3]Research Institute, KAGOME CO. LTD, NasuShiobara, Tochigi 329-2762, Japan. (*Corresponding author: shiina@affrc.go.jp)

Physical stresses such as shock and vibration deteriorate the quality of fresh fruits and vegetables not only by direct physical damage but also through the endogenous biological activities of the produce. Ethylene seems to be the most important factor of the stress response in the produce. In this paper we study the effects of slight dropping (1, 3, 10 times) on the ethylene and CO_2 production of the whole tomato fruits and accumulation of the ACC synthase (ACS) mRNA in the tissue. The mature green tomato (*Lycopersicon esculentum*) fruits which were screened by ethylene production rate (less than 0.2 $\mu l \cdot kg^{-1} \cdot h^{-1}$) were used to avoid a contamination of climacteric phase fruit.

Dropping of the fruits from the height of 5 cm on the concrete floor dramatically enhanced the ethylene and CO_2 production rate without any external or internal damage on the fruits. The more the dropping times, the greater the enhancement level was. The maximum value of relative CO_2 production rate of the tomato with 1-, 3- and 10-time dropping were 1.1-, 1.3- and 1.5-folds of the value for reference sample, respectively. After that, relative CO_2 production rate of the tomato for 1-, 3- and 10-time dropping showed gradual decrease and reached to the same level with reference tomato. Ethylene production of tomato fruits with dropping stimuli showed clear increases compared with the reference samples. Similar trend was observed as in the case of respiration rate. However, the peak values of relative ethylene production rate were much higher than that of respiration rate. The peak values for 1-, 3- and 10-time dropping were 2.8-, 3.1- and 8.1-folds of reference tomato, respectively. *LeACS2* gene is known to be wound inducible in fruits (Yip *et al.*, 1992; Lincoln *et al.*, 1993). In this

A. Ramina et al. (eds.), Advances in Plant Ethylene Research: Proceedings of the 7th International Symposium on the Plant Hormone Ethylene, 393–394.

study, *LeACS2* expression measured by RT-PCR method in the tomato with ten dropping showed remarkable increase (relative value was 9.6-folds at the peak after 5 hours from dropping stimuli) and it was closely related with the ethylene production rate among the *LeACS* gene family as shown in Fig. 1.

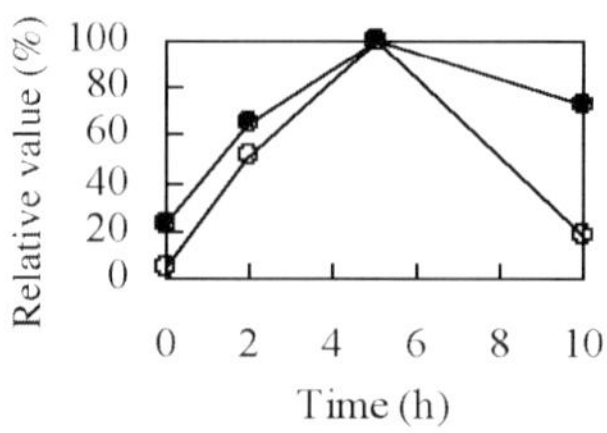

Fig. 1. Comparison of ethylene production and accumulation of LeACS2 mRNA after dropping (•: Ethylene; o: *LeACS2*).

It suggests that LeACS2 worked as the key enzyme for response to the stress by dropping similarly in the case of wound-induced enhancement of ACS activity and ethylene production (Lincoln *et al.*, 1993). Tatsuki and Mori (1999) showed the surprisingly enhanced ethylene production rate and expression of *LeACS1A* and *LeACS6* of mature green tomatoes between 30 min and 2 h after touch stimuli (no increase in *LeACS2*). Contrarily in our study, no enhancement of *LeACS1A* and *LeACS6* was observed in the fruit with dropping stimuli. The differences among studies suggest that different *LeACS*s are responsive to different kinds of the physical stress and responsive genes are differently controlled in transcript level, whereas the similarity of *LeACS2* gene expression shows the possibility to group the stresses into several categories. It seems that dropping falls into the stress category of wounding.

References

W. K. Yip, T. Moore, S. F. Yang (1992) Differential accumulation of transcripts for four tomato 1-aminocyclopropane-1-carboxylate synthase homologs under various conditions, Proc. Natl. Acad. Sci. USA, 89, 2475–2479.

J. E. Lincoln, A. D. Campbell, J. Oetiker, W. H. Rottmann, P. W. Oeller, N. F. Shen, A. Theologis (1993) *LeACS4*, a fruit ripening and wound-induced 1-aminocyclopropane-1-carboxylate synthase gene of tomato (*Lycopersicon esculentum*), J. Biol. Chem., 268, 19422–19434.

M. Tatsuki, H. Mori (1999) Rapid and transient expression of 1-aminocyclopropane-1-carboxylate synthase isogene by touch and wound stimuli in tomato, Plant Cell Physiol. 40(7), 709–715.

Ethylene production of *Botrytis cinerea in vitro* and during *in planta* infection of tomato fruits

Cristescu, S.M.[1]*, Harren, F.J.M.[1] and Woltering, E.J.[2]

[1]Department of Molecular and Laser Physics, Institute for Molecules and Materials, Radboud University Nijmegen, Toernooiveld 1, 6525 ED Nijmegen, the Netherlands. [2]Wageningen University and Research Center, Agrotechnology & Food Sciences Group(ASFG), PO Box 17, 6700 AA Wageningen, the Netherlands. (*Corresponding author: simona@science.ru.nl)

We present a new approach to study the pathogen-host interaction by online monitoring of ethylene release using a laser-based photoacoustic detector. The instrument allows detection of ethylene emission in a continuous flow system down to 10 pptv (pptv = parts-per-trillion volume, $1{:}10^{12}$) (Bijnen *et al.*, 1996) and has relatively high time resolution for measuring the dynamics of ethylene production by *B. cinerea in vitro* and by infected tomatoes. Moreover, it has been proved to be a powerful tool to study the relationship between ethylene released by the fungus *in vitro* and the enhanced ethylene production in *B. cinerea* infected tomato with respect to disease development.

Infection of tomato fruits with *B. cinerea* resulted in enhanced ethylene production which started to rise before visible decay development, indicating that ethylene can be a sensitive marker for early infection in harvested fresh products. The observed dynamics of ethylene production both from *B. cinerea in vitro* and from *B. cinerea* infected tomato fruit, showed that ethylene emission by the tomato-fungus system is not triggered directly by ethylene production of *B. cinerea*, although it is strongly 'synchronised' with the development of the fungus in the tomato tissue (Cristescu *et al.*, 2002).

B. cinerea does not use the plant ethylene biosynthesis pathway but produces ethylene from L-methionine via the KMBA pathway (Cristescu *et al.*, 2002, Chague *et al.*, 2002). Ethylene production *in vitro* is partly dependent on light. Using the photoacoustic ethylene detector we were able to monitor the dynamics of ethylene released by *B. cinerea* while interchanging the light and dark regimes (2 min timescale) (Fig. 1). We

A. Ramina et al. (eds.), Advances in Plant Ethylene Research: Proceedings of the 7th International Symposium on the Plant Hormone Ethylene, 395–397.

found that KMBA conversion to ethylene is considerably lower in the dark than in light. Similar results were reported by Chague *et al.* (2002).

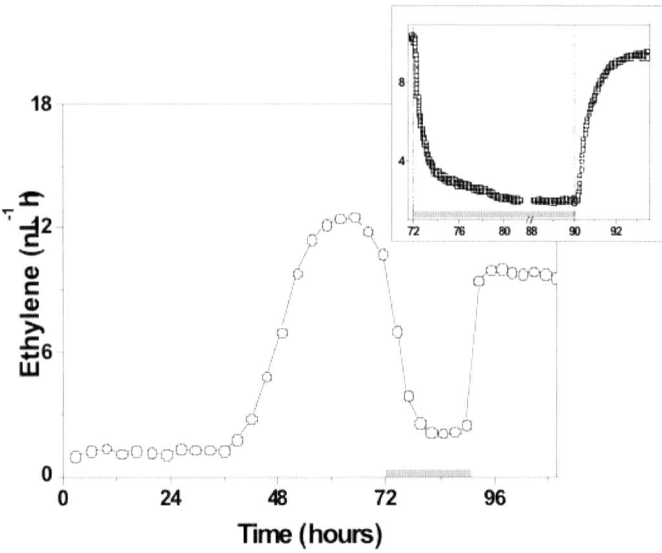

Fig. 1. Ethylene released by *B. cinerea in vitro* at 2×10^5 conidia ml^{-1} (160 μl) plated on PDA containing 25 mM L-methionine in light and dark (grey bar), respectively. Inset: the KMBA conversion to ethylene when switching from light to dark and back to light for 2×10^5 (o) conidia ml^{-1} starts within 2 min.

The mechanism for ethylene production by *B. cinerea in planta* is rather complex due to the multiple ways to convert the KMBA; possible ways for this conversion are discussed.

A major role in the process of KMBA conversion to ethylene is played by the free radicals which are generated in both dark and light conditions. In the dark, the radicals are formed by respiratory processes (chemical). In the light, additional radicals will be generated due to photochemical processes. In *B. cinerea* other sources may be considered for the formation of free radicals (e.g. pigments and/or the reaction of light directly with chemical compounds present in the fungus cells, etc.) (Cristescu *et al.* in press).

In 1-MCP pretreated tomatoes that were later inoculated with *B. cinerea*, the ethylene levels were comparable to those from non-treated infected tomatoes. This indicates that ethylene produced by tomato in response to *B. cinerea* infection is not due to autocatalysis but directly elicited by other mechanisms initiated by the fungus (e.g. production necrosis- and ethylene-inducing protein, NEP). The lack of significant levels of ethylene production from *B. cinerea* during the infection process indicates that *B. cinerea* ethylene does not play a role in the infection process.

References

Bijnen FGC, Reuss J, Harren FJM (1996) Geometrical optimization of a longitudinal resonant photoacoustic cell for sensitive and fast trace gas detection. Review of Scientific Instruments 67:2914–2923

Chague V, Elad Y, Barakat R, Tudzynski P, Sharon A (2002) Ethylene biosynthesis in *Botrytis cinerea*. FEMS Microbiology Ecology 40:143–149

Cristescu SM, de Martinis D, te Lintel Hekkert S, Parker DH, Harren FJM (2002) Ethylene production by *Botrytis cinerea in vitro* and in tomatoes. Applied and Environmental Microbiology 68:5342–5350

Cristescu SM, Ernst J. Woltering EJ, Harren FJM (in press) Real time monitoring of ethylene during fungal-plant interaction by laser-based photoacoustic spectroscopy in: Food Mycology. Eds. J. Dijksterhuis and R.A. Samson, Traylor and Francis CRC Press, Boca Raton, USA

Ethylene production of two durum wheat cultivars exposed to cadmium

Mensuali-Sodi, A. and Ercoli, L.*

Scuola Superiore di Studi Universitari e di Perfezionamento Sant'Anna, piazza Martiri della Libertà 33, Pisa, Italy. (*Corresponding author: ercoli@sssup.it)

Cadmium is a highly toxic heavy metal for all living organisms and causes a reduction of several physiological processes in plants. Cadmium tends to accumulate in several crops grown in soils containing Cd (Grant *et al.*, 1998). Grain and cereal products are primary sources of Cd, being able to accumulate Cd in grain (Greger and Löfstedt, 2004). Ethylene production has been often associated with response to Cd exposure in several species (Rodecap and Tingery, 1981; Bhattacharjee, 1997).

In this study, ethylene emission of two cvs. of durum wheat exposed to environmentally realistic concentrations of Cd was investigated. Svevo and Creso, respectively high and low for Cd accumulation in grain, were grown hydroponically in Clark's solution until 2.5 Haun stage, when were exposed to Cd concentrations (Cd as $CdNO_3$) of 5.2, 10.4 and 15.6 μM, the former representing a slight Cd contamination, the latter a moderately elevated one. Ethylene emission was gas chromatographically determined after 24-, 48- and 96-h of Cd exposure in presence or in absence of 2 ml Clark's solution.

The two wheat cultivars reacted to the stress condition caused by the absence of nutrient solution with an increase of ethylene synthesis (data not shown). Similar results were observed in other wheat cultivars with insufficient water supply (Balota *et al.*, 2004). Ethylene production was higher from Svevo than from Creso at all Cd rates and durations of exposure (results not shown). Cadmium induced an inhibition of ethylene release in Svevo and a stimulus in Creso (Fig. 1). Cadmium ions are known to interfere with ethylene biosynthesis in many plant species (Fuher, 1982; Pennazio and Roggero, 1992; Sanità di Toppi and Gabbrielli, 1999) Cd-induced stress ethylene might however represent in cv. Creso an early general adaptation syndrome, while we can suppose in Svevo plant a mechanism of Cd detoxification, as observed in other tolerant species as *Amaranthus lividus* (Bhattacharjee 1977).

A. Ramina et al. (eds.), Advances in Plant Ethylene Research: Proceedings of the 7th International Symposium on the Plant Hormone Ethylene, 399–400.

At the end of the experiment, although visual effects, such as chlorosis, were observed in plants treated at the higher Cd level, no necrosis or tissue damage were observed, indicating that severe stress conditions did not occur. Thus, it is possible to exclude the indirect effect of tissue damage on ethylene production.

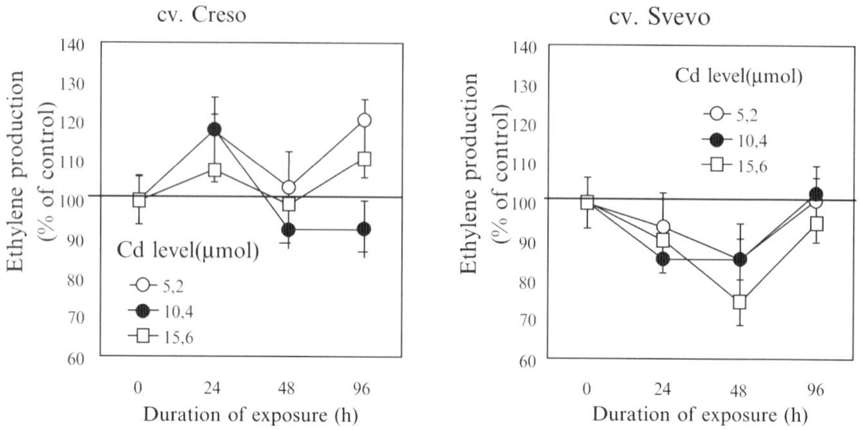

Fig. 1. Effect of Cd level and duration of exposure on ethylene production of cv. Creso (left) and Svevo (right). Vertical bars represent ± standard error of the mean; where not shown, the error bar lies within the symbol.

References

Balota M, Cristescu S, Payne WA, Lintel Hekkert S, Laarhoven LJJ, Harren FJM, (2004) Ethylene production of two wheat cultivars exposed to dessication, heat, and paraquat-induced oxidation. Crop Sci 44:812–818

Bhattacharjee S (1997) Membrane lipid peroxidation free radical scavengers and ethylene evolution in Amarantus as affected by lead and cadmium. Biol Plant 40:131–135

Fuher J (1982) Ethylene biosynthesis and cadmium toxicity in leaf tissue of beans (*Phaseolus vulgaris* L.). Plant Physiol 70:162–167

Grant CA, Buckley WT, Bailey LD, Selles F (1998) Cadmium accumulation in crops. Can J Plant Sci 78:1–17

Greger M, Löfstedt M (2004) Comparison of uptake and distribution of cadmium in different cultivars of bread and durum wheat. Crop Sci 44:501–507

Pennazio S, Roggero P (1992) Effect of cadmium and nickel on ethylene biosynthesis in soybean. Biol Plant 34:345–349

Rodecap K, Tingery D (1981) Stress ethylene: a bioassay for rhizosphere-applied phytotoxicants. Environ Monit Assess 1:119–127

Sanità di Toppi L, Gabbrielli R (1999) Response to cadmium in higher plants. Env Exp Bot 41:105–130

Ferric reductase and iron transporter gene expression in different *Arabidopsis* ethylene mutants

García, M.J.[1], Romera, F.J.[2]*, Pérez-Vicente, R.[1], Lucena, C.[2] and Alcántara, E.[2]

[1]Department of Ecology, Botany and Plant Physiology, Edificio Celestino Mutis (C-4), Campus de Rabanales, University of Córdoba, 14014-Córdoba (Spain). [2]Department of Agronomy, Edificio Celestino Mutis (C-4), Campus de Rabanales, University of Córdoba, 14014-Córdoba, (Spain). (*Corresponding author: ag1roruf@uco.es)

1. Introduction

Dicotyledonous plants need to reduce Fe(III) to Fe(II), mediated by a ferric reductase (encoded by the *FRO2* gene in *Arabidopsis*), prior to its uptake through an Fe(II) transporter (encoded by the *IRT1* gene in *Arabidopsis*). The expression of both genes is up-regulated under Fe deficiency. In the last years evidence has been presented to support a role for ethylene as activator of the ferric reductase activity but it is not known whether ethylene affects *FRO2* transcription or other aspects of this activity (Romera and Alcántara, 2004). In relation to the *IRT1* gene, it is not known whether ethylene affects it or not. The objective of this work was to study the expression of the *FRO2* and *IRT1* genes in response to Fe deficiency, and ethylene inhibitors and precursors, in the *Arabidopsis thaliana* wild-type *Columbia* and in some of its ethylene mutants (*ctr1*, *etr1*, *ein2*, *ein3eil1*). In addition, we have studied the expression of the basic chitinase gene (*CHIT*), already known by its response to ethylene. To carry out these experiments, seeds were germinated in black peat and, when appropriate, seedlings were transferred to plastic vessels with complete nutrient solution. After some days in this complete nutrient solution, plants were transferred to the different treatments: nutrient solution with Fe; with Fe plus ACC (or ethephon); without Fe; and without Fe plus ethylene inhibitors (cobalt, silver thiosulfate, AOA or AVG).

A. Ramina et al. (eds.), *Advances in Plant Ethylene Research: Proceedings of the 7th International Symposium on the Plant Hormone Ethylene*, 401–403.

2. Results and Discussion

The results obtained showed that all the ethylene mutants studied increased the expression of *FRO2* and *IRT1* under Fe deficiency (Fig. 1). All the ethylene- insensitive mutants studied (*etr1*, *ein2*, *ein3eil1*) enhanced *FRO2* and *IRT1* expression upon ACC treatment (or ethephon) (Fig. 1). *FRO2* and *IRT1* expression were negatively affected by ethylene inhibitors in all the mutants (Fig. 1). The basic chitinase gene was induced under Fe deficiency in all the mutants (which agrees with the higher production of ethylene by Fe-deficient plants (Romera and Alcántara, 2004), except in the *ein3eil1* mutant. In the *ein2* mutant, the basic chitinase gene expression was enhanced by the ethylene inhibitors (Fig. 1).

3. Conclusion

Our results suggest a role for ethylene in the regulation of *FRO2* and *IRT1* expression. In addition, the results suggest that both *FRO2* and *IRT1* genes are not regulated by ethylene in exactly the same way as the basic chitinase gene, since this last gene presents a different regulation in both *ein2* and *ein3eil1* mutants.

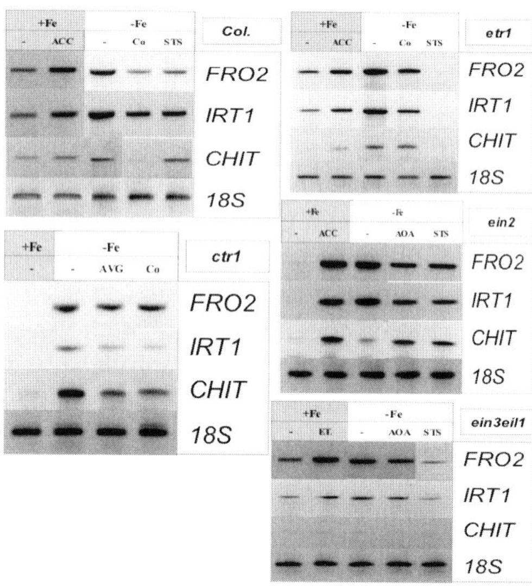

Fig. 1. Influence of ethylene precursors [ACC or ethephon (ET)] and ethylene inhibitors (AVG, Co, STS, or AOA) on the expression of *FRO2, IRT1* and *CHIT* genes in Fe-sufficient (+Fe) and Fe-deficient (–Fe) *Columbia* and ethylene mutants.

Acknowledgements

This work was supported by the 'Ministerio de Educación y Ciencia' (Project AGL2004-07630) and the 'Junta de Andalucía' (Research Group AGR115).

References

Romera FJ, Alcántara E. 2004. *Functional Plant Biology* 31, 315–328.

7. BIOTECHNOLOGY AND APPLIED ASPECTS

Modulating effects of ethylene and ethylene inhibitors in the control of fruit ripening

Tonutti, P.[a]*, Bonghi, C. and Ramina, A.

Department of Environmental Agronomy and Crop Science. University of Padova. 35020 Legnaro (Padova). Italy. [a]Present address: Sant'Anna School of Advanced Studies. 56127 Pisa. Italy. (*Corresponding author: pietro.tonutti@unipd.it)

1. Introduction

The control of the ripening process in climacteric fruit mainly relies on the possibility of affecting ethylene biosynthesis and action. This can be achieved through genetic manipulation, by modulating environmental parameters (temperature, atmosphere composition) during storage, or using specific inhibitors of ethylene biosynthesis/perception. The effects of treatments with ethylene inhibitors on climacteric fruit, however, differ according to several factors including genotype. For example, applications of 1-methylcyclopropene (1-MCP), an antagonist of ethylene for the receptor binding sites, induce variable responses in different fruit species (Blankenship and Dole, 2003; http://www.hort.cornell.edu/mcp). In a recent paper, Dal Cin *et al.* (2006) compared at physiological and molecular level the responses of peaches and apples to 1-MCP and concluded that the different behaviour of the two species to the ethylene action inhibitor might be related to differences in terms receptor type (ETR and ERS) expression and/or turnover. Exogenous ethylene applications enhance ripening in climacteric but not in non-climacteric fruit, although not all the ripening-related processes in climacteric fruit are ethylene dependent (Lelievre *et al.*, 1997). In addition, the same process can be modulated at different extent by ethylene in climacteric fruit belonging to different species: for example, ripening-associated colour changes appear to be differently affected by reduced levels of ethylene in tomato and melon (Klee, 1993; Guis *et al.*, 1997) Also in non-climacteric fruits some specific processes are influenced by the gaseous hormone. In citrus fruit, green colour is bleached out following ethylene fumigation (Abeles *et al.*, 1992); in grape berries, ethephon treatments performed at veraison (the beginning of colour change) increase anthocyanin synthesis and decrease juice acidity,

A. Ramina et al. (eds.), Advances in Plant Ethylene Research: Proceedings of the 7th International Symposium on the Plant Hormone Ethylene, 407–415.

whereas opposite effects are induced by 1-MCP [literature on ethylene and grape berry ripening has been recently reviewed by Mailhac and Chervin, (2006)]. Taken together, these data would indicate that similar regulatory mechanisms of ethylene action are shared by climacteric and nonclimacteric fruits, while divergent mechanisms might be present in different climacteric fruit species. Microarrays are powerful tools to study large-scale transcriptome profiles, to perform comparative analyses and possibly to identify structural and regulatory genes involved in specific metabolic pathways.

2. Materials and Methods

Peaches (*Prunus persica* Batsch cv Summer Rich) and grape berries (*Vitis vinifera* L. cv Raboso Piave) were harvested at commercial ripening stage. Peaches were incubated at 20°C in a sealed jar containing 1-MCP (1 μL/L) and KOH (as CO_2 absorber) for 24 h, and then transferred to air for an additional period of 14 days at 20°C. Grape berries were maintained for 7 days at 20°C in a jar flowed with ethylene-enriched (1000 ppm) humidified air (6.0 L/h). At the end of the treatment, grape berries were transferred to air at 20°C for 7 days. At the end of 1-MCP treatment and in correspondence of different post-treatment dates, flesh firmness and ethylene evolution were measured in peaches. Grape berries were sampled at the end of the ethylene treatment period and 7 days later. Peach mesocarp and grape skins were collected in correspondence of the sampling dates and stored at −80°C for molecular analyses. Peach (μPEACH 1.0) and grape (AROS v1) microarrays were constructed using the oligo sets provided by Operon (http://www.operon.com/arrays/oligosets_plants_overview.php). Considering peaches, microarray analyses compared cDNAs obtained from RNAs extracted from mesocarp of fruit maintained for 24 h either in 1-MCP or in air. For grapes, cDNAs targets were retro-transcribed from total RNA extracted from skins of grape berries at the end of the ethylene treatment or air. Microarray construction, hybridization, data analysis and validation were performed as described by Trainotti *et al.* (2006). Northern analysis was carried out on RNA extracted from control and ethylene-treated berry skins using grape phenylalanine ammonia lyase (PAL) and stilbene synthase (STS) cDNA probes as described by Versari *et al.* (2001).

2. Results and Discussion

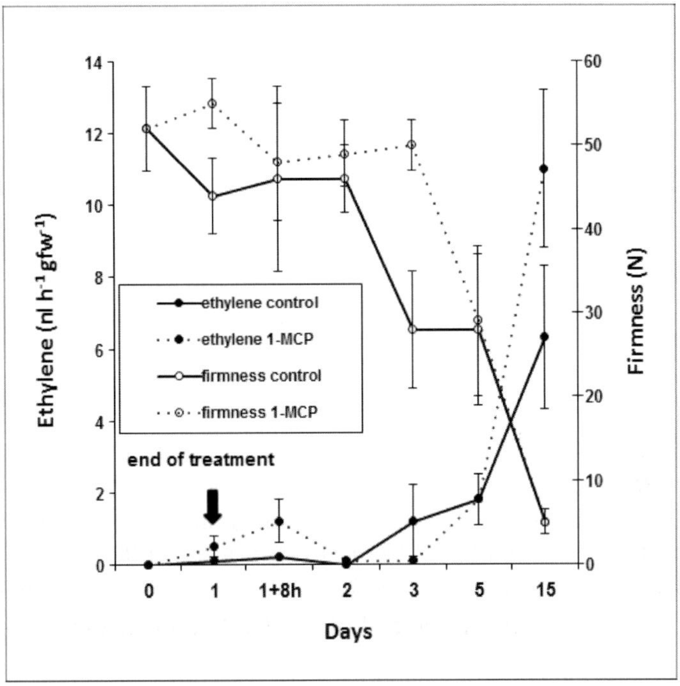

Fig. 1. Ethylene evolution (•) and firmness (**O**) in mesocarp of 1-MCP-treated (dotted line) and control (solid line) peach fruits. Arrow indicates the end of the incubation period. Vertical bars represent SD.

The application of 1-MCP induced only a limited effect in terms of firmness retention. In fact, 1-MCP-treated fruit were firmer than those of the control at the end of the incubation period and 3 days later (Fig. 1). The melting process rapidly occurred thereafter and no difference was observed between samples. Considering ethylene production (Fig. 1), a similar trend was observed in control and 1-MCP-treated fruits. Small amounts of ethylene were produced from treated fruit 8 h after the end of the incubation period. After 3 days, ethylene was detectable in control but not in 1-MCP-treated fruit. No difference was observed at the following sampling dates, even though higher values of ethylene biosynthesis were registered in treated fruit at the end of the experiment. Microarray analyses comparison of transcriptome profiles was performed of peach mesocarp sampled at the end of the incubation period with 1-MCP and in air (control). The Venn diagram shown in Fig. 2 reports the number of genes induced and repressed in 1-MCP-treated (8 and 7, respectively) and control (55 and

36, respectively) fruits. Considering the up-regulated genes, six are in common, whereas only three down-regulated are shared by both 1-MCP-treated and control fruits.

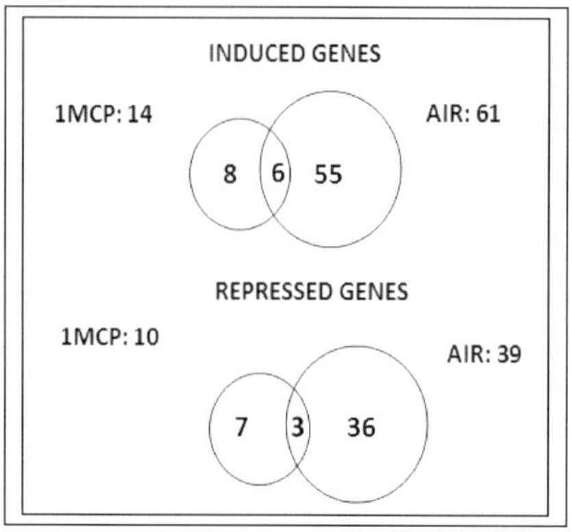

Fig. 2. The Venn diagram reports the number of induced and repressed genes in peach fruit treated with 1-MCP or maintained in air. Genes have been identi fied following μPEACH 1.0 microarray hybridization. In the overlapping areas the number of genes induced and repressed in common by both treatments are reported.

Considering the specific functions and categories of genes differently regulated in 1-MCP-treated and untreated peaches, among genes involved in ethylene physiology, only Pp-*ACO1* appeared to be significantly down-regulated by 1-MCP, even though a decrease of *Pp-ETR2* (an EIN4-like receptor) transcription resulted in treated fruit (Fig. 3A). Expression of some cell wall related genes as *PG* (contig 420), *PMEs* (contigs 4533 and 4499) and an expansin (contig 938) was negatively affected by 1-MCP. (Fig. 3B). On the contrary, an expansin gene (contig 941), down-regulated at ripening (Trainotti *et al.*, 2006), appeared to be induced by the chemical. Among genes involved in isoprenoid pathways, only β-carotene hydroxylase (contig 711) showed a significant down-regulation by 1_MCP. (Fig. 3C).

These preliminary results using microarray on 1-MCP-treated peaches, confirms a slight effect of the chemical on the expression of Pp-*ERS* and Pp-*ETR1* (Dal Cin *et al.*, 2006). On the contrary, *Pp-ETR2* appears to be strongly down-regulated by 1-MCP demonstrating that it is more sensitive to ethylene than *Pp-ETR1* and *Pp-ERS1* (Begheldo *et al.*, see this issue). These results pointed out a strong down-regulation induced by the chemical on ethylene-dependent genes, in spite of the limited transient effect on parameters defining the ripening syndrome.

In a second set of experiments, the microarray approach has been used to evaluate the effects of a saturating ethylene treatment (1000 ppm for 7 days) performed on wine grape berries after harvest. At the end of the treatment, 15 genes appeared down-regulated whereas 68 showed an increased expression (data not shown).

3A. Ethylene biosynthesis, perception and transduction

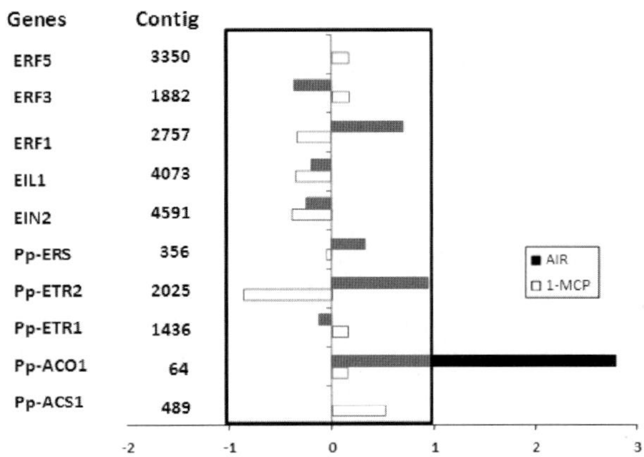

3B. Cell wall hydrolases

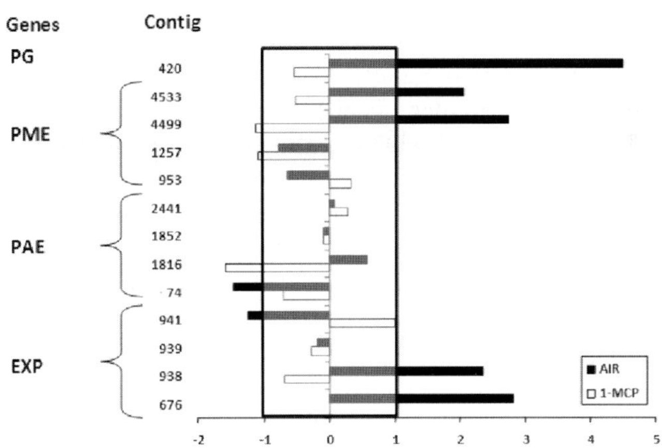

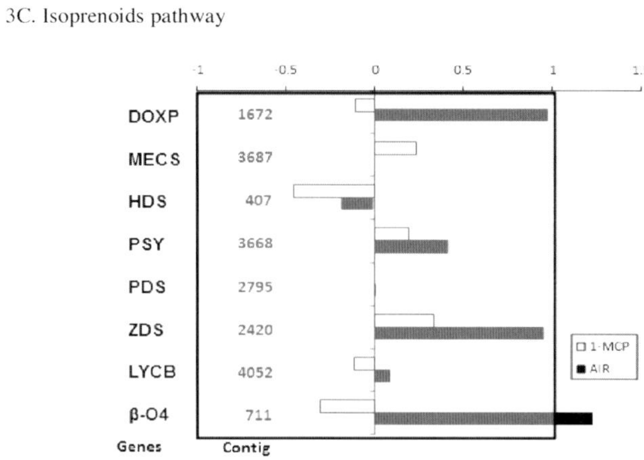

Fig. 3. Hybridization intensity ratio, reported as log(2), of probes corresponding to genes involved in ethylene biosynthesis, perception and transduction **(A)**, cell wall metabolism **(B)**, and isoprenoid pathway **(C)**, spotted on μPEACH 1.0 microarray. The ratio has been calculated by comparing hybridization signal of cDNAs from mesocarp of fruit at harvest with those of fruits maintained for 24 h in 1-MCP (white bars) or air (black bars). The ratio threshold value has been arbitrarily fixed in −1 and 1. Abbreviations: DOXP: 1-deoxy-D-xilulose 5-phosphate synthase; MECS: 2C-methyl-D-erythritol 2,4-cyclodiphosphate synthase; HDS: 1-hydroxy-2-methyl-2-(E)-butenyl 4-diphosphate synthase; PSY: phytoene synthase, ZDS: ζ-carotene desaturase; LCYB: lycopene β-cyclase and β-O4: β-carotene hydroxylase.

Among the up-regulated genes, 11 appeared involved in polyphenol metabolism and, in particular, in the biosynthetic pathway (Fig. 4). A Northern analysis using PAL cDNA validated the microarray experiments (Fig. 5) showing a marked accumulation of specific transcripts at the end of the treatment. An additional hybridization of the same RNAs has been performed using a grape stilbene synthase (STS) probe (not present in the microarray oligo set). The strong increase of STS gene expression following exposure to ethylene, the corresponding increase of resveratrol content in the skins (data not shown) together with the microarray expression data clearly indicate the presence of ethylene-controlling mechanisms active not only at veraison (Maihlac *et al.*, 2006), but also during the late developmental stages of grape berry.

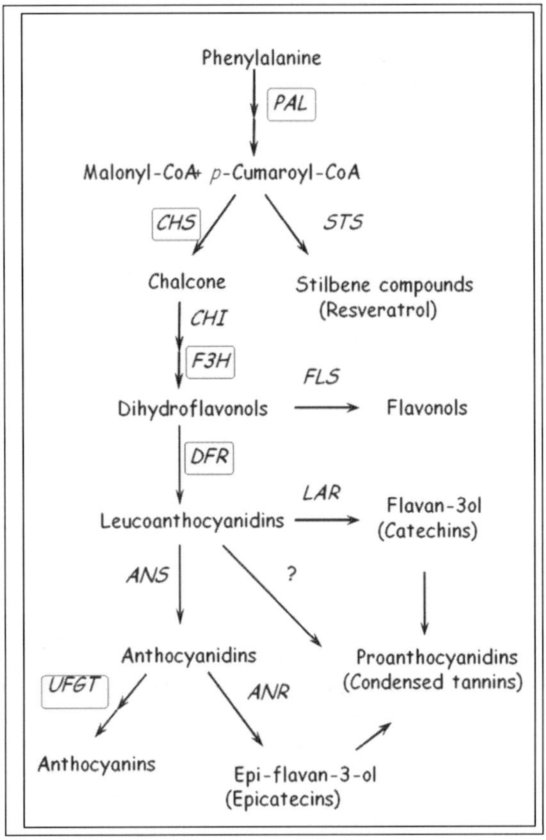

Fig. 4. Simplified scheme of polyphenol biosynthetic pathway. In a microarray experiment using a grape oligo set from Operon, expression of grey-boxed genes appeared to be induced in grape berry skin by a postharvest ethylene treatment (1000 ppm for 7 days). Abbreviations: PAL: Phenylammonialiase, CHS: chalcone synthase, STS: stilbene synthase, F3H: Flavanol 3-hydrolase, FLS: Flavonol synthase, DFR; Dihydroflavonol 4-reductase, LAR: Leucoanthocyanidin reductase, ANS: anthocyanidin synthase, UFTG: UDP-flavonol 3-O-glucosyltransferase, ANR: anthocyanidin reductase.

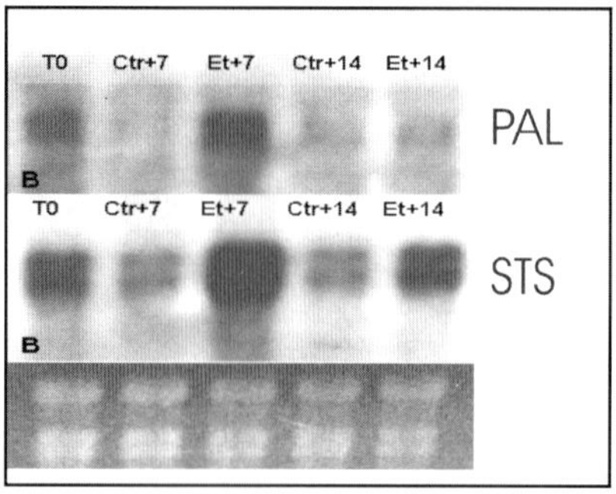

Fig. 5. Northern analysis carried out on total RNA extracted from skins of grape berries after 7 days of postharvest ethylene treatment (Et + 7) and 7 days after the end of the treatment (Et + 14). Control fruits (Ctr + 7 and Ctr + 14) have been maintained in air. cDNAs encoding for phenylalanine ammonia lyase (PAL) and stilbene synthase (STS) have been used as probes. Equal RNA loading have been checked by ethidium bromide staining.

References

Abeles B, Morgan P.W., Saltveit M.E., Jr. 1992. Ethylene in plant biology. Academic Press, Inc.

Begheldo, M., Ziliotto F., Rasori A., Bonghi C. The use of μPEACH 1.0 to investigate the role of ethylene in the initiation of peach fruit ripening, 265–267.

Blankenship, S.M., Dole, J.M. 2003. 1-Methylcyclopropene: a review. Postharvest Biology and Technology 28: 1–25.

Dal Cin V., Rizzini F.M., Botton A., Tonutti P. 2006. The ethylene biosynthetic and signal transduction pathways are differently affected by 1-MCP in apple and peach fruit. Postharvest Biology and Technology 42: 125–133.

Guis M., Rotondi R., Ben-Amor M., Ayub R., Bouzayen M., Pech J.C. 1997. Ripening-associated biochemical traits of cantaloupe Charentais melons expressing an ACC oxidase transgene. Journal of the American Society for Horticultural Science 122: 748–751.

Klee H. 1993. Ripening physiology of fruit from transgenic tomato (*Lycopersicon esculentum*) plants with reduced ethylene synthesis. Plant Physiology 102: 911–916.

Lelievre J.M., Latché A., Jones B., Bouzayen M., Pech J.C. 1997. Ethylene and fruit ripening. Physiology of Plant 101: 727–739.

Mailhac N. and Chervin C. 2006 Ethylene and grape berry ripening. Stewart Postharvest Review (online). Vol n 2, Issue 2. Available from: http://www.stewartpostharvest.com/April_2006/Chervin.htm

Versari A., Parpinello G.P., Tornielli G.B., Ferrarini R., Giulivo C. 2001. Stilbene compunds and stilbene synthase expression during ripening, wilting and uv treatment in grape cv corvine. Journal of Agricultural and Food Chemistry, 11: 5531–5536.

Trainotti L., Bonghi C., Ziliotto F., Zanin D., Rasori A., Casadoro G., Ramina A., Tonutti P. 2006. The use of microarray mu PEACH1.0 to investigate transcriptome changes during transition from pre-climacteric to climacteric phase in peach fruit. Plant Science 170: 606–613.

Issues with commercialization of 1-methylcyclopropene (1-MCP) for apples

Watkins, C.B.*, Jung, S.K., Razafimbelo, F. and Nock, J.F.

Department of Horticulture, Cornell University, Ithaca, NY 14853, USA, (phone 001 607 2551784). (*Corresponding author: cbw3@cornell.edu)

1. Introduction

1-Methylcyclopropene (1-MCP) is an inhibitor of ethylene binding that can prevent ethylene-dependent responses in many fruits and vegetables (Blankenship and Dole, 2003; Watkins, 2006). 1-MCP has been commercialized as SmartFresh and food use registration has been obtained for a wide range of crops in several countries (Watkins, 2006). The apple was the first crop to be registered and 1-MCP has been rapidly incorporated into storage and handling protocols around the world, especially because it maintains fruit quality throughout the marketing chain. However, the 'apple' is a fruit with many cultivars, each with unique ripening rates, susceptibility to physiological and pathological disorders, and storage potential. As such the apple has become an excellent model to illustrate the opportunities and limitations of 1-MCP-based technologies. Here we report on issues related to 1-MCP treatment of apples from three aspects: (1) the influence of harvest date on responses of fruit to 1-MCP treatment; (2) the effects of 1-MCP on internal ethylene concentrations (IEC) during air storage and (3) the effects of 1-MCP on susceptibility of 'Empire' apple fruit to CO_2 injury.

2. Materials and Methods

2.1 The influence of harvest date on responses of fruit to 1-MCP treatment

'McIntosh' (Rogers) apple fruit were harvested at 3–4 day intervals from four sets of replicate trees from 10 Sept. through 1 Oct., 2002. A minimum of 200 fruit (100 per apple crate) were harvested from each replicate group of trees. Ten fruit per replicate were used for analysis of IEC and flesh

A. Ramina et al. (eds.), Advances in Plant Ethylene Research: Proceedings of the 7th International Symposium on the Plant Hormone Ethylene, 417–422.

firmness as described by Watkins and Nock (Watkins and Nock, 2005). Crates of fruit from each replicate were either untreated or treated with 1 μL L^{-1} 1-MCP for 24 h in hermetically sealed chambers at 20°C (Watkins and Nock, 2005). The fruit were then stored in air at 0.5°C for 4 months. Each month, 20 fruit samples from each replicate box were transferred to a room kept at 20°C, and IEC and firmness of ten fruit measured after 1 and 7 days.

2.2 The effects of 1-MCP on IEC during air storage

'McIntosh' (RedMax) apple fruit were harvested at intervals from 19 Sept. to 10 Oct., 2005. A minimum of 100 fruit were harvested on each occasion and the IEC of each fruit measured within 2 h of harvest by gas chromatography (Watkins and Nock, 2005). The fruit were then treated with 1 μL L^{-1} 1-MCP for 24 h at 20°C, and then kept in air at 0.5°C. Fruit were separated into categories based on their IEC: <0.5, 0.5–1, 1–10, 11–50, 50–100 and >100 μL L^{-1}. Fruit were removed at 4 week intervals from storage and IEC were measured after 24 h at 20°C. Only categories with sufficient fruit to provide at least three apples per removal were analysed.

2.3 The effects of 1-MCP on susceptibility of 'Empire' apple fruit to CO_2 injury

'Empire' apple fruit were harvested and divided into jars, each containing 40–45 fruit. Fruit were cooled overnight and then untreated or treated with 1 μL L^{-1} 1-MCP for 24 h at 2°C. In Experiment 1, fruit from both treatments were then exposed to 1, 2.5 and 5% CO_2 (in 2% O_2) at 2°C for 20 weeks using a flow through system as described by Watkins *et al.* (1997). In Experiment 2, fruit were kept in air for 1, 2, 7 and 14 days before being exposed to 5% CO_2 for 10 weeks in the flow through system. The injury incidence was assessed after 7 days at 20°C. In each experiment, four replicate jars were used for each treatment.

3. Results and Discussion

3.1 The influence of harvest date on responses of fruit to 1-MCP treatment

Limited research is available on the effect of harvest date on the response of apple fruit to 1-MCP, but in general late harvest is associated with declining effectiveness of treatment (Watkins and Nock, 2005; Mir *et al.*, 2001). We are testing the effects of harvest date using several cultivars; our data on 'McIntosh' are presented here because it is a relatively rapid ripening cultivar that can vary in responsiveness to 1-MCP treatment. Fruit showed normal changes in harvest indices over time, with the IEC increasing from 0.3 to 39.7 $\mu L\ L^{-1}$, and firmness decreasing from 72.2 to 54.5N, from 10 Sept. through 1 Oct., respectively. Fruit were untreated or treated with 1-MCP, and firmness and IEC evaluated at monthly intervals during air storage (Fig. 1). Rates of softening and IEC increase were rapid in fruit without 1-MCP treatment, irrespective of harvest date. In contrast, the IEC of 1-MCP-treated fruit from earlier harvests increased much more slowly than in those from later harvests. These data confirm that control of ripening by 1-MCP is greater in fruit with lower IEC at harvest.

The firmness changes are more difficult to interpret however, as fruit softened both during the harvest period and during storage. Also, changes in rates of softening over time are also affected by harvest date during storage; for example, firmness of fruit from the last two harvests was maintained only for the first 2 months. When the data were expressed as percentage of the harvest firmness, a clearer picture of the effects of 1-MCP emerge (Fig. 2). Untreated fruit usually softened to 90% of harvest firmness between 1 and 2 months, while 1-MCP-treated fruit usually softened to similar levels between 2 and 3 months. Thus within the time period of this experiment, 1-MCP delayed softening of fruit from all harvest dates, irrespective of its effects on IEC. However, from a commercial point of view, fruit have to meet market requirements for firmness, and therefore firmness at harvest remains a key factor regardless of whether 1-MCP is used or not.

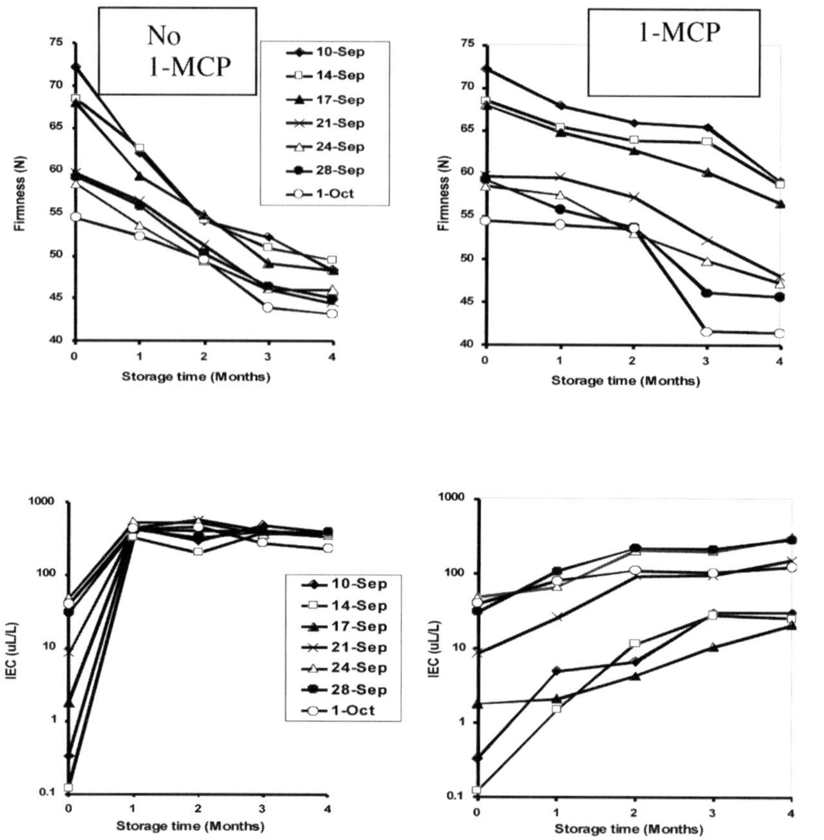

Fig. 1. Firmness and IEC of 'McIntosh' apples harvested from 10 Sept. to 1 Oct., 2002, untreated or treated with 1-MCP and stored at 0.5°C for up to 4 months.

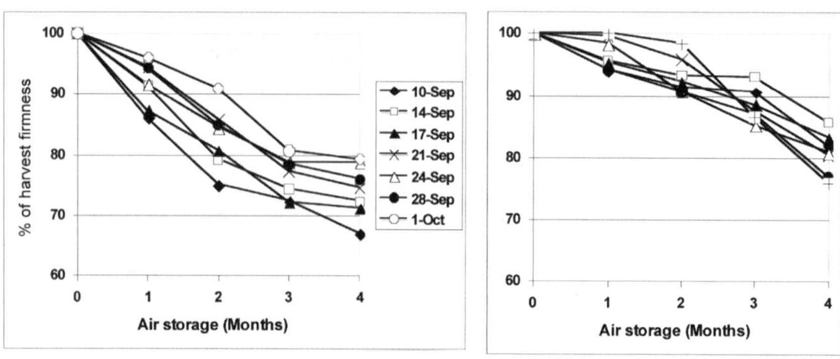

Fig. 2. Firmness of untreated and 1-MCP-treated fruit recalculated as percentage of harvest firmness.

3.2 The effects of 1-MCP on IEC during air storage

Variability of fruit in response to 1-MCP is a significant concern for cultivars such as 'McIntosh'. To better understand how fruit respond to 1-MCP, we separated them by IEC at harvest and followed the changes of IEC during air storage. A single harvest date has been selected to illustrate patterns of response that were consistent throughout the experiment (Fig. 3). Our expectation was that fruit with high IEC, e.g. 50–100 μL L^{-1}, would not respond to 1-MCP, but these fruit showed an initial decline before increasing. Fruit with low IEC at harvest remained low during the period of this experiment. The data indicate that even fruit that are more mature at harvest will respond, but that the delays of ripening will be shorter.

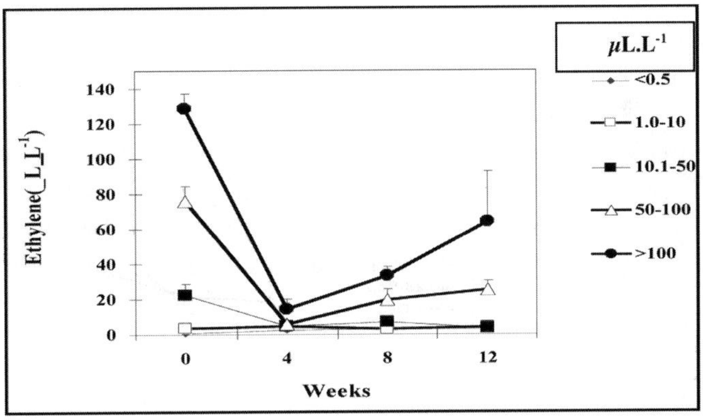

Fig. 3. IEC of 1-MCP-treated 'McIntosh' apple fruit separated into IEC categories (μL L^{-1}) at harvest (see legend).

2.3 The effects of 1-MCP on susceptibility of 'Empire' apple fruit to CO_2 injury

1-MCP-treated 'Empire' apple fruit have increased susceptibility to external CO_2 injury. When exposed to 1, 2.5 and 5%, the injury incidence was 4, 25 and 31%, respectively, for untreated fruit, and 8, 38 and 57%, respectively, for 1-MCP-treated fruit (effects of CO_2 and 1-MCP significant at $P < 0.001$ and $P = 0.011$, respectively). We have been investigating the reasons for this increased susceptibility. One possibility is that 1-MCP prevents the adaptation period that occurs after harvest; delays in air before exposure to elevated CO_2 can markedly reduce injury development (Watkins *et al.*, 1997). The incidence and severity of external CO_2 injury of fruit without 1-MCP treatment was not affected by a 2-day delay before exposure to 5% CO_2 compared with 1 day, but by 7 days was reduced by almost 50% of those measured in fruit treated on day 1 (Table 1). A 14-day delay resulted in almost complete control of injury. In fruit that were treated with 1-MCP, however, the incidence of external CO_2 injury remained statistically unchanged by all delays.

The results indicate that special care has to be taken to avoid CO_2 injury under commercial conditions. Using low CO_2 (< 1%) in the storage atmosphere for the first 4 weeks when susceptibility to injury is highest has been used with some success. Use of diphenylamine (DPA) has been shown to totally prevent risk of injury (Watkins *et al.*, 1997), but has the disadvantage of being an additional expense and is chemically based.

Table 1. External CO_2 injury (%) of 'Empire' apples fruit untreated or treated with 1-MCP and exposed to 5% CO_2 (in 2% O_2) for 10 weeks.

	Day 2	Day 7	Day 14
Untreated	68.7a	34.1b	0.6c
1-MCP-treated	71.1a	70.7a	58.9a

In summary, research to understand the variable responses of apples to 1-MCP in terms of quality and susceptibility to disorders remains an important issue for industry. Much remains to be learned as we adopt this new technology to ensure that we maximize quality of fruit in the marketplace.

References

Blankenship SM, Dole JM (2003) 1-Methylcyclopropene: A review. Postharvest Biol Technol 28: 1–25

Mir NA, Curell E, Khan N, Whitaker M, Beaudry RM (2001) Harvest maturity, storage temperature, and 1-MCP application frequency alter firmness retention and chlorophyll fluorescence of 'Redchief Delicious' apples. J Am Soc Hortic Sci 126: 618–624

Watkins CB, Silsby KJ, Goffinet MC (1997) Controlled atmosphere and antioxidant effects on external CO2 injury of 'Empire' apples. HortScience 32: 1242–1246

Watkins CB, Nock JF (2005) Effects of delays between harvest and 1-methylcyclopropene (1-MCP) treatment, and temperature of treatment, on ripening of air- and controlled atmosphere-stored apples. HortScience 40: 2096–2101

Watkins CB (2006) The use of 1-methylcyclopropene (1-MCP) on fruits and vegetables. Biotechnol Adv 24: 389–409

Comprehensive analysis of candidate genes involved in ethylene production and perception during apple ripening: phenotypic dissection and functional profiling

Costa, F.[1]*, Stella, S.[1], Soglio, V.[2], Gianfranceschi, L.[2], Schouten, H.[3], Van de Weg, W.E.[3], Guerra, W.[4], Serra, S.[1], Musacchi, S.[1] and Sansavini, S.[1]

[1]Dipartimento di Colture Arboree, Universita' degli Studi di Bologna, 40127 Bologna, Italy. [2]Dipartimento di Scienze Biomolecolari e Biotecnologie, Università degli Studi di Milano, Via Celoria 26, 20133 Milano, Italy. [3]Plant Research International, Wageningen, PO Box 16, 6700A, The Netherlands. [4]Centro per la Sperimentazione Agraria e Forestale Laimburg, Posta Ora, 39040 Vadena (BZ), Italy. (*Corresponding author: fcosta@agrsci.unibo.it)

1. Introduction

The ripening of fleshy fruit is generally distinguished into two categories, climacteric and non-climacteric, based on the fruit's ability to produce or ethylene in high concentrations during maturation. A gaseous hormone produced by higher plants, ethylene regulates a number of physiological processes ranging from fruit development to maturation and senescence (Yang and Hoffmann 1984; Giovannoni 2004). At the onset of ripening, climacteric fruits are characterized by an ethylene burst due to the transition from ethylene system 1 to system 2-physiological stages controlled by the expression of different *1-aminocyclopropane-1-carboxylic acid synthase (ACS)* genes that change the pre-climacterium fruit physiology into climacetric (Barry *et al.*, 2000). Susequent to the action of the two main pathway genes (*ACS* and *1-aminocyclopropane-1-carboxylic acid oxidase, ACO*), ethylene is perceived by a series of receptors that, together with transcription factors and a signaling cascade, regulate the functional expression of several ethylene target genes (Klee, 2002).

Given its complex mode of action, ethylene controls several traits related to fruit quality, one of the main objectives of modern breeding programs. We investigated allele diversity and trait association of three candidate genes, *Md-ACS1* and *Md-ACO1* involved in ethylene production and *Md-ERS1* in ethylene perception. The haplotype diversity of these genes

A. Ramina et al. (eds.), Advances in Plant Ethylene Research: Proceedings of the 7th International Symposium on the Plant Hormone Ethylene, 423–429.

was explored to specify their map position, QTL effect and transferability in four different mapping populations and in a suite of 30 separate apple cultivars.

The investigated genes belong to complex families in which each member is specifically expressed in different physiological stages or tissues. To improve the confidence about gene ripening specificity and to analyze their ethylene dependence, their gene expression profiles were assessed via real time PCR technique. Expression profiles of these candidates were also run in a functional comparison with 1-MCP treated vs. control fruit samples to further elucidate the gene-ethylene relationship. The ripening samples were thus used to identify new candidate genes active in the late ripening step using two novel approaches: microarray wide genome profiling and total proteome investigation.

2. Material and Methods

Four apple populations and a suite of 30 apple cultivars were employed for trait association mapping and QTL identification. The four progeny considered are located in different districts in Europe: Prima x Fiesta (The Netherlands) was used as the reference map due to its marker density (Van de Weg, unpublished data); Fuji x Braeburn and Golden Delicious x Fuji are in northern Italy (Laimburg Research Station in Bolzano) and were used to analyze the haplotype effect and phenotypic trait dissection; and Fuji x Mondial Gala (University of Bologna's experiment station) was used for functional marker (candidate gene-based approach) positioning and QTL mapping.

At harvest, the date being set at the starch degradation value of 7 (1–10 scale), the 1-MCP treatment (1 ppm of concentration) was applied to a sub-sample of Mondial Gala apples for 12 h. Ethylene evolution of control and treated fruits was monitored for 30 days by measuring ethylene production three times a week. Fruit softening data were calculated as the difference between flesh firmness assessed with an Effegi penetrometer at harvest and after 2 months' cold storage at 4°C. Allele variability was explored by analyzing the sequence polymorphism of three functional markers related to the genes *Md-ACS1* and *Md-ACO1* involved in the ethylene pathway and *Md-ERS1* in ethylene reception. Molecular mapping was performed with JoinMap 3.0 (Van Ooijen *et al.*, 2001) and QTL mapping with MapQTL 4.0 (Van Ooijen *et al.*, 2000) softwares.

3. Results and Discussion

The Fuji x Breaburn progeny was used to assess phenotype segregation and natural ethylene variability. Twenty-six seedlings showed a 1:1 ethylene segregation ratio, around a threshold of 30 μl Kg^{-1} h^{-1} (Costa *et al.*, 2005; Fig. 1). Even if there is trait variability, it is clear in this distribution that ethylene tends to be qualitatively inherited, thereby suggesting the involvement of a restricted number of genes, some with a major effect on the phenotypic control. *Md-ACS1* and *Md-ACO1* haplotype variability showed a direct impact on ethylene distribution in this population, the former being the gene with a greater impact than the latter (Harada *et al.*, 2000; Costa *et al.*, 2005).

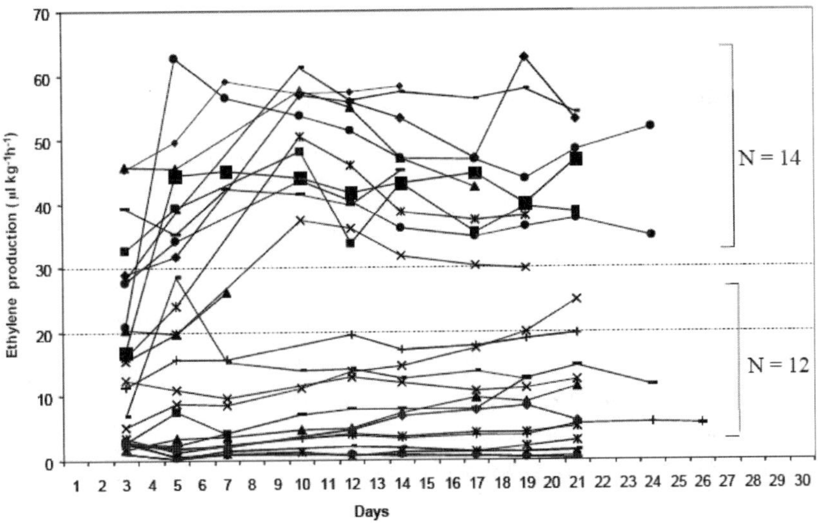

Fig. 1. Ethylene evolution of 26 seedlings of the Fuji x Breaburn population over 30 days after harvest (Costa *et al.*, 2005).

In Prima x Fiesta *Md-ACS1* was positioned in the LG15 of both parental cultivar maps (Fig. 2). The alignment of five LG15 (Prima, Fiesta, Fuji. Mondial Gala, Discovery) enabled the assembly of a microsatellite fluorescent multiplex system covering linkage group 15 in *Md*-ACS1's surrounding genomic area. This system was set up to amplify the LG 15 in Breaburn, which has to date gone unmapped. Combining the marker segregation and the ethylene data enabled the identification of a putative QTL for ethylene in LG15. The characterization of a possible QTL with the highest LOD value close to the genomic position of *Md-ACS1* highlights the importance of this gene in controlling ethylene production and indicates that ethylene is a quantitative trait but its variability is tightly controlled by *Md-ACS1*.

The second gene involved in the ethylene pathway, *Md-ACO1*, mapped in LG10 of the Prima x Fiesta maps, within the 5% interval border of a known major QTL for fruit firmness (Fig. 2; Costa *et al.*, 2005; Maliepaard *et al.*, 2001). *Md-ACO1* was also mapped for Mondial Gala in LG10, and its allele diversity affects phenotype variability in the Fuji x Mondial Gala population, thus highlighting QTL transferability over different environments. The QTL LOD significance initially detected in Prima x Fiesta population was in fact also observed in this second progeny, confirming the reliability of this functional marker in phenotypic control of fruit softening.

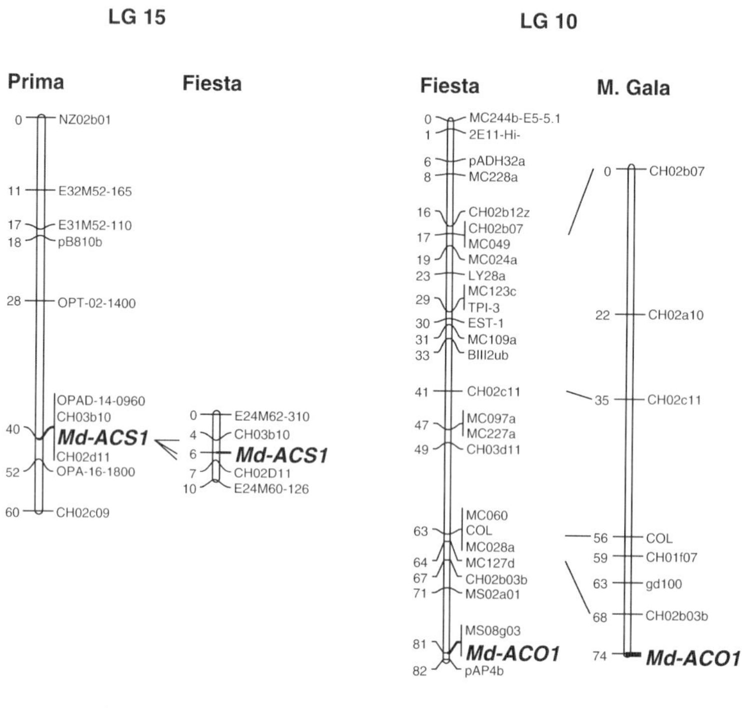

Fig. 2. *Md-ACS1* and *Md-ACO1* map position on LG15 of Prima and Fiesta and LG10 of Fiesta and Mondial Gala, respectively (Costa *et al.*, 2005).

The QTL transferability of these two candidate genes for fruit firmness was further assessed in a third population, Golden Delicious x Fuji. In this progeny *Md-ACO1* showed a major effect with respect to the *Md-ACS1* haplotyope variability; a similar effect was also observed in allele association with fruit firmness in the apple cultivar suite. These results thus confirm that *Md-ACS1* is the limiting step in the control of ethylene production in apple, and that *Md-ACO1* has the major impact on fruit firmness control. These findings are also supported by the genetic mapping of *Md-ACO1*, a gene involved in the ethylene pathway, within a QTL region for fruit firmness in apple (Fig. 3).

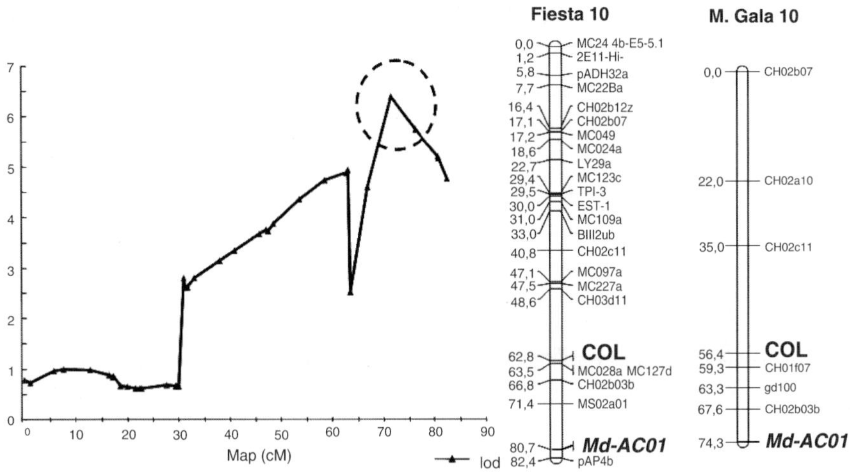

Fig. 3. Fruit firmness QTL mapping on LG10 of Fiesta and Mondial Gala.

Md-ERS1 is a novel putative gene involved in the ethylene reception system identified in apple and reveals a good sequence identity to the tomato *Nr* gene. *Never ripe* in tomato is a semi-dominant mutation (receptor-like to the *ETR1* class) that shows an altered ethylene perception determining distorted or failed ripening (Wilkinson *et al.*, 1995). The *Md-ERS1* functional marker in apple was positioned in LG3 in the Prima x Fiesta mapping population. Overall fruit softening phenotype data (Maliepaard *et al.*, 2001) show that the functional marker based on this gene was coincident with an increased LOD value in a genomic region presenting a minor QTL for fruit firmness.

Md-ACO1 and *Md-ERS1* were both spotted on an apple microarray chip (EU-funded HiDRAS Project, data unpublished) and their relative expression proved to be up-regulated throughout fruit maturation, their transcript being highly abundant in the sample that corresponded to full ripening. The microarray expression data were further confirmed via real time PCR by assessing the transcriptome profile of these genes over fruit ripening. To elucidate the ethylene dependence of gene expression, the functional profile was also performed in a physiological comparison of control against the 1-MCP-treated sample. *Md-ACO1* and *Md-ERS1* were highly expressed in the control sample at the beginning of the ethylene burst but were notably down-regulated by the 1-MCP treatment (Fig. 4). *Md-ACS1* was also highly expressed at the ethylene peak and, like the other two, down-regulated by 1-MCP. Map positions and the expression profiles support the involvement of these gene members in the late stage of fruit ripening, and validate the use of the functional marker technique in molecular-assisted selection (MAS) and breeding programs for ethylene production and fruit

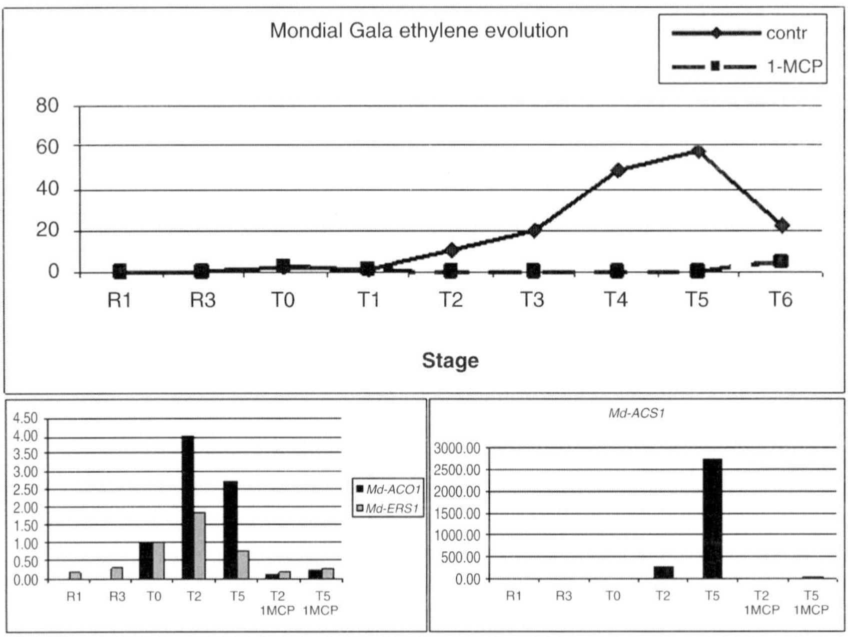

Fig. 4. Expression profiles of *Md-ACS1*, *Md-ACO1* and *Md-ERS1* during fruit ripening and comparison with 1-MCP treatment.

Two novel strategies were employed to explore the transcription dynamics over fruit development and ripening. The first is a heterologous genomic comparison with tomato, a model species for fruit ripening. In studies carried out using TOM1 array. This extensively characterized and annotated genomic tool has been used to target differentially expressed novel genes throughout fruit ripening (Alba *et al.*, 2004). The second strategy, used for candidate gene identification, is a total proteomic profiling carried out between two samples, one characterized by normal (control) and the other by modified ripening (1-MCP). Preliminary data indicate a different content in the proteome pattern, revealing that 1-MCP causes a notable variation in the transcriptome and in fruit proteome activity.

References

Alba R., Fei Z., Payton P., Liu Y., Moore S., Debbie P., Cohn J., Dascenzo M., Gordon J., Rose J., Martin G., Tanksley S., Bouzayen M., Jahn M. and Giovannoni J. ESTs, cDNA microarray, and gene expression profiling: tool for dissecting plant physiology and development. Plant J (2004) 39: 697–714.

Barry S.C., Llop-Tous M.I., and Grierson D. The regulation of 1-Aminocyclopropane-1-Carboxylic Acid Synthase gene expression during the transition from system-1 to system-2 ethylene synthesis in tomato. Plant Physiology (2000), 123: 979–986.

Costa F., Stella S., Van de Weg W.E., Guerra W., Cecchinel M., Dallavia J., Koller B. and Sansavini S. Role of the genes Md-ACO1 and Md-ACS1 in ethylene production and shelf life of apple (*Malus domestica* Borkh). Euphytica (2005) 141: 181–190.

Giovannoni J.J. Genetic regulation of fruit development and ripening. The Plant Cell (2004) 16: S170–S180.

Harada T., Sunako T., Wakasa Y., Soejima J., Satoh T. and Niizeki M. An allele of the 1-aminocyclopropane-1-carboxylate synthase gene (Md-ACS1) accounts for the low level of ethylene production in climacteric fruits of some apple cultivars. Theor. Appl. Genet (2000) 101: 742–746.

Klee H.J. Control of ethylene – mediated processes in tomato at the level of receptors. J Exp Bot (2002), 53: 2057–2063.

Maliepaard C., Silanpää M.J., Van Ooijen J.W., Jansen R.C. and Arjas E. Bayesian versus frequentist analysis of multiple quantitative trait loci with an application to an outbred apple cross. Theor Appl Genet (2001) 103: 1243–1253.

Van Ooijen J.W., Boer M.P., Jansen R.C., Maliepaard C, 2000. MapQTL 4.0: software for the calculation of QTL positions on genetic maps. Plant Research International, Wageningen, the Netherlands.

Van Ooijen J.W. and R.E. Voorrips, 2001. JoinMapÆ 3.0, Software for the calculation of genetic linkage maps. Plant Research International, Wageningen, the Netherlands.

Wilkinson J.Q., Lanahan M.B., Yen H-C., Giovannoni J.J. and Klee H.J. An ethylene-inducible component of signal transduction encoded by Never ripe. Science (1995), 270: 1807–1809.

Yang S.F. and Hoffmann N.E. Ethylene biosynthesis and its regulation in higher plants. Annu Rev Plant Physiol (1984), 35: 155–189.

Chilling injury as related to climacteric behaviour in plums

Candan, A.P.[1], Graell, J.[2] and Larrigaudière, C. [2]*

[1]Postharvest, INTA Alto Valle, CC 782 (8338), General Roca, Río Negro, Argentine. [2]Postharvest, UdL-IRTA Center, CeRTA, Rovira Roure 191, 25198 Lleida, Spain. (*Corresponding author: christian.larrigaudiere@irta.es)

1. Introduction

Plums have a limited postharvest life due to their elevated rate of ripening and because they can develop chilling injury after long cold storage, even when stored at optimal temperatures (0°C ± 0.5°C). Flesh browning and flesh translucency are the main symptoms and the susceptibility depends on the cultivar and storage temperature (Crisosto *et al.* 1999). Flesh browning is a brown discolouration of the flesh that starts below the epidermis. Flesh translucency manifests as a translucent gelatinous breakdown of the mesocarp tissue around the stone. Although these symptoms mainly develop during shelf life following cold storage, changes in membrane permeability have been observed before the occurrence of symptoms. These changes in membrane permeability were associated with the development of chilling injury in 'Songold' plums (Taylor *et al.* 1993).

In climacteric fruits, ethylene production at the onset of ripening controls the changes in colour, aroma, texture, flavour and other biochemical and physiological attributes. In contrast, the ripening of non-climacteric fruits is considered as an ethylene-independent process (Lelièvre *et al.* 1997). Plum cultivars can be classified into climacteric or suppressed climacteric according to their ripening behaviour. Suppressed climacteric phenotype results from an inability of the fruit to produce enough ethylene to coordinate ripening (Abdi *et al.* 1997). Many trials have been carried out to study the relationship between ethylene production, ripening and ethylene inhibitors such as 1-MCP in plums (Martínez-Romero *et al.* 2003; Candan *et al.* 2006). However, further studies are necessary to explain the role that ethylene plays in the occurrence of chilling injury in plums.

A. Ramina et al. (eds.), Advances in Plant Ethylene Research: Proceedings of the 7th International Symposium on the Plant Hormone Ethylene, 431–436.

2. Material and Methods

'Larry Ann' and 'Angeleno' plums were harvested at the pre-climacteric stage from a commercial orchard in Rio Negro, Argentina. On the day of harvest, maturity parameters were measured, and fruits were divided into two lots and treated with 0 (control) or 400 ppb 1-MCP during 24 h at 0°C and then stored at 0°C for up 60 days. After 1, 3, 7, 11, 15, 30, 40, 50 and 60 days at 0°C, fruits of each treatment was removed from cold storage and ethylene production and ion leakage were determined. After 30, 40, 50 and 60 days at 0°C, plus 3, 7 and 11 days of shelf life at 20°C maturity and chilling injury symptoms were also assessed. Ethylene production (nl/g/h) by enclosing six fruits in sealed jars, taking 1 ml sample from the head-space and injecting it in a GC (Shimadzu 14-A). Ion leakage (%) was calculated in six replicates per treatment by determination of electrical conductivity of 2 g of fruit flesh submerged in 40 ml of Mannitol 0.4 M, after and before frozen each sample. Maturity parameters: firmness (N) was measured by an electronic penetrometer (GÜSS) with an 8 mm plunger. Pieces of each fruit were juiced to determine soluble solids content (%) by a refractometer (Atago) and titratable acidity (%) by titration of 10 ml of juice with NaOH 0.1 N to an end point of pH 8.2. Epidermis colour was measured with a Minolta CR300. Flesh colour was measured visually after cutting each fruit in half along the equatorial axis. Chilling injury symptoms were assessed visually by cutting each fruit in half along the equatorial axis.

3. Results

3.1 Changes in maximal ethylene production

Each data of Fig. 1 represent the value of the ethylene peak of a kinetic carried out at 20°C after different time of storage at 0°C. The value of maximal ethylene production was cultivar dependent and clearly affected by exposure to low temperatures (0°C) and treatments (Fig. 1). At each point of analysis, the magnitude of the ethylene peak was higher in the climacteric cultivar than in the suppressed climacteric. A twofold increase in maximal ethylene production was observed in climacteric fruits after 1 and 3 days of exposure at 0°C. Later, the magnitude of the climacteric peak decreases during storage at 0°C, while it steadily increases in the suppressed climacteric cultivar. A sharp increase in ethylene production was observed in the climacteric fruits on a short-term basis, but not when the fruits were treated with 1-MCP or in the suppressed climacteric fruits.

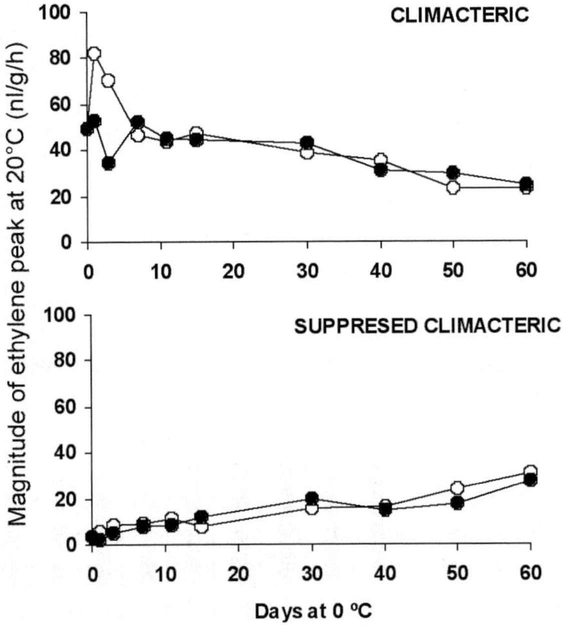

Fig. 1. Changes in the magnitude of the ethylene peak (nl/g/h) at 20°C in climacteric ('Larry Ann') and suppressed climacteric ('Angeleno') plums stored for 60 days at 0°C. (○): control fruits; (●): 1-MCP-treated fruits.

3.2 Maturity parameters

Fruits from the climacteric cultivar showed softening and losses of titratable acidity and epidermis colour (hue). 1-MCP-treated fruits maintained higher firmness than control fruits after 3 and 7 days of shelf life and exhibited more acidity and hue values than control fruits (data not shown).

During storage and shelf life periods, the suppressed climacteric fruits did not show important changes in firmness, titratable acidity, soluble solids content and epidermis colour (hue), and no differences were observed between control and 1-MCP-treated fruits (data not shown). Even in the last evaluation (60 days plus 11 days), fruit maintained 60% of the initial firmness and 80% of initial acidity.

3.3 Ion leakage and chilling injury

Ion leakage (%) was affected by 1-MCP treatment mainly in 'Larry Ann' cultivar. Higher ion leakage values were found in control fruits in comparison to 1-MCP-treated fruits after 3, 7, 11, 15 and 60 days of exposure at 0°C (Fig. 2).

'Larry Ann' developed the first flesh translucency symptoms during shelf life after 30 days of storage (data not shown). 1-MCP treatment reduces the severity of these symptoms and the percentage of affected fruits (Fig. 3). After 30, 40, 50 and 60 days plus 3 days of shelf life, the 1-MCP treatment reduced the percentage by 7%, 20%, 32% and 62%, and by 3%, 37%, 28% and 42% when shelf life was 7 days. 'Angeleno' fruits remained healthy until the end of the experimental period, with no development of translucency even after 60 days of storage plus 11 days of shelf life irrespective of treatment (Fig. 3).

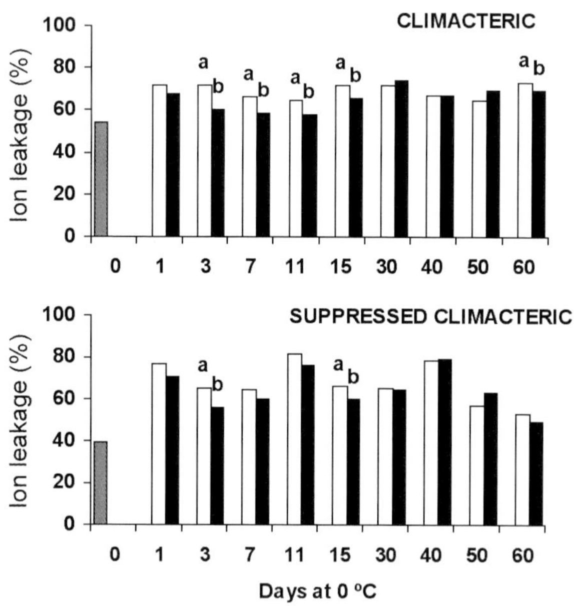

Fig. 2. Effect of time of storage at 0°C on ion leakage in climacteric ('Larry Ann') and suppressed climacteric ('Angeleno') plums cultivars stored for 60 days at 0°C. (○): control fruits; (●):1-MCP-treated fruits.

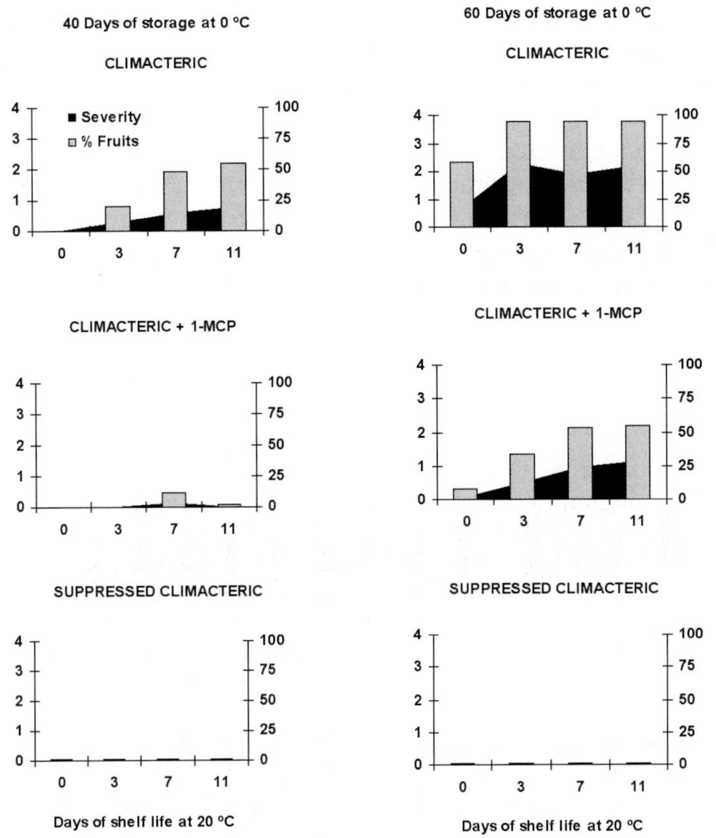

Fig. 3. Percentage of fruits affected by flesh translucency (grey bars) and severity of this symptom (black areas) in climacteric cultivar ('Larry Ann'), climacteric cultivar + 1-MCP and suppressed climacteric cultivar ('Angeleno') irrespective of treatment, after 40 and 60 days of storage at 0°C + 3, 7 and 11 days of shelf life at 20°C.

4. Discussion

The increase in maximal ethylene production observed in 'Larry Ann' control fruits after a short time exposure to cold was not observed in 1-MCP-treated fruits. In accordance with this result, control fruits showed higher ion leakage than 1-MCP-treated fruits. This result may be due to the action of ethylene on membrane integrity.

The first step of chilling injury is recognized as a loss in membrane integrity. The second stage corresponds to the development of the symptoms (Saltveit and Morris 1990). In agreement with this model, untreated climacteric fruits showed high severity and a great number of fruits affected by chilling injury. 1-MCP treatment clearly reduced these chilling symptoms. This last result highlighted the putative role that ethylene plays in the occurrence of chilling injury disorder. In accordance with this result, the suppressed climacteric cultivar, that produces low amounts of ethylene, was less susceptible to chilling injury. Collectively, these results show that chilling injury in plums is related to the climacteric behaviour of the fruit and more precisely to the changes in maximal ethylene production on a short-term basis. 1-MCP as an inhibitor of ethylene production effectively reduces the occurrence of chilling disorders.

References

Abdi N, Holford P, Mc Glasson WB, Mizrahi Y (1997) Ripening behaviour and responses to propylene in four cultivars of Japanese type plums. Postharvest Biology and Technology 12: 21–34.

Candan AP, Graell J, Crisosto C, Larrigaudiere C (2006) Improvement of storability and shelf life of 'Blackamber' plums treated with 1-methylcyclopropene. Food Science and Technology International 12: 437–443.

Crisosto CH, Mitchel FG, Ju Z (1999) Susceptibility to chilling injury of peach, nectarine, and plum cultivars grown in California. HortScience 34 (6): 1116–1118.

Lelièvre JM, Latché A, Jones B, Bouzayen M, Pech JC (1997) Ethylene and fruit ripening. Physiologia plantarum 101: 727–739.

Martínez-Romero D, Dupille E, Guillén F, Valverde JM, Serrano M, Valero DJ (2003) 1-Methylcyclopropene increases storability and shelf life in climacteric and nonclimacteric plums. Agricultural and Food Chemistry 51: 4680–4686.

Saltveit ME, Morris LL (1990) Overview on chilling injury of horticultural crops. In: Wang ChL (ed) Chilling injury of horticultural crops. Chapter 1: 3–16.

Taylor MA, Jacobs G, Rabe E, Dodd MC (1993) Physiological factors associated with over-ripeness, internal breakdown and gel breakdown in plums stored at low temperature. Journal of Horticultural Science. 68 (5): 825–830.

Effect of combining 1-MCP treatment and heat treatment on the storage capability of 'blanquilla' pear

Vilaplana, R.*, Soria, Y., Valentines, M.C., Reig, G. and Larrigaudière, C.

Postharvest, UdL-IRTA Center, CeRTA, Rovira Roure 191, 25198 Lleida, Spain. (*Corresponding author: rosa.vilaplana@irta.es)

'Blanquilla' pear (*Pyrus communis*) has delicious fragrance and flavour and represents an important produce for Spanish retailers. Although refrigeration can be used to prolong its storage, fruit ripens very quickly, even during cold storage, and therefore has a limited commercial life. This fruit is also very sensitive to mechanical bruising and can develop scald during long-term storage. In order to eliminate these specific problems, fruit were treated with 1-MCP and kept in air storage during 4 months. In general, in these conditions the fruit remain evergreen and the ripening process is blocked (Bai 2006, Trinchero *et al.* 2004). To solve this problem, combined treatments that mixed 1-MCP treatment and heat treatment to induce ripening were assayed.

'Blanquilla' pear was obtained from an experimental orchard in Lleida (Spain), harvested following local recommendations. Fruit was divided in five sets and four of them were treated with 1-MCP (for 300 nL L^{-1}: 0.48 g m^{-3} of chamber) following the procedure of Vilaplana *et al.* (2006). Fruit was then stored at −0.5°C for 4 months and approximately after 3 months of storage, three sets of 1-MCP-treated fruit were heat treated (5, 10 or 15 days at 15°C noted as E5, E10 and E15, respectively). The changes in ripening and ethylene production were determined after storage and during shelf life at 20°C, according to Vilaplana *et al.* (2006).

After 4 months of storage in air, untreated fruits were over-ripened, showing a significant reduction of firmness and high incidence of scald and mechanical bruising, especially after shelf life (Table 1). In contrast, 1-MCP-treated fruits did not ripen and remained green (data not shown) and excessively firm even during shelf life (Table 1). A synergistic effect was observed when the fruit was treated with 1-MCP and heat. Fruit remained free of scald and mechanical bruising but ripened during shelf life. The intensity by which the ripening process was restored depended on the duration of the heat treatment. A significant improvement in the general

A. Ramina et al. (eds.), Advances in Plant Ethylene Research: Proceedings of the 7th International Symposium on the Plant Hormone Ethylene, 437–439.

fruit quality (lower loss of colour, firmness, lack of disorders) was observed, especially when the fruit was treated 5 days at 15°C.

		Control	1-MCP	E5	E10	E15
Firmness (Kg)	Harvest	6.7 a				
	Storage	3.0 e	5.7 d	6.3 b	5.6 cd	5.9 c
	3 d S. life	2.1 b	5.9 a	6.1 a	5.1 a	4.5 a
	7 d S. life	1.0 d	5.8 a	4.0 b	2.7 c	1.3 d
SSC (%)	Harvest	10.5 b				
	Storage	11.9 a	12.0 a	11.9 a	11.9 a	12.2 a
	3 d S. life	11.9 a	11.8 a	11.8 a	11.3 a	11.8 a
	7 d S. life	11.6 b	12.0 a	11.7 ab	11.5 b	12.1 a
Acidity ($g.L^{-1}$)	Harvest	2.11 a				
	Storage	1.45 b	1.48 b	1.37 b	1.44 b	1.42 b
	3 d S. life	1.28 a	1.55 a	1.45 a	1.33 a	1.72 a
	7 d S. life	1.61 a	1.57 a	1.76 a	1.49 a	1.55 a
Scald incidence (%)	Harvest	0 a				
	Storage	0 a	0 a	0 a	0 a	0 a
	3 d S. life	63 a	0 b	0 b	0 b	0 b
	7 d S. life	80 a	0 b	0 b	0 b	0 b

Table 1. Effect of 1-MCP and combinations with heat treatment on physico-chemical attributes and scald incidence after 4 months of storage and during shelf life at 20°C of 'Blanquilla' pear. Means in a row not followed by a common letter are significantly different at $P \leq 0.05$, using LSD test (harvest data are compared to storage data).

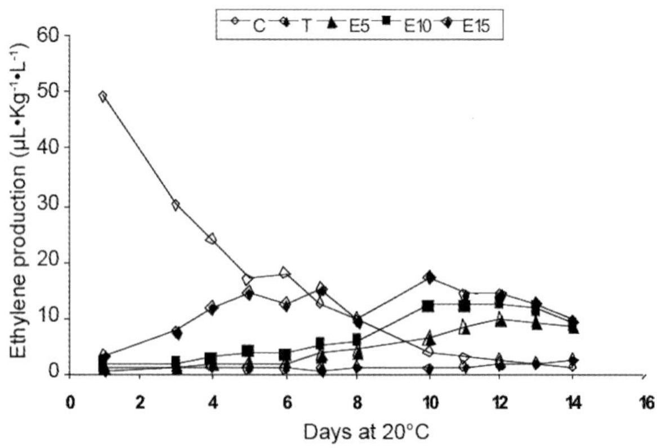

Fig. 1. Effect of 1-MCP treatment and combination of heat and 1-MCP treatments on ethylene production during shelf life of 'Blanquilla' pear after 4 months of storage.

Maximal ethylene production in control was found at the beginning of the shelf life period but decreased later (Fig. 1). On the contrary, 1-MCP-treated fruit did not produce ethylene during shelf life. The effects of heat treatment on ethylene production depended on the duration of this treatment. In E5 and E10, 1-MCP blocked the ethylene production until the eighth day of measurements. For E15, significant ethylene rates were observed after 3 days and this production remained stable until the end of the experiment. Changes in quality parameters as described above were directly related to the changes in ethylene production. Collectively these results showed that 1-MCP when combined to heat treatment may be an interesting mean to improve the commercial quality of 'Blanquilla' pear.

References

Bai J (2006) Comparison of CA storage and pretreatments of 1-MCP and ethoxyquin in altering α-farnesene metabolism and reducing superficial scald of 'd'Anjou' pears. Acta Horticulturae (in press).

Trinchero GD, Sozzi GO, Covatta F, Fraschina AA (2004) Inhibition of ethylene action by 1-methylcyclopropene extends postharvest life of 'Bartlett' pears. Postharvest Biol. Technol. 32:193–204.

Vilaplana R, Valentines MC, Toivonen PMA, Larrigaudière C (2006) Antioxidant potential and peroxidative state of 'Golden Smoothee' apples treated with 1-methylcyclopropene. J. Am. Soc. Hortic. Sci. 131:104–109.

Life science trace gas facility: a way towards top-research on biological systems

Cristescu, S.M. and Harren, F.J.M.

Department of Molecular and Laser Physics, Institute for Molecules and Materials, Radboud University Nijmegen, Toernooiveld 1, 6525 ED Nijmegen, the Netherlands. (*Corresponding author: simona@science.ru.nl)

Ethylene (C_2H_4) is a plant hormone involved in many aspects of the plant life cycle, as well as in the plants response to many environmental stimuli. Over the years the laser-based detectors have been used for online monitoring of ethylene production in various processes in plants and microorganisms, such as germination (Petruzzelli *et al.*, 1995; Thuring *et al.*, 1994), growth, flower and fruit development, senescence of plant organs (Woltering *et al.*, 1993; Wagstaff *et al.*, 2005), programmed cell death, plant defense, biotic stress (e.g. pathogen attack) (Cristescu *et al.* 2002) and abiotic stress conditions (e.g. wounding, hypoxia, drought, heat, chilling and freezing) (Leprince *et. al.*, 2000), interaction with auxin (Van der Bussche *et al.*, 2003), circadian rhythm (Thain *et al*, 2004), nitrogen fixation by cyanobacteria (Staal *et al.*, 2003), etc.

The strength of these instruments resides in the possibility to perform non-invasive, fast (seconds timescale) and online detection of ultra low gas concentrations at and below the ppbv-level (1 ppbv = 1 part per billion volume = $1:10^9$) under rapidly changing external conditions (Harren and Reuss, 1997).

With more than 20 years of experience with applications in life science, the Trace Gas Facility from Nijmegen is offering free access, training and full support to researchers engaged in experiments where high sensitivity measurements are important to determine the character and the timing of the observed processes. With our unique state-of-the-art detectors, traces of ethylene and also other biological interesting gases from plants, fruits, algae, bacteria, fungi, insects, breath, human skin, culture media, etc. can be monitored under rapidly changing conditions (e.g. temperature, O_2-, CO_2-levels, biotic stress, etc.) in a seconds timescale and without incubation periods.

Among ethylene, other components of biological interest can be monitored in real time with the laser-based photoacoustic detectors and proton

A. Ramina et al. (eds.), Advances in Plant Ethylene Research: Proceedings of the 7th International Symposium on the Plant Hormone Ethylene, 441–442.

transfer reaction mass spectrometer (PTR-MS), such as nitric oxide, ethane, higher aldehydes (e.g. formaldehyde, acetaldehyde, propanal, etc.), ketones, alcohols (e.g. methanol, ethanol, propanol, iso-propanol, *n*-propanol, etc.), acids (e.g. formic acid, acetic acid, propanoic acid, etc.) and esters as well as many unsaturated, aromatic and N or S substituted hydrocarbons. Examples of applications are numerous and can be found on the web site www.ru.nl/tracegasfacility).

References

Cristescu SM, de Martinis D, te Lintel Hekkert S, Parker DH, Harren FJM (2002) Ethylene production by *Botrytis cinerea in vitro* and in tomatoes. Applied and Environmental Microbiology 68:5342–5350

Harren FJM and Reuss J (1997) Photoacoustic spectroscopy, p. 413–435. *In* Trigg GL (ed.), Encyclopedia of applied physics, vol. 19. VCH Publishers, Inc., Weinheim, Germany

Leprince O, Harren FJ, Bultink J, Alberda M, Hoekstra FA (2000) Metabolic disfunction and unabated respiration precede the loss of membrane integrity during dehydration of germinating redicles. Plant Physiol 122:597–608

Petruzzelli L, Harren FJM, Perrone C, Reuss J (1995) On the role of ethylene in seed germination and early root growth of *Pisum Sativum.* Journal of Plant Physiology 145:83–86

Staal M, te Lintel-Hekkert S, Harren FJM, Stal LJ (2003) Light action spectra of nitrogenase activity in Baltic Sea cyanobacteria. Journal of Phycology 39 (4):668–677

Thain SC, Vandenbussche F, Laarhoven LJ, Dowson-Day MJ, Wang Zy, Tobin EM, Harren FJ, Millar AJ, Van der Straeten D (2004) Circadian rhythms of ethylene emission in Arabidopsis. Plant Physiol 136:3751–3761

Thuring JWJF, Harren FJM, Nefkens GHL, Reuss J, Titulaer GTM, de Vries HSM, Zwanenburg B (1994) Ethene production by seeds of Striga hermonthica induced by germination stimulants. *In* Pieterse AH, Verkleij JAC and ter Borg SJ (eds),Biology and management of Orobanche, Royal Tropical Institute, Amsterdam, 225–236

Van der Bussche F, Vriezen WH, Smalle J, Laarhoven LJJ, Harren FJM, Van der Straeten D (2003) Ethylene and auxin control the Arabidopsis response to decreased light intensity. Plant Physiology 133:517–527

Wagstaff C, Chanasut U, Harren FJM, Laarhoven LJ, Thomas B, Rogers HJ, Stead AD (2005) Ethylene and flower longevity in Alstromeria: relationship between tepal senescence, abscission and ethylene biosynthesis. Journal of Experimental Botany 56:1007–1016

Woltering EJ, Van Hout M, Somhorst D, Harren FJM (1993) Roles of pollination and short-chain saturated fatty acids in flower senescence. Plant Growth Regulation 12:1–10

Biochemical and physiological characteristics of transgenic *CaMV 35S::iaaM* tomato

Vioque, B.[1], Castellano, J.M.[1]* and Chamarro, J.[2]

[1]Instituto de la Grasa-CSIC, Dept. Fisiología y Tecnología de Productos Vegetales. Avda. Padre García Tejero 4, E-41012 Sevilla, Spain; [2]Instituto de Biología Molecular y Celular de Plantas-UPV-CSIC, Dept. Biología del Desarrollo, Avda. de los Naranjos s/n, E-46022 Valencia, Spain. (*Corresponding author: jmcas@ig.csic.es)

Plants expressing the bacterial *iaaM* gene from promoters of different strength and tissue specificity can be used as a good genetic tool to manipulate the endogenous auxin levels and study the involvement of the hormone in different physiological responses (Romano *et al.*, 1993; Mezzeti *et al.*, 2004). Indole-3-acetic acid (IAA, auxin) plays an active role during fruit development and ripening interacting with ethylene to regulate its biosynthesis and/or perception. The expression in tomato plants (*Lycopersicum esculentum* Mill. cv Ailsa Craig) of the *iaaM* gene from *Agrobacterium tumefaciens* under the control of the *CaMV 35S* promoter originates fruits that overproduce both auxin and ethylene (Castellano *et al.*, 1999). This altered hormonal status induces substantial changes in many aspects of the fruit physiology that might have important effects on the quality and shelf life.

We first confirmed by Northern analysis, the correct insertion and expression of the *35S::iaaM* construction. Despite that in transformed and non-transformed plants, the first flowers appeared simultaneously, the maximal rate of flowers production was advanced 1–2 weeks in transgenic plants. These were able to produce parthenocarpic fruits in absence of pollination with a very high index of fruit setting. Under favourable conditions for pollination, transgenic fruits presented a reduced number of seeds, being a significant percentage (45%) strictly parthenocarpic. Genetic transformation did not modify the number of locular cavities, but it was noticeable the low abundance of the jelly-like parenchyma of the fruit with 73% of them in which it was scarce or non-existent. The length of the development period, measured as time since anthesis to *Breaker*, was typically 60 days in WT and slightly longer (62 days) in transgenic fruit. On the

A. Ramina et al. (eds.), Advances in Plant Ethylene Research: Proceedings of the 7th International Symposium on the Plant Hormone Ethylene, 443–444.

other hand, the weight achieved by the fruit at the onset of ripening was significantly higher (121%) in *35S::iaaM* fruit.

Fruit ripening culminates with dramatic changes in the colour, texture, flavour and aroma of its flesh. A study of the evolution of these parameters indicated that in *35S::iaaM* fruit the ripening process is slowed down. Transgenic fruit exhibited a delay in the acquisition of the red colour, softening, and in the accumulation of soluble solids. It also displayed lower titratable acidity. This behaviour is in agreement with our data indicating that *35S::iaaM* tomato shows a significant delay in the autocatalytic phase of the ethylene production (Vioque *et al.*, 2006). In summary, the expression in tomato of the *35S::iaaM* transgene induces phenotypic modifications that resemble those observed after exogenous applications of auxins. The results confirm the viability of the experimental strategy used to study the auxin/ethylene interaction in fruits. Our data also support that high levels of auxin during the ripening slow down this process, and are compatible with the working hypothesis in tomato postulating that the ripening process is initiated or signalled by low levels of free IAA (Normanly, 1997). This work was supported by the research project AGL2001-2288 (Plan Nacional I + D + I 2000–2003, Spain).

References

Castellano JM, Chamarro J, Vioque B (1999) Ethylene biosynthesis in transgenic auxin-overproducing tomato plants. Biology and Biotechnology of the Plant Hormone Ethylene II. Kluwer Academic Publishers. Dordrecht

Mezzetti B, Landi L, Pandolfini T, Spena A (2004) The *defH9-iaaM* auxin-synthesizing gene increases plant fecundity and fruit production in strawberry and raspberry. BMC Biotechnol 4:1–10

Normanly J (1997) Auxin metabolism. Physiol. Plant. 100:431–442

Romano CP, Cooper ML, Klee HJ (1993) Uncoupling auxin and ethylene effects in transgenic tobacco and *Arabidopsis* plants. Plant Cell 5:181–189

Vioque B, Castellano JM, Chamarro J (2006) Changes in the autocatalytic ethylene production in transgenic tomato expressing the *iaaM* gene from *Agrobacterium tumefaciens*. 7th International Symposium on the Biology and Biotechnology of the Plant Hormone Ethylene. P 4–13

Synergistic actions of ethylene with gibberellin in the growth of rice

Watanabe, H.* Hase, S. and Saigusa, M.

Field Science Center, Grad. School of Agric. Sci., Tohoku University, Miyagi 989-6711, Japan. (*Corresponding author: watanabe@bios.tohoku.ac.jp)

1. Introduction

Seedling establishment is one of the most important agronomic traits in direct seeding rice cultivation. It has been considered that plant growth regulators (PGRs), such as gibberellin (GA) and ethylene, have promotive effects on rice seedling organs (Takahashi and Kaufman, 1983). However, the effects of these plant hormones are occasionally divers due to various environmental factors, including temperature and flooding depth. As well, synergistic or counteracting plant hormone interaction can be found in several growth systems of plants (Davies, 1995). A notable case of the former interaction is internode elongation in deep-water rice (Kende *et al.*, 1998). Ethylene promotes the growth of internodal tissue of deep-water rice, which responds to flooding by rapid elongation induced by ethylene formation. Ethylene promotes growth in part by increasing the responsiveness to the internodal tissue to GA. Here, we investigated the effects of single or combined applications of ethylene and GA on the growth of different organs of rice seedlings growing under growth chamber and flooding soil conditions.

2. Material and Methods

The cultivar used in the study was Hitomebore (*Oryza sativa* L.). Sterilized seeds were immersed in water and then soaked in the test solution. Ethephon was used as an ethylene-releasing agent, and GA_3 was used for gibberellin. The components of the test solution were as follows: (1) water alone (control), (2) 50 ppm ethephon (ET), (3) 100 ppm GA_3, (4) 50 ppm ET + 100 ppm GA_3 (ET + GA). After treatment with the plant hormone solution, the seeds were again immersed in water to remove any excess test

A. Ramina et al. (eds.), Advances in Plant Ethylene Research: Proceedings of the 7th International Symposium on the Plant Hormone Ethylene, 445–447.

solution. The imbibed seeds were geminated in water at 30°C in the dark, and the germinated seeds sown at 0.5 cm of seeding depth flooding soil.

2. Results and Discussion

The mesocotyl growth was increased by a single treatment of GA or ET over that of the control. However, effects of combined applications of GA and ET on seedling growth, including mesocotyls, coleoptiles and 1st to 3rd leaves were more pronounced than those of GA or Et alone. This shows that the pair, GA and ET, has synergistic effects on rice seedling growth (Fig. 1).

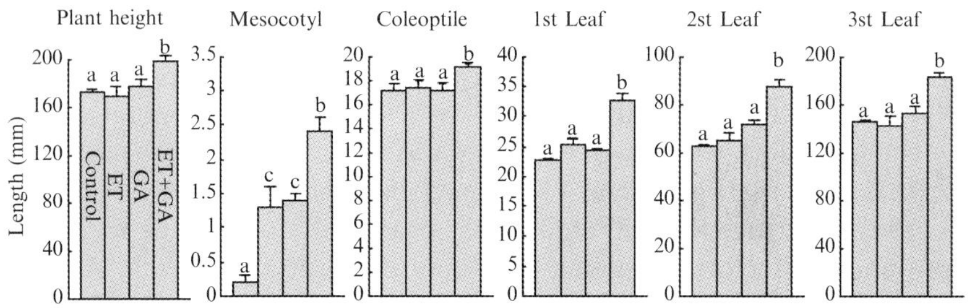

Fig. 1. Effect of ET and GA on the elongation of organs of rice seedlings. Different letters indicate significant differences by Fisher's protected LSD ($P < 0.05$).

No significant differences were found in RWL (the ration of shoot dry weight to shoot length of the seedlings) among the PGRs treatments, indicating that the growth-promoting effects of PGRs treatments on rice seedlings are not merely a spindly growth; an increase in shoot growth by PGRs treatments were accompanied by the enhancement of dry weights (Table 1).

Table 1. Effects of ET and GA on DW and RWL of rice seedlings.

Treatment	DW[1] (mg/plant)	RWL[2] (mg/cm)
Control	16.0a	0.92a
ET	14.8a	0.88a
GA	17.4a	0.98a
ET+GA	16.9a	0.85a

Same letter do not differ at $P < 0.05$, by Fisher's LSD test. [1]Dry weight. [2]Dry weight/plant height ratio of shoot dry weight to shoot length.

In the field experiment, a pair of ET and GA treatment showed a highest seedling emergence rate among the treatments. This might be due to both high elongation rate and maximum length of mesocotyls plus coleoptiles (Fig. 2). As is the same case with seedling emergence at the coleoptiles elongation stage, occurrence of the 1st leaf in ET + GA treatments is also the fastest compared with other treatments. This also might be due to the maximum length and elongation rate of 1st leaves in this treatment (Fig. 2). Taken together, the growth regulation of target organs of rice seedlings will be possible by using proper combinations (ex. ET + GA treatment) of plant growth regulators.

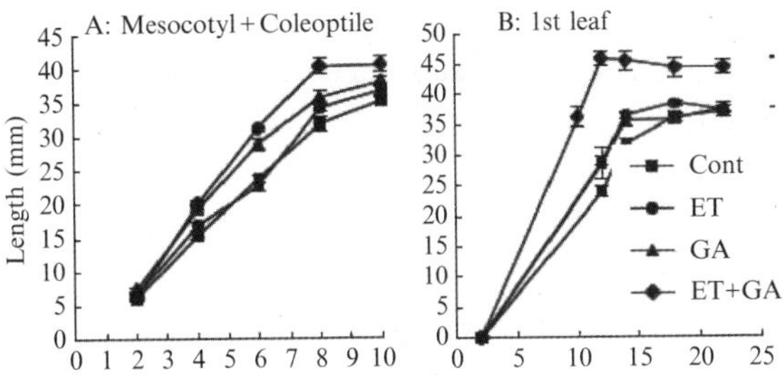

Fig. 2. Effect of ET and GA on early growth of rice seedling.

References

Davies, P.J. (1995). Plant Hormones: Physiology, Biochemistry and Molecular Biology. Kluwer Academic Publishers, Dordrecht, Pages 1–35.

Kende H., Van der Knaap and Cho, H.T. (1998). Deep water rice: A model plant to study stem elongation. Plant Physiol. 118, 1105–1110.

Takahashi, K. and Kaufman, P.B. (1983). Growth regulation of rice seedlings. Proc. Plant Growth Regul. Soc. Am. 10, 229–234.

ANNEX – COLOR FIGURES

A novel membrane protein conserved in plants and animals is important for ethylene receptor function in *Arabidopsis thaliana* (Fig. 1). Resnick *et al.*

A

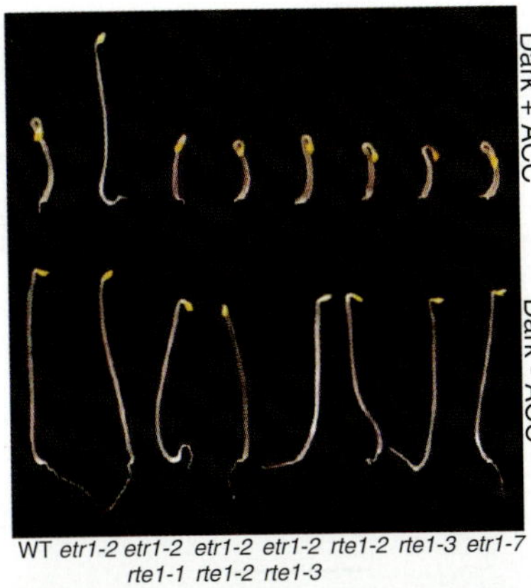

B

The melon ethylene receptor CmERS1 is localized to the endoplasmic reticulum with an N_{lum}-C_{cyt} membrane orientation (Fig. 1). Ma *et al.*

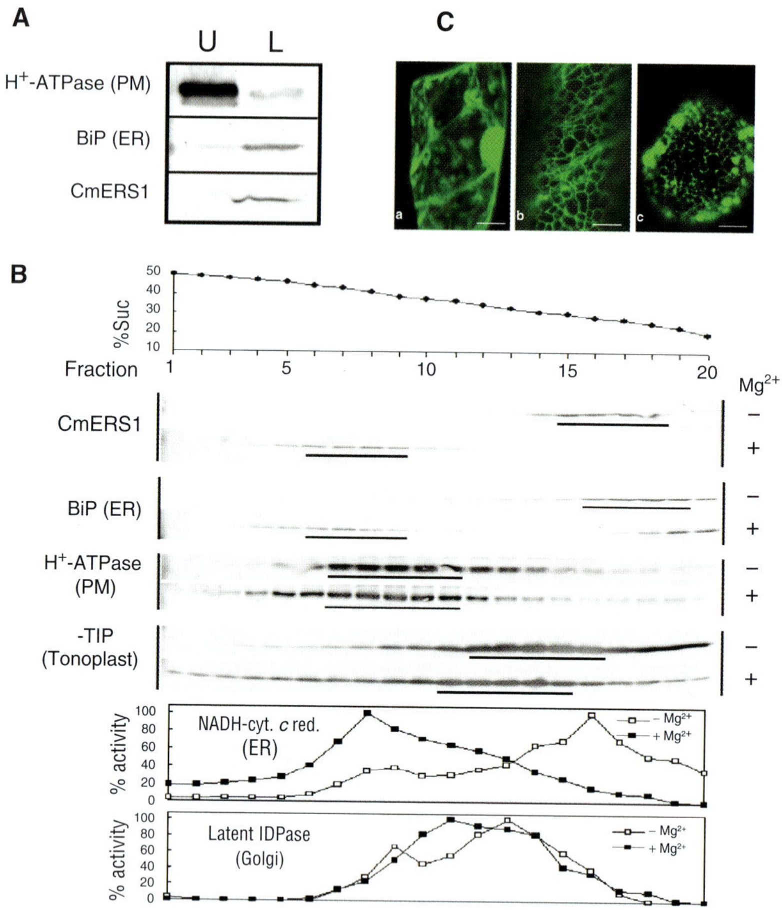

Cloning and differential expression of banana genes coding for EIN3-LIKE proteins involved in ethylene action (Fig. 1). Mbéguié-A-Mbéguié *et al.*

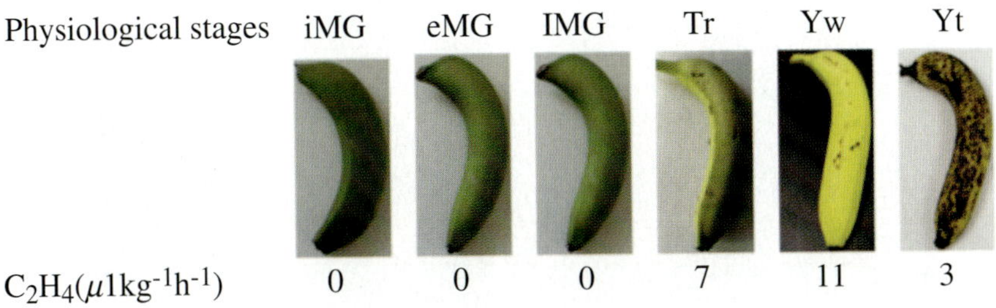

Characterization of three ethylene receptor genes in *Coffea canephora* Pierre (Fig. 2). Bustamante *et al.*

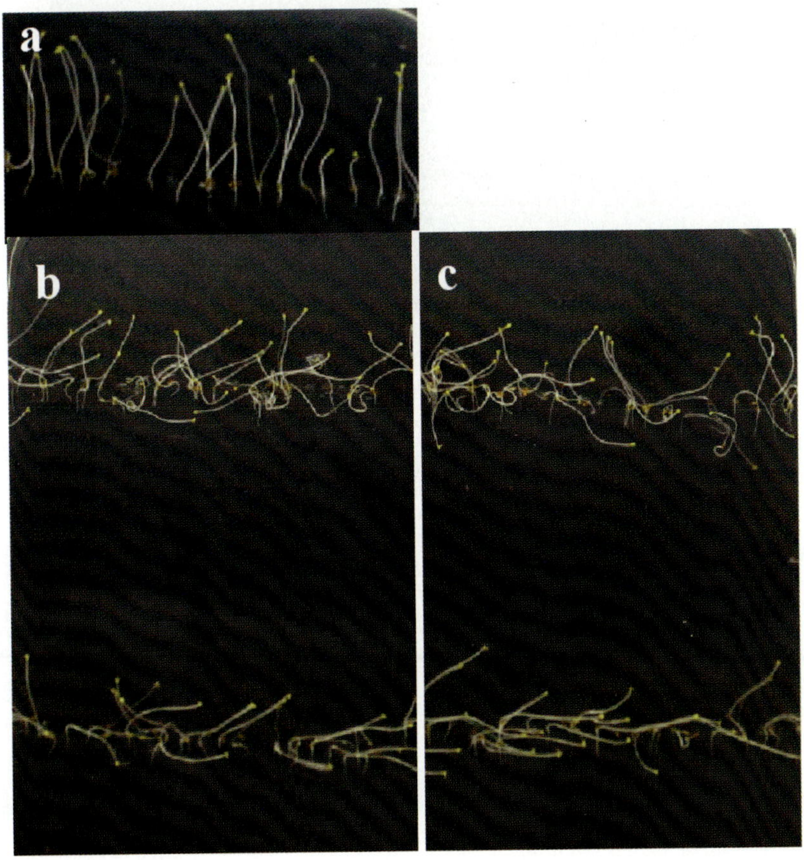

Blue light dependence of Arabidopsis seedling ethylene responses (Fig. 1). Vandenbussche *et al.*

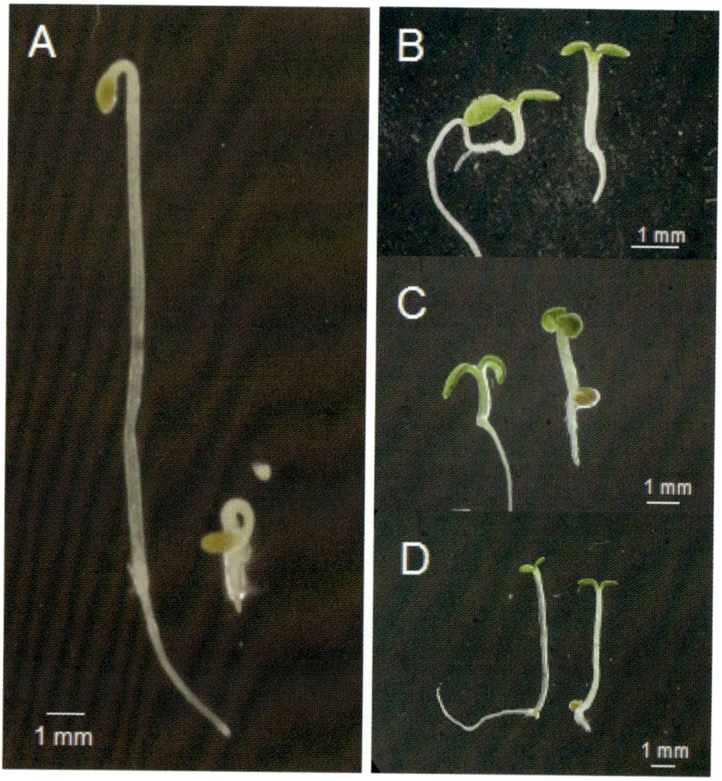

Unravelling ethylene biosynthesis and its role during tracheary element formation in *Zinnia elegans* (Fig. 1). Pesquet and Tuominen

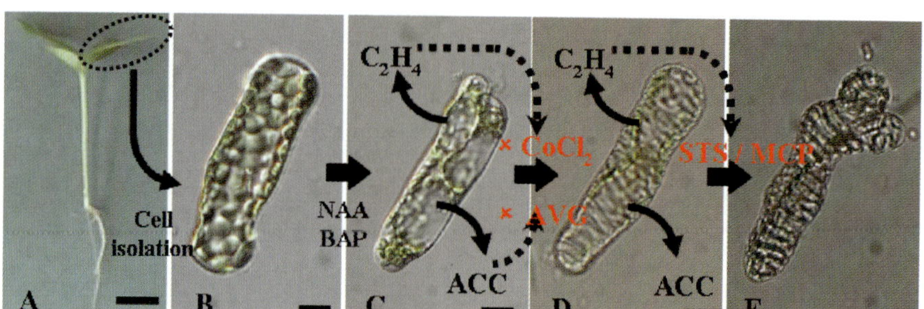

Synergistic effect of kinetin and benzyl adenine improves the regeneration of cotyledon explants of Bottle gourd (*Lagenaria siceraria*) on ethylene production (Fig. 1). Shyamali and Kazumi

a b c

Ethylene Responses in Abscission and other processes of cell separation in Arabidopsis (Fig. 5). Patterson *et al.*

Circadian rhythm of ethylene related genes under postharvest conditions in transgenic Arabidopsis plants (Fig. 2). Kenigsbuch *et al.*

Ethylene receptor signaling and plant salt-stress responses (Fig.1). Cao *et al.*

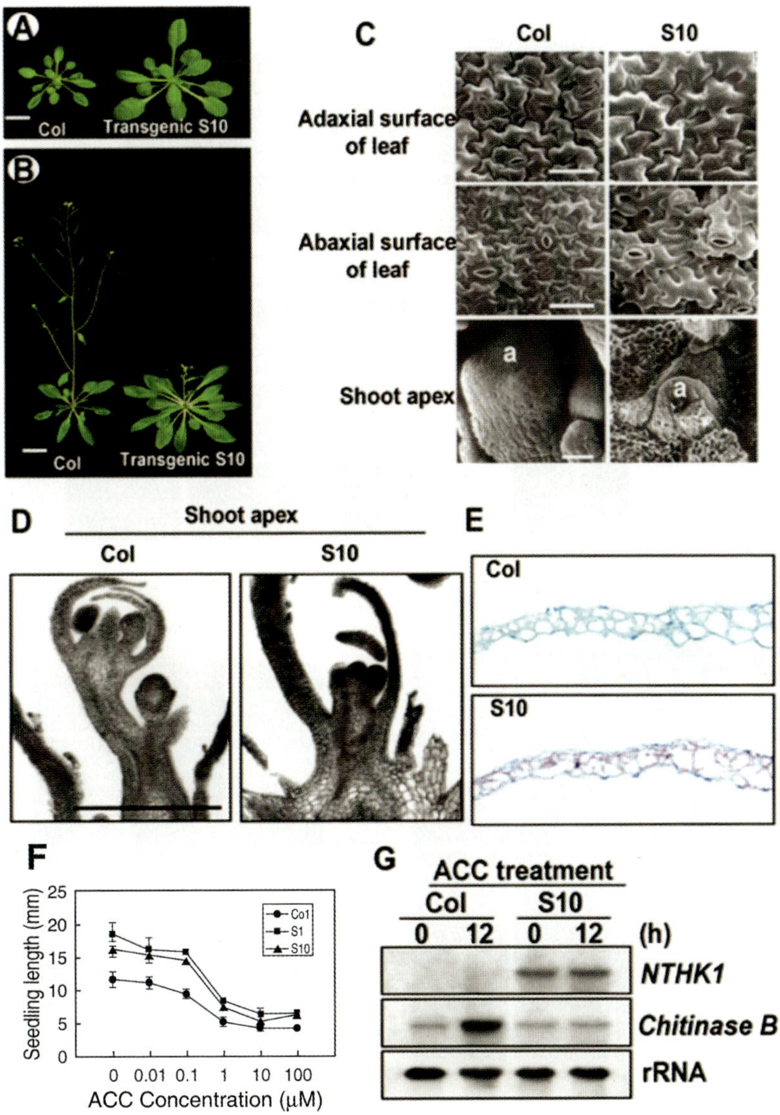

Ethylene receptor signaling and plant salt-stress responses (Fig.2). Cao *et al.*

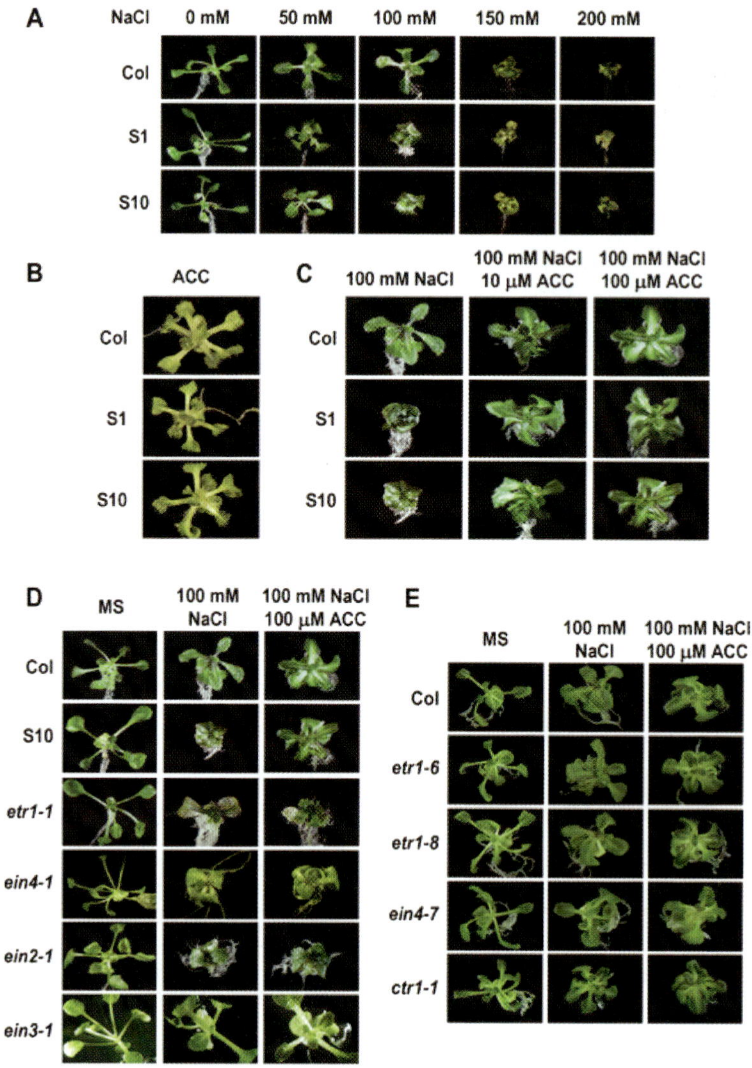

Metabolomic approaches to understand ethylene mediated defenses in *Arabidopsis thaliana* against *Botrytis cinerea* (Fig. 1). Lloyd *et al.*

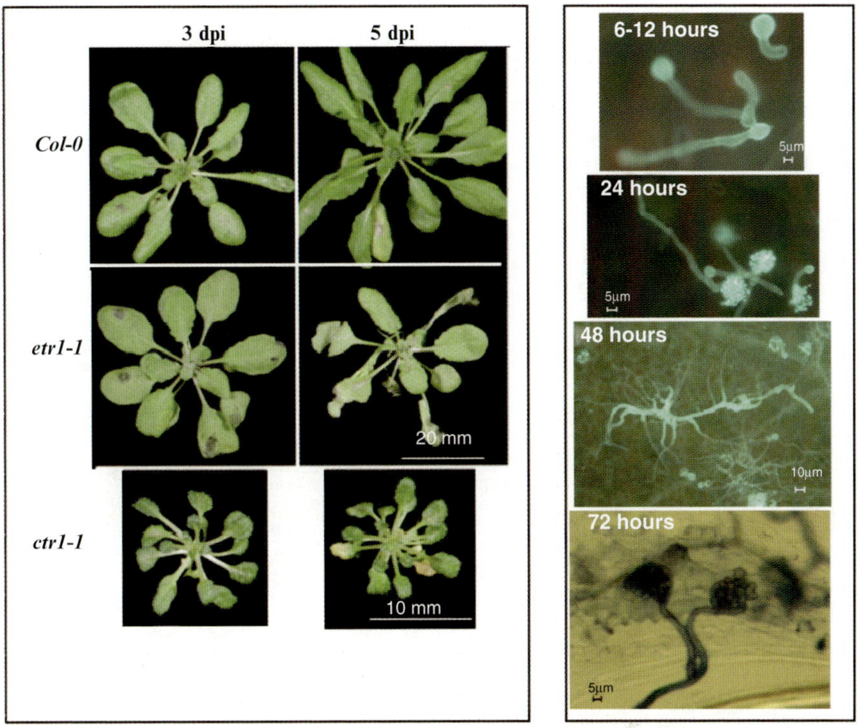

Metabolomic approaches to understand ethylene mediated defenses in *Arabidopsis thaliana* against *Botrytis cinerea* (Fig. 2). Lloyd *et al.*

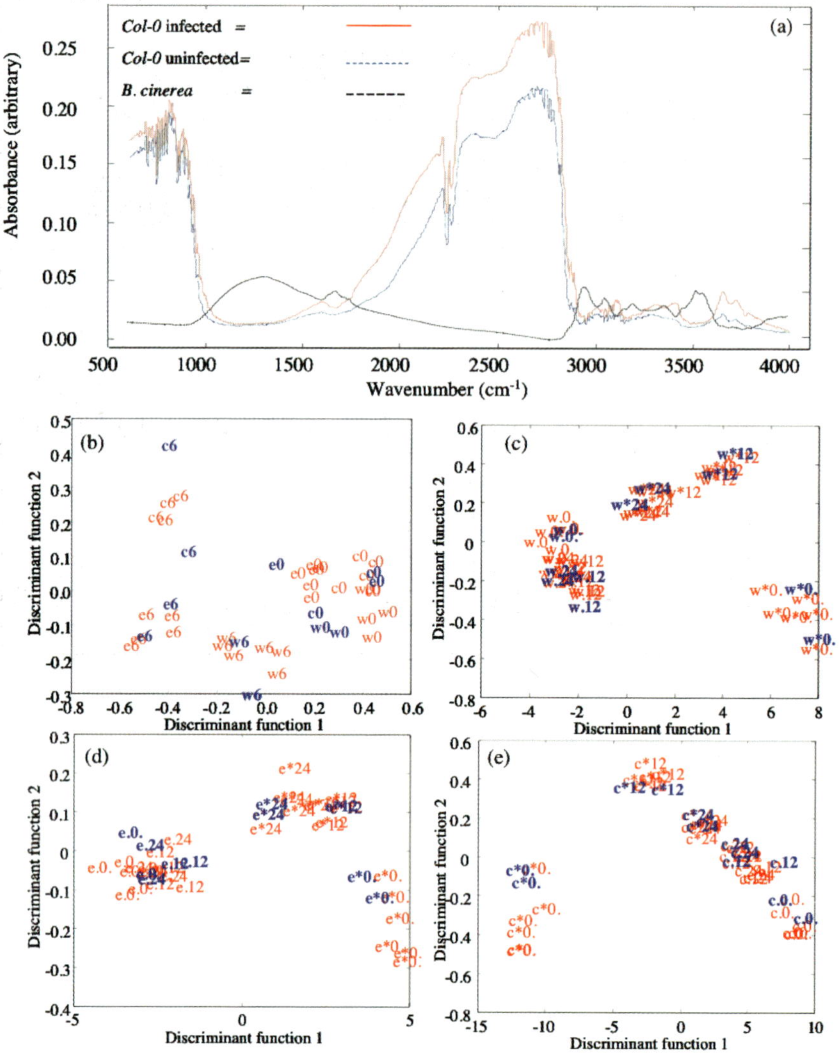

Role of ethylene in triggering ROS production in the tomato mutant *Nr* subjected to acute ozone treatment (Fig 1). Castagna *et al.*

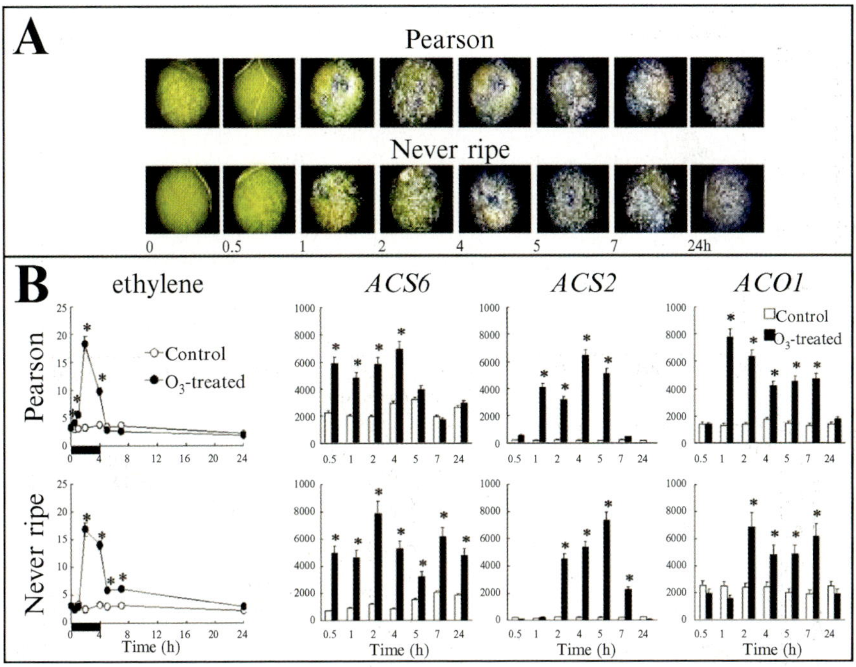